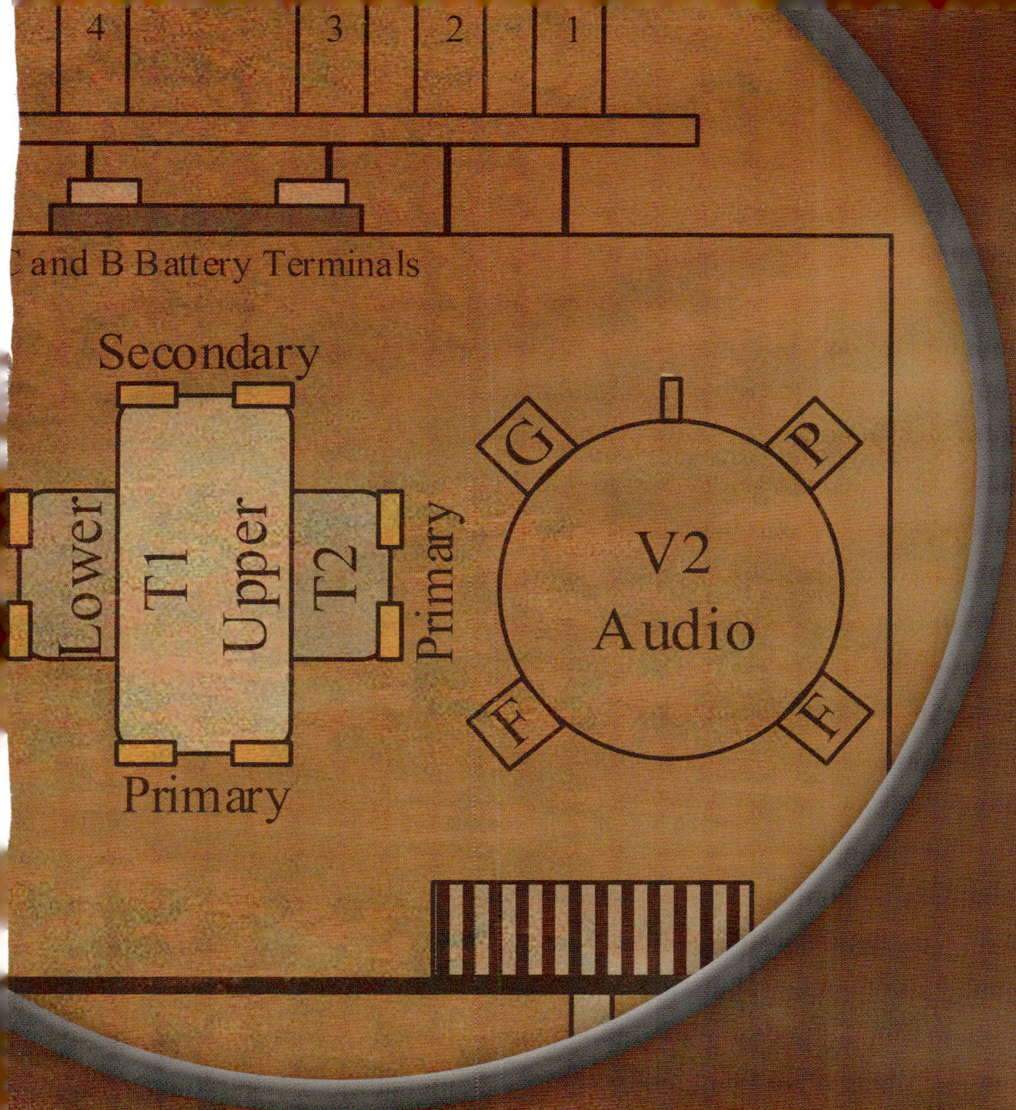

The Early Zenith Radios

The Battery Powered Table Sets
1922–1927

Gilbert M. Hedge with Durell M. Roth

Schiffer Publishing Ltd
4880 Lower Valley Road • Atglen, PA 19310

Library of Congress Control Number: 2014942302

Designed by RoS
Type set in Agency FB/GoudyOlSt BT

ISBN: 978-0-7643-4674-3
Printed in China

Published by Schiffer Publishing, Ltd.
4880 Lower Valley Road
Atglen, PA 19310
Phone: (610) 593-1777; Fax: (610) 593-2002
E-mail: Info@schifferbooks.com

For our complete selection of fine books on this and related subjects, please visit our website at www.schifferbooks.com. You may also write for a free catalog.

This book may be purchased from the publisher. Please try your bookstore first.

We are always looking for people to write books on new and related subjects. If you have an idea for a book, please contact us at proposals@schifferbooks.com.

Schiffer Publishing's titles are available at special discounts for bulk purchases for sales promotions or premiums. Special editions, including personalized covers, corporate imprints, and excerpts can be created in large quantities for special needs. For more information, contact the publisher.

*To my mother Ursula,
my dad Buford,
my brothers Bill and Hugh,
my sister Coral Lee,
and especially my older brother Kerry,
who was always there
for all of us.*

Acknowledgments

My appreciation and gratitude go to Durell Roth for his assistance in compiling this book: for his editing expertise, antique radio knowledge, and electronics background, and for encouraging me, for believing in this project, and for providing suggestions that significantly improved the readability and the depth of information, for which the reader will undoubtedly be thankful.

To fellow Houston Vintage Radio Association (HVRA) members Kathy Botto, Robert Botto, David Moore, John Schmitt, William Werzner, Sheila Wyborny, and Wendell Wyborny who were kind enough to review the first draft and provide valuable comments.

To fellow HVRA member and fellow Boeing employee Lance Borden who always had time, in his busy schedule, to answer my technical questions. He also provided guidance on the discussion related to antennas. I guess I owe you another scotch.

To my daughter-in-law Kacie (a school teacher) who reviewed the very first draft, kept me straight on the grammar, and helped me make some early decisions on format.

To my sons Shawn and Ryan and my grandchildren Zariah, Connor, and Cayden for putting up with me while I worked on this project.

To my relatives, out I-10, Barbara, Darryl, Barbie, William, Robin, Scott, Kerry John, Lorraine and their families, who put up with those very short visits. I promise to stay longer from now on.

To my nephews John, Danny, and Tommy and my niece Tammy for their patience.

To my wife Chere, who would answer the phone, talk a few minutes and then say to the caller, "Oh, he's working on the book" and would tell everyone, lovingly, that she's a book widow. Thanks for your support and for putting up with me during this endeavor.

My heartfelt thanks! Any errors are all mine.

Contents

Introduction

There is a broad range of published information available to the collector and restorer of antique radios and associated equipment. Many books have been written about the major and minor manufacturers of radios, the development of radio, and the pioneers who made it work. Much of this documentation includes alphabetical listings of the equipment produced by numerous manufacturers that contain model numbers, dates of production, etc. Still other related documents contain schematic diagrams, discussions of basic electronics, trouble-shooting and repairing, and cabinet refinishing.

With all that information available to the antique radio collector, there is still something missing; specific information about the individual radios themselves. *The Early Zenith Radios* fills that gap by providing a single comprehensive source of information for each table top, battery powered radio that Zenith manufactured in the 1920s. Excluded are the Phonograph Set, the model 15 and the model 31.

With the exception of some information (articles, advertisements, notes, etc.) that is directly related to the radios herein, no attempt has been made to discuss the history of the Zenith Radio Corporation, which is adequately covered in other texts.

During the research and development phases of this document only one of each radio and/or amplifier model was available to be examined, photographed and documented. Because of this limited availability, engineering changes to any given model, such as the 1-R, 3-R, 2-M, etc., may not be reflected in this documentation and your set may not be an exact match to the one discussed. Such changes were frequently made during and between production runs and may or may not have been documented by the manufacturer. These changes, however, are very important because they can help determine technological advances and help identify differences between production runs.

Zenith manufactured two different versions of the 3-R and 4-R receiver but did not change the model numbers that would differentiate between the two versions. Over the years radio collectors have adopted a suffix letter system to identify a specific version of the set: "E" for the early version and "L" for the later version. In order to eliminate the possibility of confusion, this book will use this suffix method when discussing a particular version of the model 3-R and 4-R radios.

The reader does not have to know how a radio works or have extensive electronic experience in order to understand and benefit from most of the information presented in this book, however, a minimal level of electronic knowledge and radio theory is helpful when interpreting radio schematics, understanding the circuit explanations, or attempting to perform basic troubleshooting and testing.

The Early Zenith Radios is a unique book in terms of its arrangement, its extensive coverage of information related to these particular radios, and its in-depth photographic and textual documentation used to explain each topic. Because of this uniqueness, it is important here to provide an overview of the layout and philosophy of *The Early Zenith Radios*.

The text is divided into three distinct sections: Each of these sections is further subdivided into chapters that provide detailed photographs and/or discussions on a given radio or related topic as outlined below.

Section I: Radio Descriptions includes a physical and functional description of each radio and audio amplifier covered in the text. The physical description includes multiple views of the cabinet, chassis and assemblies. The details of the mechanisms and sub-assemblies are further clarified with explanations and expanded views, diagrams or drawings. The functional section includes a schematic; multiple color coded diagrams (used to describe how the set works); plus instructions on how to install and operate the radio. Each physical and functional description section is preceded with a quick reference guide, which allows the reader to quickly locate specific information.

Section II: Useful Details for Understanding Zenith Models and Marketing presents some facts and conclusions that have been realized over the years about these radios. In addition, Chapter 13 is a discussion of Successes and Failures had by Zenith as they apply to specific sets discussed throughout the text. This is followed in Chapter 14 by various examples of publicity material used by Zenith to promote their equipment to the general public and to communicate product-related information to their distributors, dealers and sales personnel.

Section III: Radio Restoration uses step-by-step instructions that guide you through the five stages of radio restoration. Each topic includes detailed text, diagrams, drawing, pictures, etc., to ensure the information is clear and complete. Detailed instructions are also included for making original- style working batteries, which enable you to play that "new" radio just like the original manufacturer had intended. Plus, various other restoration techniques are included throughout the text in an effort to help broaden the reader's restoration skills.

Radio Preservation

Maintaining originality may not be as simple and straightforward as it first appears. Basically, it is a multilevel philosophy of restoration with which every collector of vintage radios and related items must deal every time she or he obtains another antique set. Maintaining originality could mean leaving it in the original "as found" condition; that is, as found by the collector. It could also mean "reconditioning" that radio to its original like-new (just off the assembly line) condition, or it could mean something in between.

Typically, antique radio collectors define originality as being the "original condition and configuration of a given radio, and related equipment, as it was when manufactured." This is a high-quality restoration standard that requires the collector to consider a series of questions such as: Were there any alterations made to the equipment, are the accessories missing (power cable, instruction manual, etc.), are parts missing or damaged? These are just a few of the more important aspects of the collecting hobby, and they allow both hobbyists and historians to compare one radio to another identical radio, in order to identify any engineering design changes made by the manufacturer during a production run or between multiple production runs for a specific model.

During the last three and a half decades, collectors have produced many excellent restorations that brought many "basket cases" back to life. We, as fellow collectors, are proud to be associated with them and are very thankful for their work. Had they not taken on the challenge of restoring these sets to their former glory and beauty, they and we would have never enjoyed hearing them play, and the sets may have been lost forever.

Not all restorations, however, are real and over-restoration is a prime example. Some radios have been restored to a condition that makes them "prettier" than when they were brand new; this adds credence to the phrase "restoration is another form of destruction." Frequently, restorations have internal modern electronic components as substitutes for the original parts. Too, on rare occasions, some restorations have been shown to be completely misrepresented in virtually every aspect of the finished repairs. As collectors of historical artifacts, we have a responsibility to repair as accurately and completely as possible and to act with integrity and honesty in documenting all such repairs.

Another practice among the general public that has evolved in recent years is the removal of the radio chassis from a console cabinet in order to install a bar, a fish bowl, or a modern AM-FM/CD radio combination. Many of these console radios could have been restored to their former grandeur, but now they are lost forever.

Restorers of early Zenith radios face a problem that restorers of other brands are rarely required to address—pot metal! Zenith used this low grade metal to make many of the internal parts such as switches, gears, etc. for many of their early sets. As time took its toll, the pot metal expanded causing a given control to bind and become inoperable. Some of these parts are repairable; however, the majority of them have crumbled into little more than dust and require replacement. There is no commercial source for any of these components, and usually a part removed from a "parts radio" will suffer from the same or worse maladies as the one being replaced. The parts, however, can be reproduced, and depending on one's abilities and the availability of shop equipment (drill press, lathe, etc.) can be made to function very much like the original component. Methods of reproducing the commonly needed Zenith parts are covered in Section III: Radio Restoration. Not all the parts can be totally replicated; therefore, to a Zenith collector, maintaining originality may mean "maintaining originality as closely as possible."

There are many positives and negatives in any discussion of restoration. Ultimately, however, it is up to the individual collector to determine what needs restoring and what should be left alone. Again, with the decision comes the responsibility of fully documenting all phases of the restoration for the benefit of future collectors.

Reproductions

Another aspect of collecting antique radios that has gained momentum in recent years is reproductions: the creation of an exact duplicate of an original radio, radio apparatus, or other related item. Many of these reproductions are "works of art" that are the result of exceptional talent, workmanship, and attention to detail. Some of these are so good that it's virtually impossible, even for the seasoned collector, to recognize the fake. Already these beautiful but fake works of art are finding their way into collections of truly vintage equipment and in some cases there is no way to distinguish the reproduction from the real thing. Eventually, maybe 50 or 100 years from now, these reproductions will have infiltrated even the most carefully guarded and documented collections of original antique radio equipment, creating gross misinformation regarding these historical artifacts. As beautiful and artful as the reproductions are, they are simply reproductions and the craftsman should boast of her or his abilities while at the same time carefully and permanently documenting, either on or in the equipment, that it is a working replica and not restored vintage equipment.

Identification of Restorations and Reproductions

Although the restoration or reproduction of antique radios or related items is wholeheartedly supported by this author, we need keep in mind that our current actions will have a profound impact, far into the future, on the whole antique radio collecting community. In acting responsibly, we need to assure that the integrity of the radios that we restore or reproduce is maintained; this means documentation and traceability. I suggest we consider physically identifying these restorations and reproductions with a label that states something similar to the following:

Restoration or Reproduction performed by:
Date of restoration or reproduction:
Manufacture Date:
Manufacturer or Original Manufacturer:
Model or Original Model:
Comments:

The identification could be in the form of a vintage-looking label that would be easily and permanently placed inside the cabinet or under the chassis of the restoration or reproduction. One example of how to identify the restoration is presented on page 251.

Remember that your collection of radios will only be in your possession for a relatively short period of time, and will eventually end up somewhere else; in another collection or maybe even in a museum. If the radios are appropriately marked, then everyone will know who restored it or reproduced it and what wonderful skills that person possessed. Your name deserves to be associated with that radio or that related item forever. Each of us, as a temporary custodian of these historical artifacts, is responsible for seeing that these radios are preserved in a manner that will assure that future generations will be able to accurately interpret the historical facts, and we must avoid clouding the evidence with unidentified restorations or reproductions.

Common Radio Attachments

Each early Zenith radio requires an antenna to pick up the signal coming from the radio station, a loudspeaker to allow you to comfortably hear the information coming out of the radio, and power for the tubes in the radio. These are collectively referred to here as Common Radio Attachments because they are common to all the sets presented. This information will make you aware of some to the options that are available when you get to the section for preparing a particular radio for operation.

Antennas

All the receivers in this book, except for the Super-Portable, require some type of antenna.

Because today's broadcasting stations have higher transmitting power than those of the 1920s, an outside antenna is not recommended and is generally not necessary. It is, therefore, recommended that an antenna be installed inside the residence. An insulated or non-insulated wire may be used; however, if the wire is non-insulated be sure that it doesn't touch anything made of metal. It should be at least 30 ft. long and the higher the better. It can be a zig-zag pattern or a square or any other pattern that lets the antenna fit a particular location. Try it in various locations. A ground is optional, but may improve reception. The installation options are endless, but make sure it is safe (regarding children, electrical outlets, tripping, etc.). I use a 40-ft. antenna that is in the attic, similar to the one shown in the Zenith Antenna Installation drawing below.

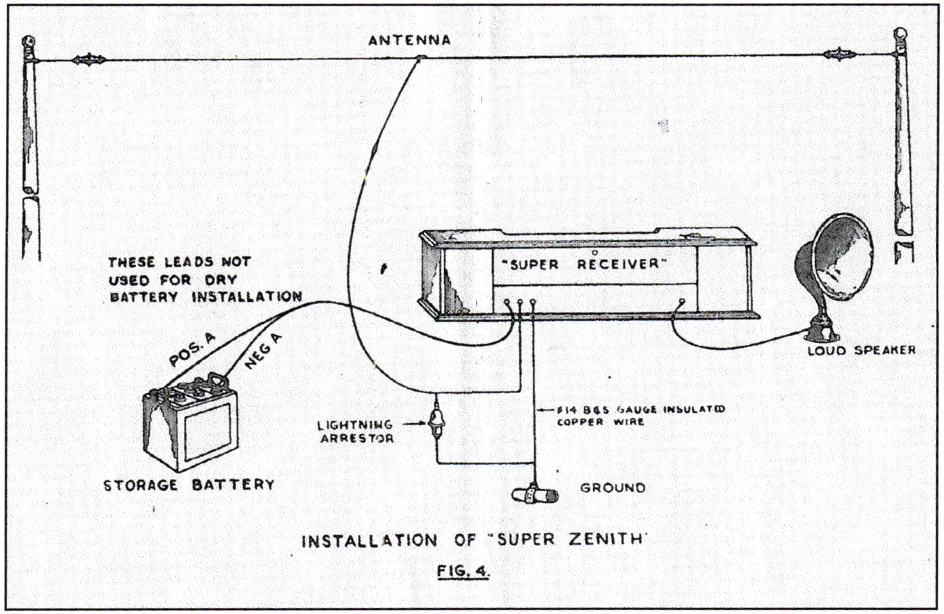

Zenith Antenna Installation

If an outside antenna is required, for long distance listening or for some other reason, use the typical Zenith Antenna Installation drawing as a guide for the installation. The lead-in wire can be attached at any location on the antenna. Zenith recommends that the outside antenna be approximately 50 ft. long. The antenna and lead-in wire should not exceed 125 ft. from end to end.

Antenna kits (antenna wire, lead-in wire, insulators, anchors, etc.) are available from Borden Radio Company (www.xtalman.com). When installing the kit, be sure to follow the installation instructions.

Loudspeakers

Any magnetic loudspeaker (horn or cone) that was manufactured in the early to mid-1920s should work with any of these early Zenith radios. The only exception is the 1-R radio, which doesn't have enough power to drive a loudspeaker. However, when a 2-M or 3-M AF amplifier is attached to the 1-R, a loudspeaker can be used with them.

Loudspeaker wires usually have small pins on the ends that can be easily attached to the terminals of the radio. In cases where there are no terminals, such as with the 4-R radio, simply attach the pins on the ends of the speaker wire to a phone plug and insert the plug into the jack on your receiver. There are numerous methods of attaching wires to plugs and the requirements become clear when looking at the plug. Several different styles of plugs are shown in the Assortment of Plugs picture.

Assortment of Plugs

Horn and Cone Loudspeakers

Zenith recommends that whenever one of the early radios is tuned, that it be tuned with the operator using headphones and after the station is located and properly tuned, a loudspeaker can be connected. See the specific chapter for your set regarding any questions that may arise with respect to connecting a speaker to that specific set.

Power Options

The radios of the early 1920s used an assortment of batteries that provided the power to the various elements of the tubes. The "A" battery heats the filament, the "B" battery provides the plate voltage, and the "C" battery applies a negative bias to the grid. Sometimes the negative grid bias is generated by the circuit itself and does not require a bias battery. Occasionally a "D" battery will be used for the negative grid bias.

In order for the hobbyist to play a particular radio, there are three basic ways to provide the power to the various circuits: option 1—use batteries, option 2—use a battery eliminator, or option 3—use a combination of batteries and a battery eliminator.

Option 1 uses all batteries, which maintains the complete originality of the radio. The "A" battery (lead-acid) would need to be charged to maintain its voltage and occasionally the "B" and "C" batteries (reproduced) would need to be replaced. This is exactly what the original owners of these radios had to do.

Option 2 uses a battery eliminator to supply the power to a particular radio. This method completely eliminates the need to replace the batteries. When using a battery eliminator, there may not be an access hole for running the wires from the eliminator to the radio circuits, so they would have to be routed through the open door. Never drill access holes into the cabinet. A battery eliminator could be installed internally, but that would impact the originality of the radio. Make sure the battery eliminator provides the exact voltages (−3 volts, −4.5 volts, 45 volts, etc.) that are required for your set.

Option 3 uses a combination of batteries and a battery eliminator. First, all the internal "A", "B", or "C" batteries, as applicable, are replaced with reproduced (working) batteries. Second, all external power is provided to the radio with a battery eliminator, which is connected to the terminals located on the back of the radio or connected with the original or reproduced power cable. The use of a battery eliminator for the external power source can offer the best in terms of battery life, and placing the eliminator out of sight will help preserve the originality of the system.

There are some variations, depending on the battery requirements for a particular radio. The option you choose depends largely on your radio, and on the resources and skills available. Weigh the advantages and disadvantages on each system and then pick the one that best satisfies your needs.

The Super-Portable and model VII radios have "D" batteries that provide a negative grid bias to the first RF tube. These older style "D" batteries had a metal case that was part of the negative terminal of the battery, but the modern "D" battery has an insulated case. The modern "D" battery can be used in these vintage sets, if an adapter is installed. Construction of the adapter ("D" Battery Adapter Modification) is described on page 244.

Any particular radio may require that jumpers be installed, i.e., A+ to B−. The placement and number of jumpers depends on the design of the radio and the design of the battery eliminator, if used. Review the schematic for the radio and the instructions that come with the battery eliminator, if used, to determine the appropriate placement of the jumpers. The placement and number of jumpers varies.

Batteries can leak and cause damage to electronic equipment (radios, audio amplifiers, reproduced batteries, toys, etc.), so don't leave batteries installed for extended periods of time.

All the radios in this book have batteries that were replaced with working reproduced batteries. These reproduced batteries are shown in the descriptions of each radio. Construction of reproduced batteries (A, B, C) is covered in Chapter 15. The Number 6 "A" battery is usually used to power the low filament voltage tubes (e.g., 99, 199) in portable radios, e.g., Zenith Super-Portable.

Should you decide to use a battery eliminator as all or part of the power source for your radio, they are available as both wired and tested units, and as kits. A battery eliminator is an excellent test-bench instrument and if you work on a lot of radios, then a good choice might be an ARBE Universal Battery Eliminator, which is available from Antique Radios Inc. (www.arbeiii.com). Should you want a battery eliminator that stays hooked up to the radio on a permanent basis, a battery eliminator kit may the way to go. Battery eliminator kits are available from Antique Electronic Supply in Tempe, Arizona (www.tubesandmore.com). When considering the purchase of a battery eliminator, always compare the battery eliminator specifications to the power requirements of the radio.

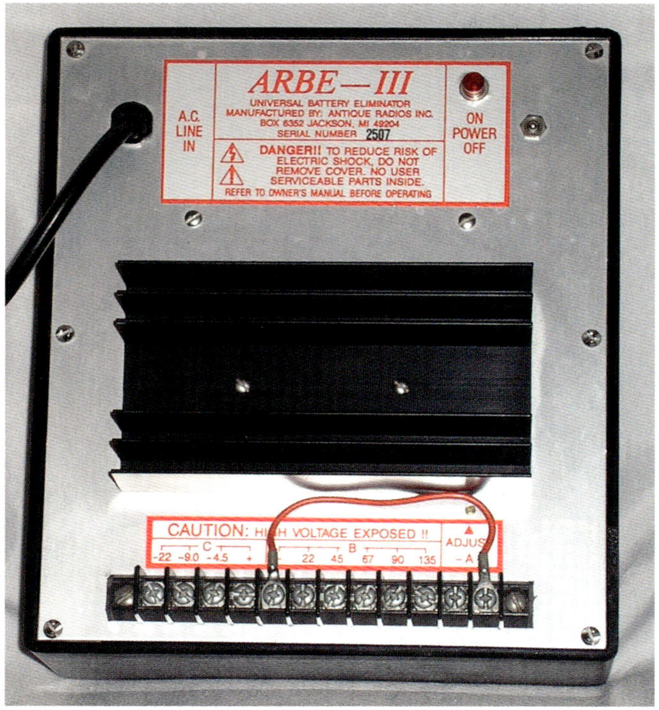

ARBE Battery Eliminator

Notable Zenith Milestones

The Notable Zenith Milestones chart is a tri-section, timeline-structured chart that provides a unique perspective of the "goings on" within Zenith between 1919 and 1929. The chart correlates the production dates of each radio with the factory in which it was most likely built, approximate production dates, as well as the production change from amateur to commercial equipment and the evolution of the corporate name—Zenith. Because some of the dates shown are approximate, this chart should not be used to determine specific dates.

Note: This chronology is provided as a reference only and should not be used to determine specific dates.

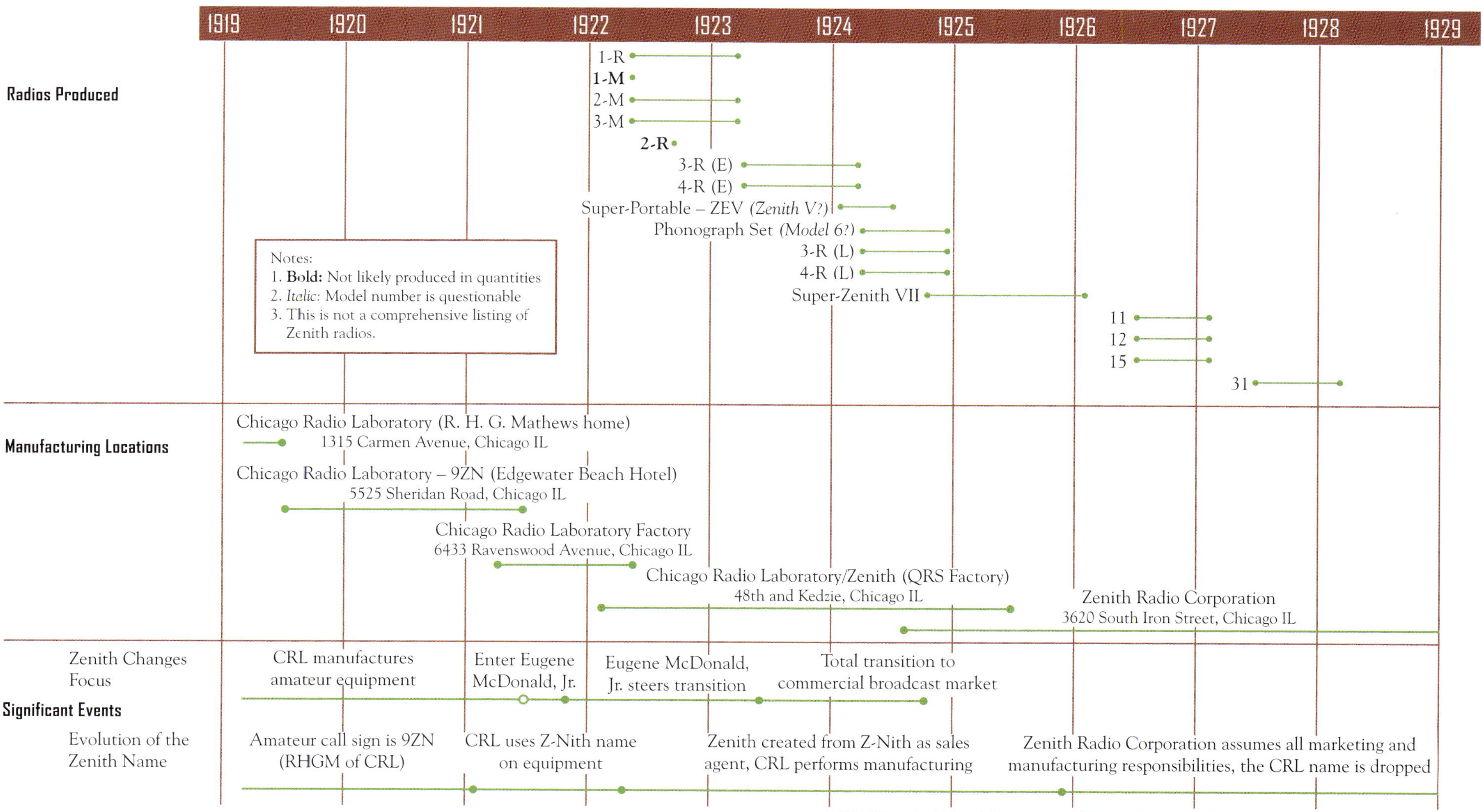

Section I

Radio Descriptions

Chapter 1: **Zenith 1-R**

1-R General Specifications

Company:	Zenith Radio Corporation
Model:	Radiophone 1-R
Year Manufactured:	1922
Manufacture Location:	48th and Kedzie, Chicago IL (QRS Factory)
Serial Number:	4183
Price:	$75
Tube:	UV-201-A (1), one detector
Circuit Type:	Regenerative

1-R General Information

The Zenith Radiophone 1-R was the first Zenith radio. With its novel design the 1-R helped Zenith make the transition from producing equipment for the limited amateur market to manufacturing radios for the fast growing commercial broadcast industry.

This radio has some unique design features. The overall outside appearance is very attractive with its sloping simulated wood panel, dark mahogany cabinet and the lack of visible tap switches, which are hidden from view on internal sub-panels. Fine tuning (vernier) is especially smooth and accomplished with gear reduction mechanisms that drive the tickler (regeneration) and variable capacitor (tuning).

Restoration details for the set that was used as the example in this chapter. *This 1-R radio (SN 4183) has been completely restored to working condition. The original gears were badly damaged and were replaced with pulleys, made of plastic, similar to the pulleys that are used in the 3-R. These plastic pulleys and some other replacement parts were provided by the late Mr. Keith Perry. Three of the seven pointers were replaced with handmade pointers.*

1-R Physical Description Quick Reference Guide

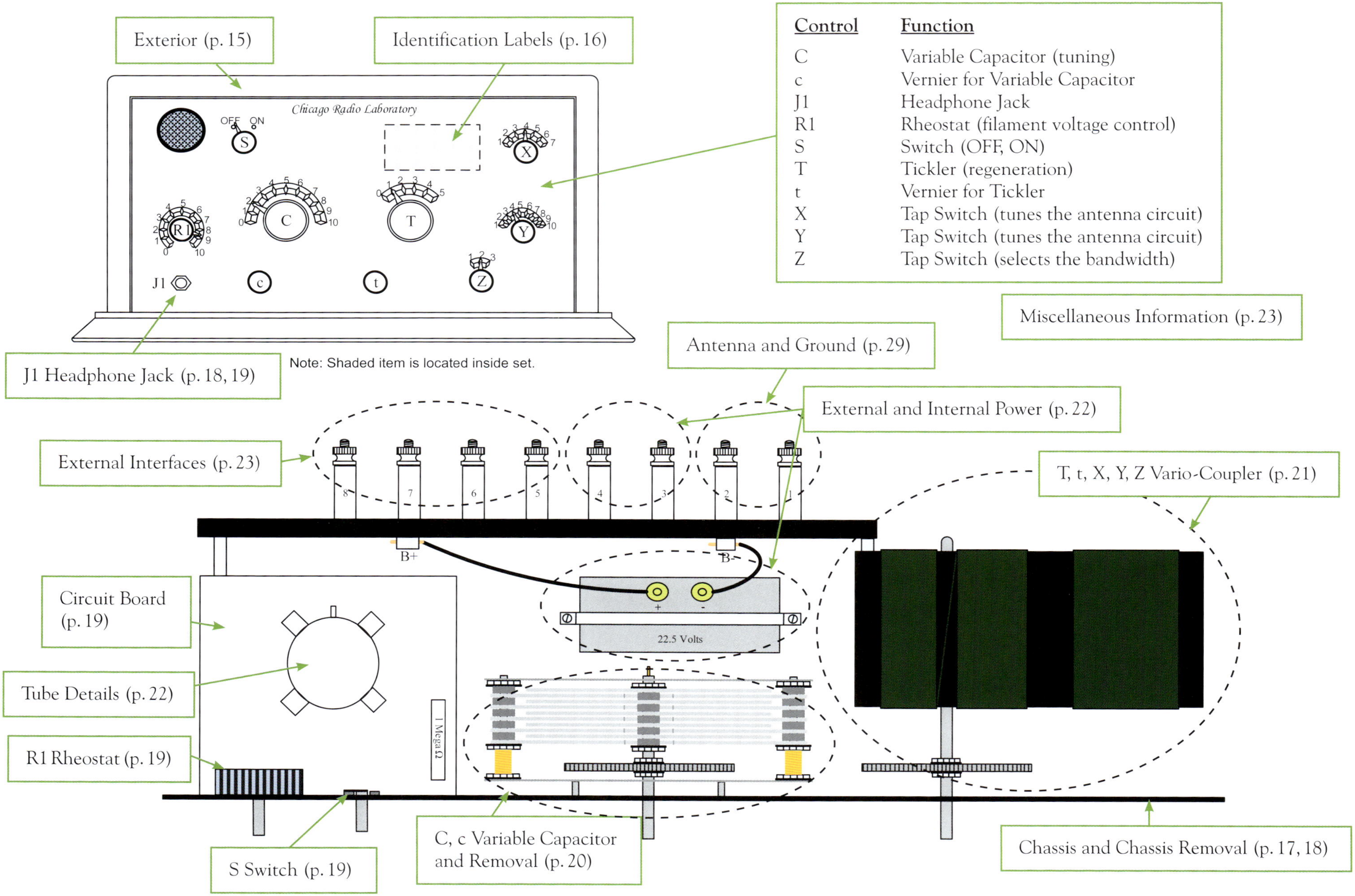

Control	Function
C	Variable Capacitor (tuning)
c	Vernier for Variable Capacitor
J1	Headphone Jack
R1	Rheostat (filament voltage control)
S	Switch (OFF, ON)
T	Tickler (regeneration)
t	Vernier for Tickler
X	Tap Switch (tunes the antenna circuit)
Y	Tap Switch (tunes the antenna circuit)
Z	Tap Switch (selects the bandwidth)

1-R Exterior

The cabinet of the 1-R receiver is made of mahogany and measures 23" by 11" by 9.75". The front panel slopes at an angle of approximately 10° and is made of brown Bakelite with faux wood grain. The panel also has a peep hole that allows the user to view the brightness of the tube.

Front View

Chicago Radio Laboratory logo and Armstrong patent information.

Top View. Zenith decal is discolored due to the fading of the lacquer overcoat.

Rear View. Posts for connecting the antenna, ground, "A" battery and optional AF amplifier.

1-R Identification Labels

Rear View (door open)

Inspection and Test Sticker

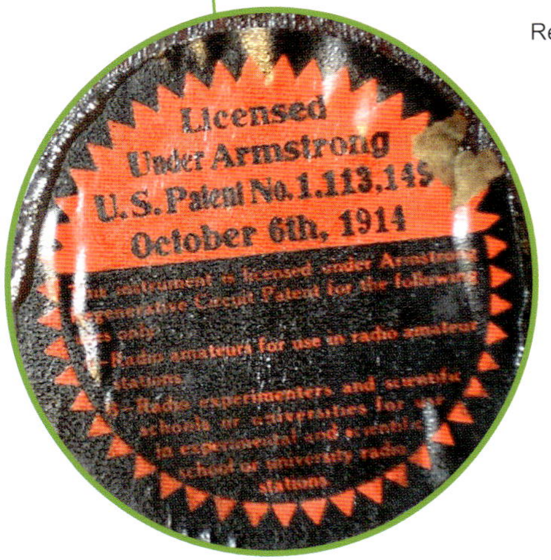

Armstrong License and Patent Label

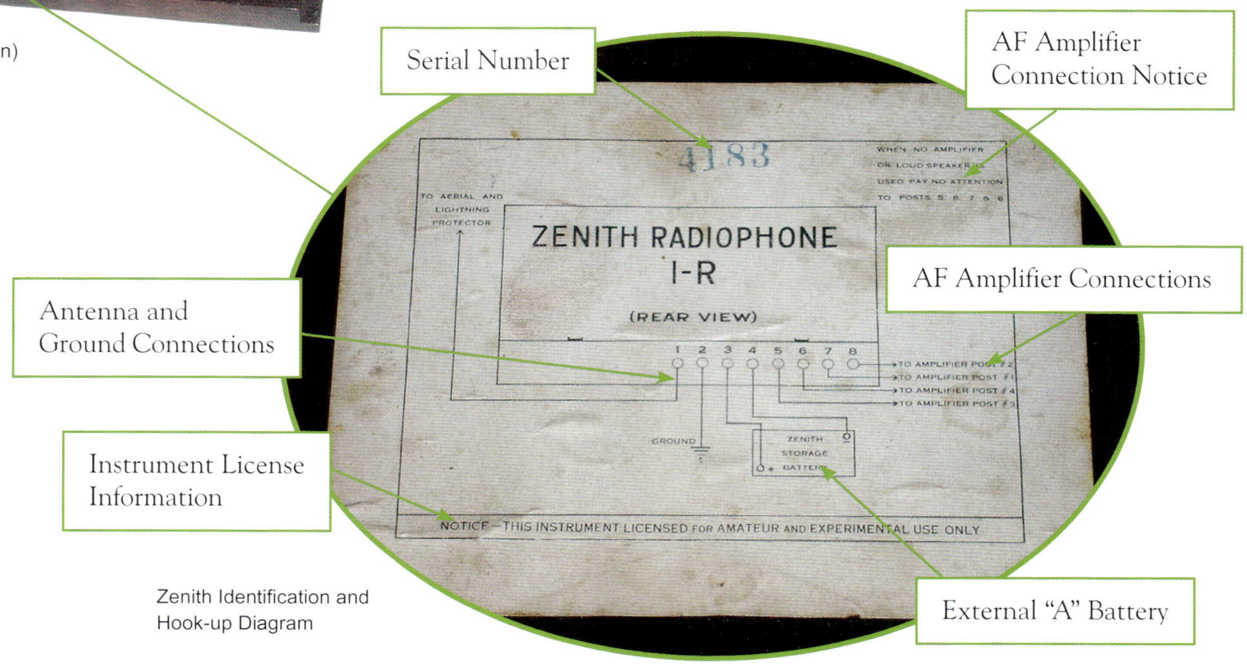

Zenith Identification and Hook-up Diagram

1-R Chassis

Rear View

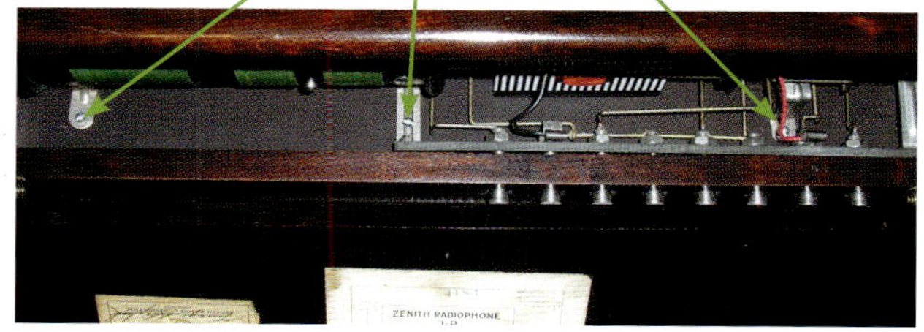

Top View (looking into cabinet with back door open)

"B" Battery

Front View

Removal of Chassis from Cabinet

1. Inside the cabinet, remove the "B" battery and bracket. Remove the three screws that hold the chassis to the bottom of the cabinet. See "B" Battery and Top View.
2. On the front panel, remove the four corner screws with the raised washers. Remove the three small screws across the top of the front panel. See Front View.
3. The whole chassis slides out the front of the cabinet. Raise the front panel above the bottom lip and gently push chassis forward from the rear.

To reinstall the chassis into the cabinet, reverse the removal process.

1-R Chassis (continued)

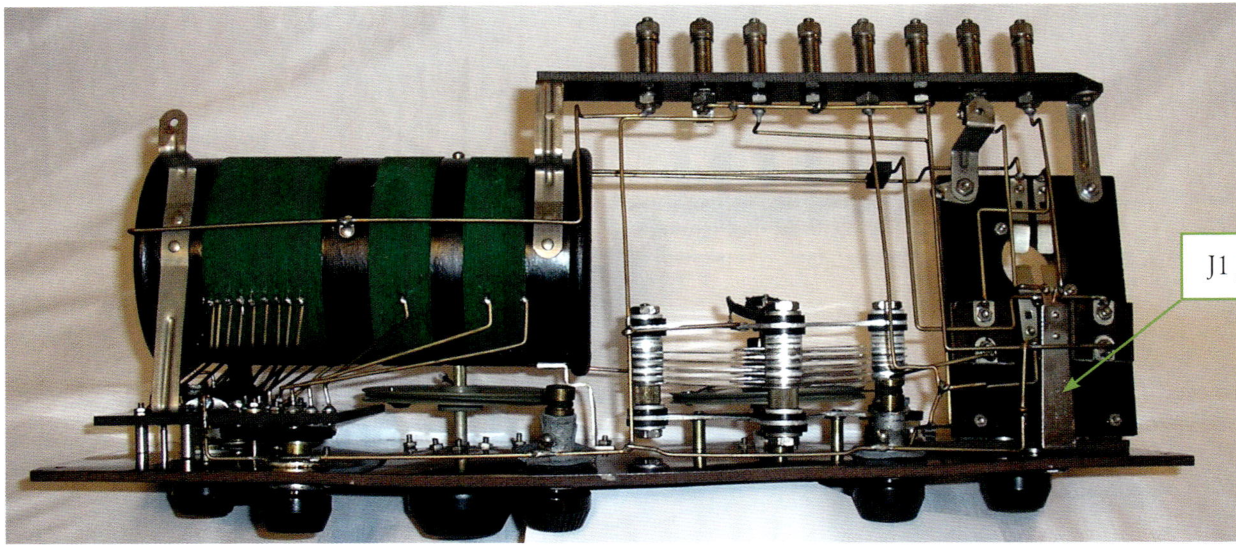

Underside View

Right Side View

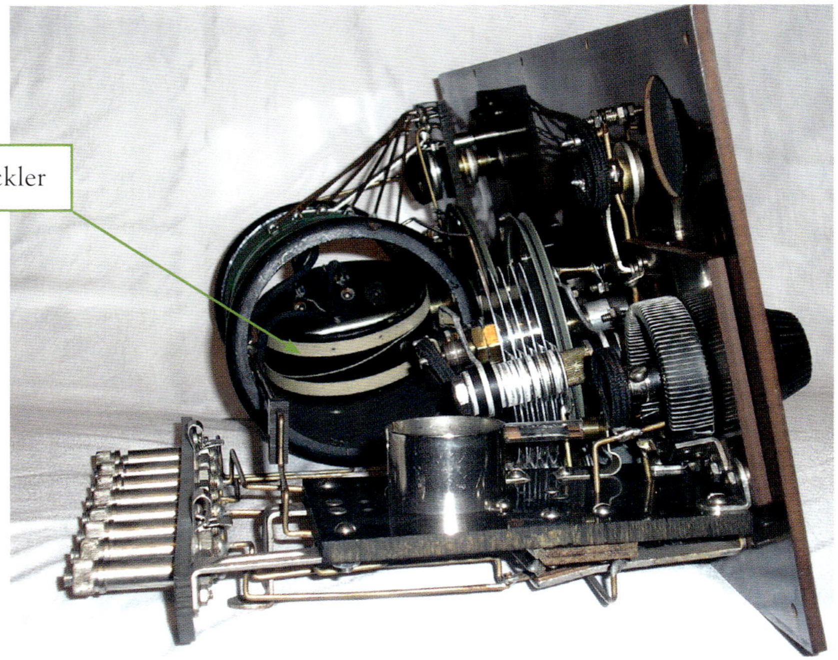

Left Side View

1-R Circuit Board

Top View

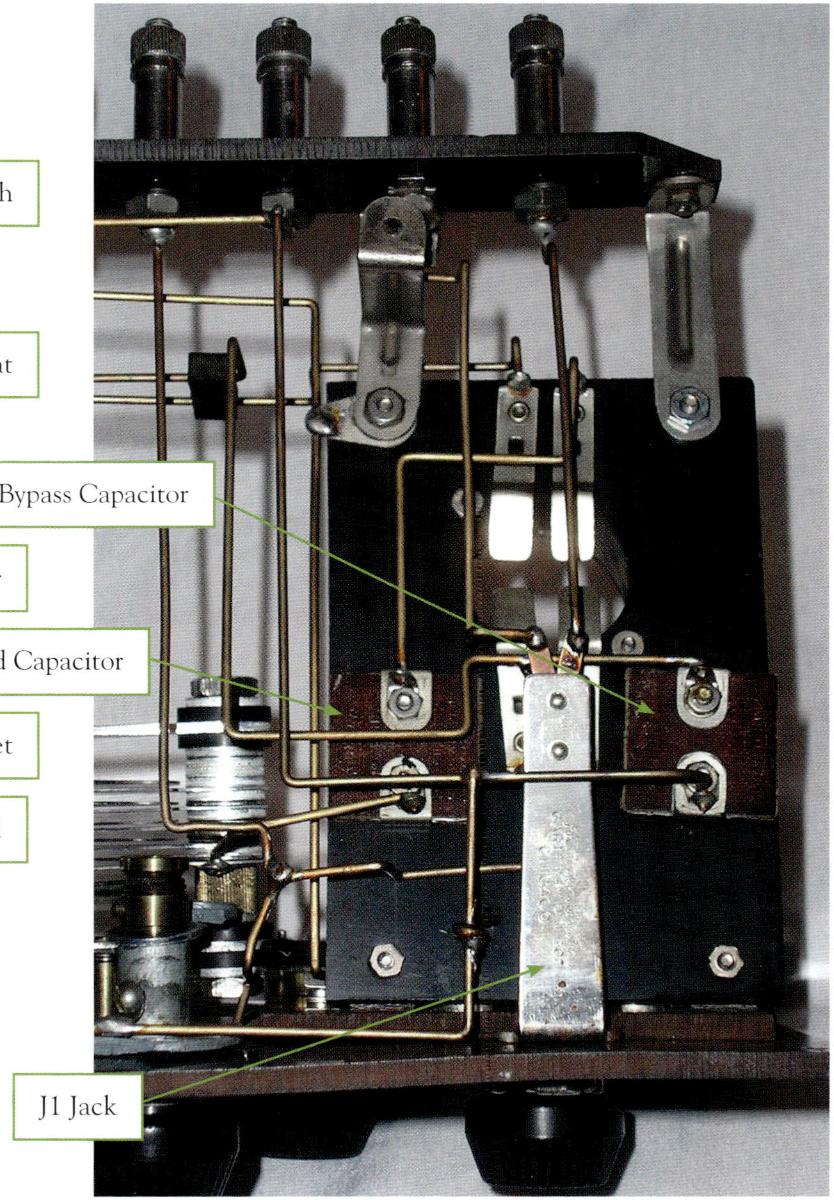

Underside View

1-R Variable Capacitor

The variable capacitor housing holds a shaft (rotor) that has 5 plates and a large gear that is driven by a smaller vernier gear. The shaft rotates the 5 plates, merging them with the 5 fixed plates (stator). The large gear, small gear and shaft are made of pot metal and usually need replacing. The original gears in this set were no exception, and were replaced with pulleys similar to those used in the 3-R receiver. The pulleys can be clearly seen in the figures titled Top View and Underside View. See 1-R Gear Replacement Modification (page 245) for instructions on how to perform this alternate repair.

Top View

Underside View

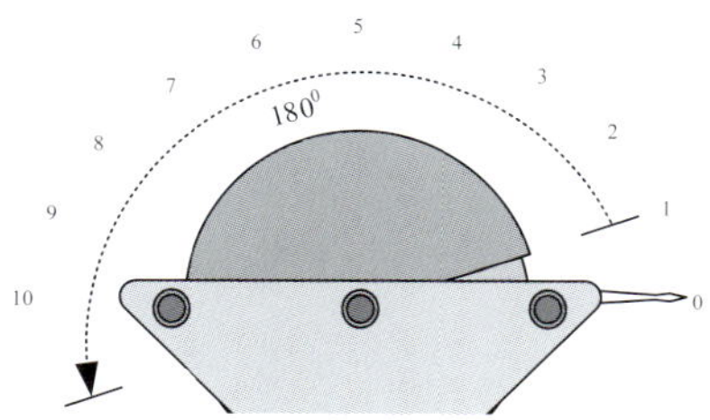

Front View

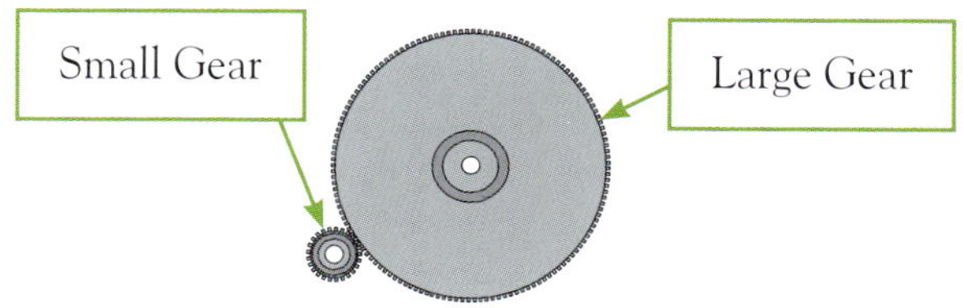

Variable Capacitor Gears

Variable Capacitor Clocking (Viewed from Rear)
Initial starting position of the rotating plates and C pointer in relation to the numbers on the front panel. As the Variable Capacitor rotates 180° right-to-left from the position shown the pointer rotates from 0 to 10, respectively. The pointer and logging numbers are located on front panel, but are shown for clarity

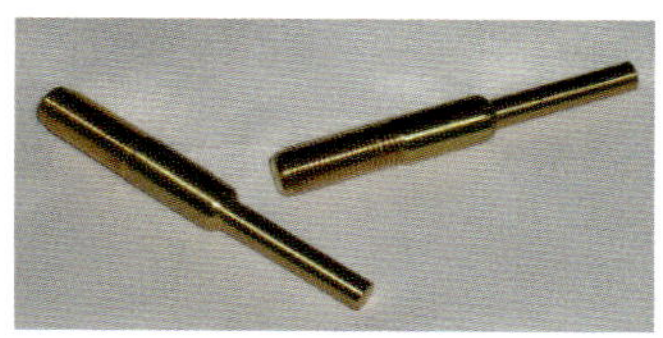

Reproduced Variable Capacitor Shafts

Removal of Variable Capacitor

1. Disconnect the 2 bus wires. See Top View and Underside View.
2. On the front panel, remove the two stops and screw. See Front View.
3. Gently remove the variable capacitor assembly from the chassis.

To reinstall the variable capacitor, reverse the removal process.

1-R Vario-Coupler

The Vario-Coupler has a shaft that rotates a small tickler coil which is mounted inside the secondary winding of the main coil. The shaft has a large gear that is driven by a smaller Vernier gear. These gears were made of pot metal, so usually require replacement, and the set used in this example was no exception. The original gears on this radio were replaced with pulleys similar to those used in the 3-R receiver. The pulleys are clearly visible in the photographs titled Top View and Underside View. See 1-R Gear Replacement Modification (page 245) for instructions on how to perform this alternate modification.

Top View

Underside View

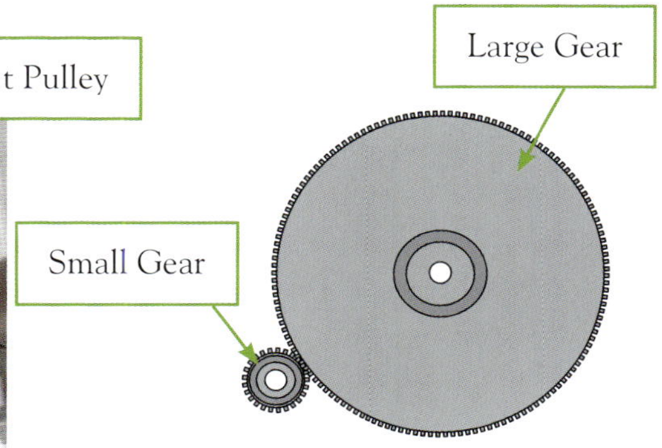

T Tickler Gears

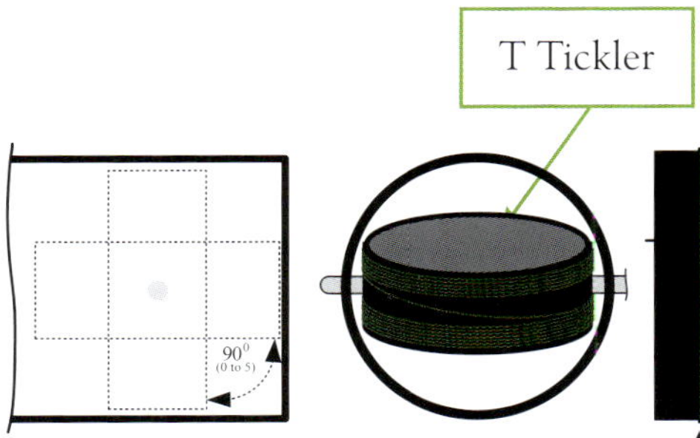

Tickler Clocking

When viewed from front the tickler rotates from horizontal, down to vertical, and the T pointer moves from 5 to 0, respectively.

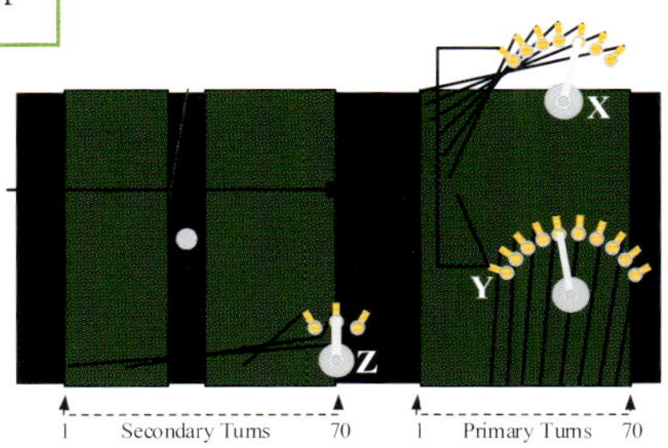

Front View

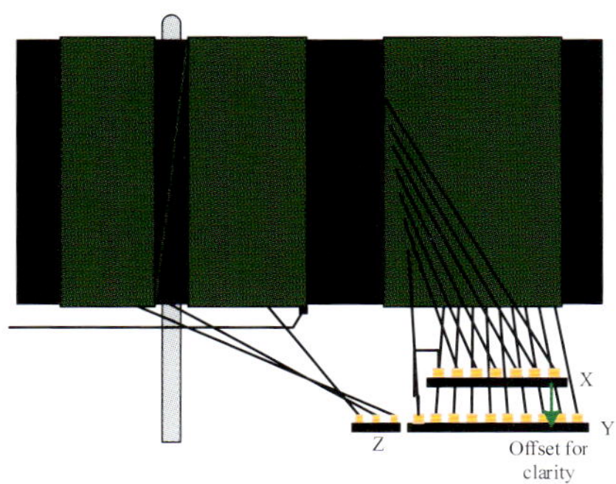

Top View

Vario-Coupler Windings and Tap Switch Connections

The main coil in the 1-R consists of a primary and secondary coil each having 70 turns. The secondary coil is physically wound as two coils separated by ½". The coil is constructed in this manner to facilitate the mounting of the small tickler coil inside the main coil. The X-switch tap points 1 through 7 are connected to turns 7 through 1 of the primary coil, respectively. In a like manner, Y-switch tap points 1 through 10 connect to turns 8, 14, 21, 28, 35, 42, 49, 56, 63, and 70 of the primary coil. Additionally, the Y-switch tap point 1 is connected to X-switch tap point 1. Tap-switch Z is connected to the turns of the secondary coil as follows: 1 (tap 3), 21 (tap 2) and 51 (tap 1). The tickler coil T has two connected sets of windings, separated by ½", with each coil section having 13 turns (no taps).

1-R Internal and External Power

Plate voltage for V1 tube is provided by an internal 22.5-volt "B" battery. The filament voltage is provided from an external "A" battery and is connected via binding posts 3 and 4. See the photographs and drawings below.

Internal "B" Battery

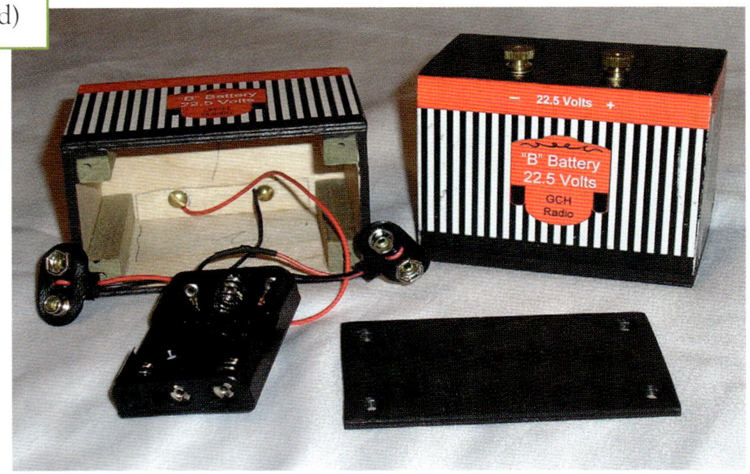

Reproduced "B" Battery. Uses 2 9-volt and 3 AAA batteries.

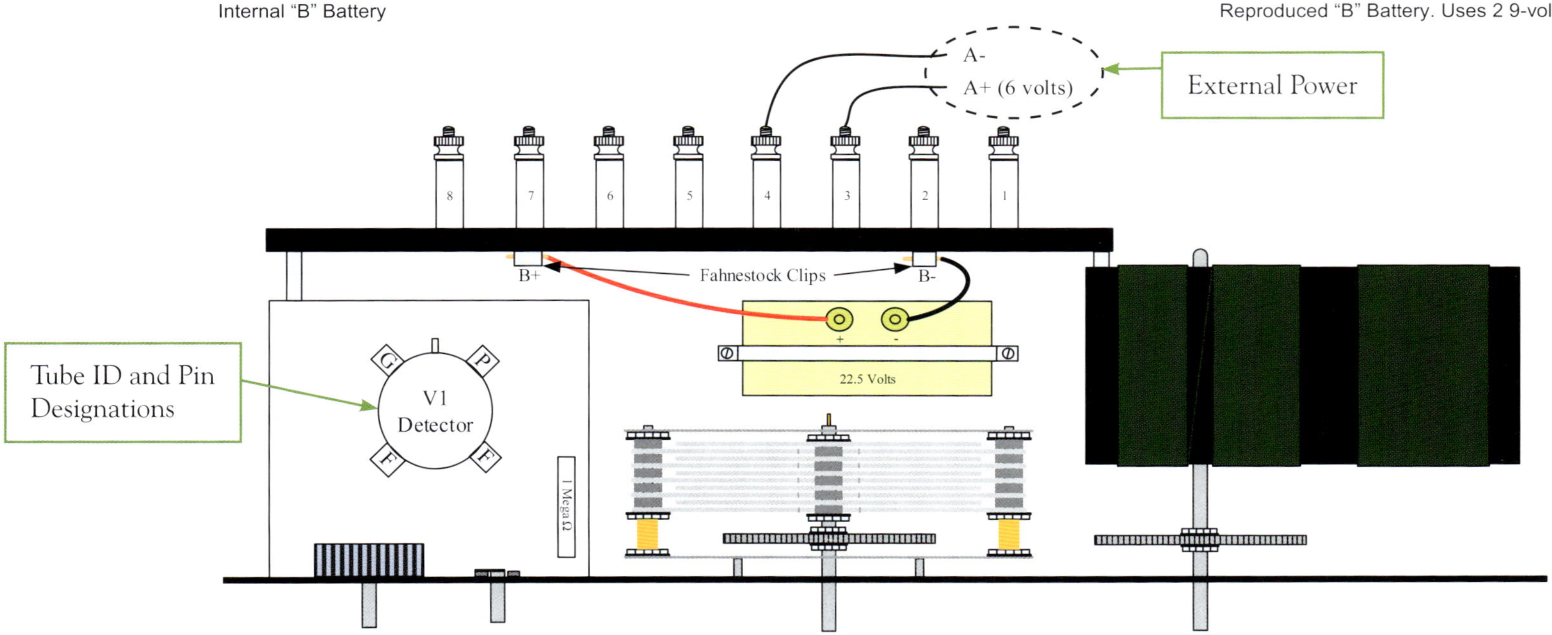

Chassis Top View. Location of the internal "B" battery, battery bracket and battery connections. The battery bracket is fastened to the bottom of the cabinet.

1-R External Interfaces

The 1-R has an internal "B" battery that provides 22.5 volts to the plate of V1. This is ample power for the standard headphones, but not enough to drive a horn speaker. To address this problem, Zenith offered two optional matching AF amplifiers. The 2-M provides two stages of AF amplification and the 3-M provides three stages of AF amplification. The connections for attaching the 2-M or 3-M AF amplifier to the 1-R radio are shown below. Also included is a table that shows what type of signal is available on each binding post. Notice that the connections are exactly the same for the 2-M or 3-M. See Chapter 2—Zenith 2-M, page 31, and Chapter 3—Zenith 3-M, page 46, for descriptions of these AF amplifiers.

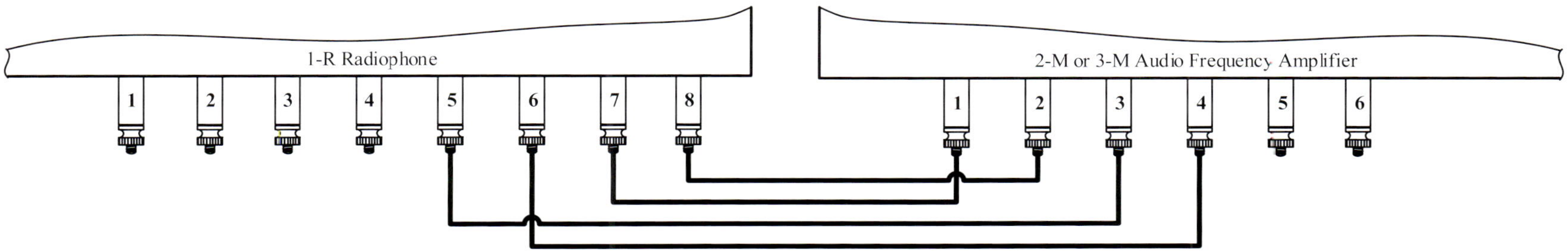

1-R to 2-M or 3-M Connections

Zenith 1-R Interface Table

1-R	Post 1	Post 2	Post 3	Post 4	Post 5	Post 6	Post 7	Post 8		
2-M or 3-M					Post 3	Post 4	Post 1	Post 2	Post 5	Post 6
Signal Type	Antenna	Ground	A+	A−	A+	A−	Audio	Audio	B+	B−

1-R Miscellaneous Information

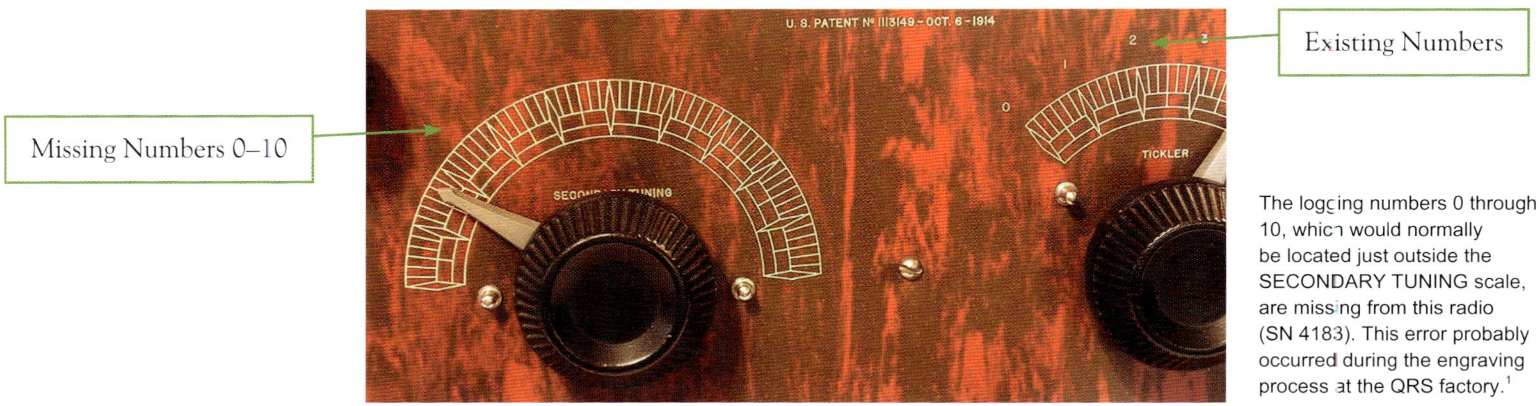

The logging numbers 0 through 10, which would normally be located just outside the SECONDARY TUNING scale, are missing from this radio (SN 4183). This error probably occurred during the engraving process at the QRS factory.[1]

1-R Functional Description Quick Reference Guide

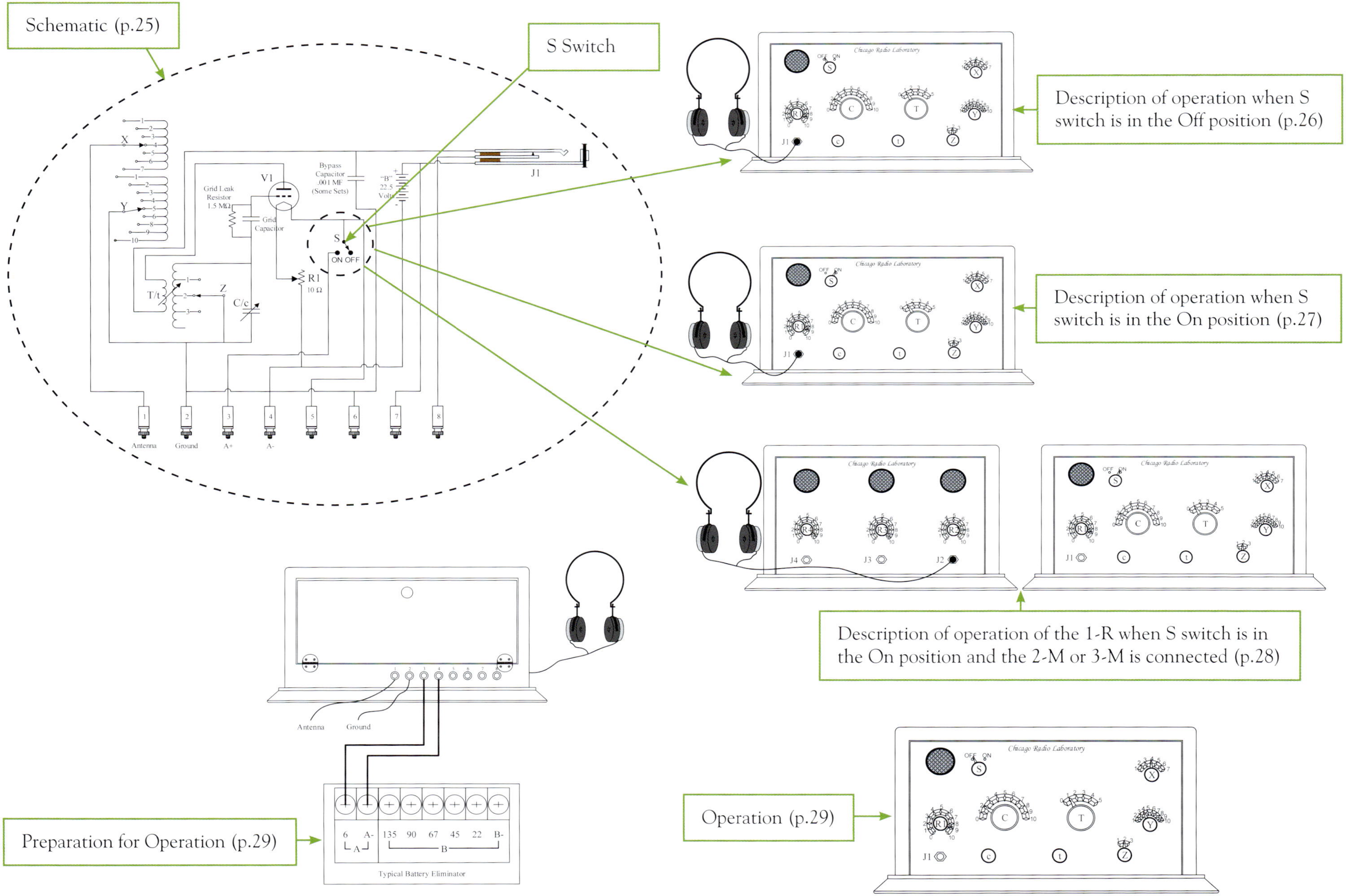

1-R Schematic

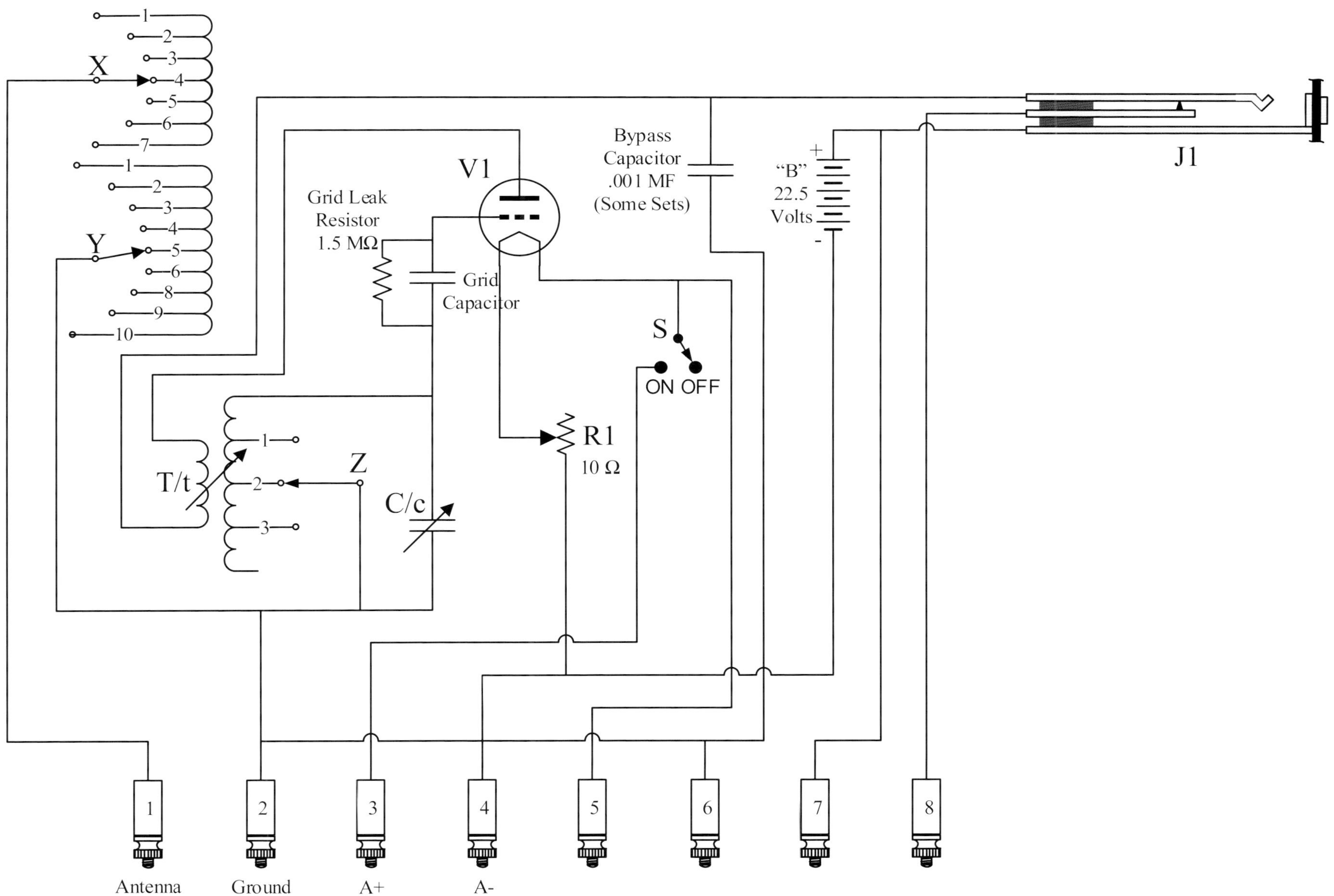

1-R Description of Operation—Switch S Off

When switch S is in the OFF position the filament circuit is open and the radio is inoperative. Plate voltage is applied to the plate of V1. This is considered the initial state of the radio for future discussions of circuit operation.

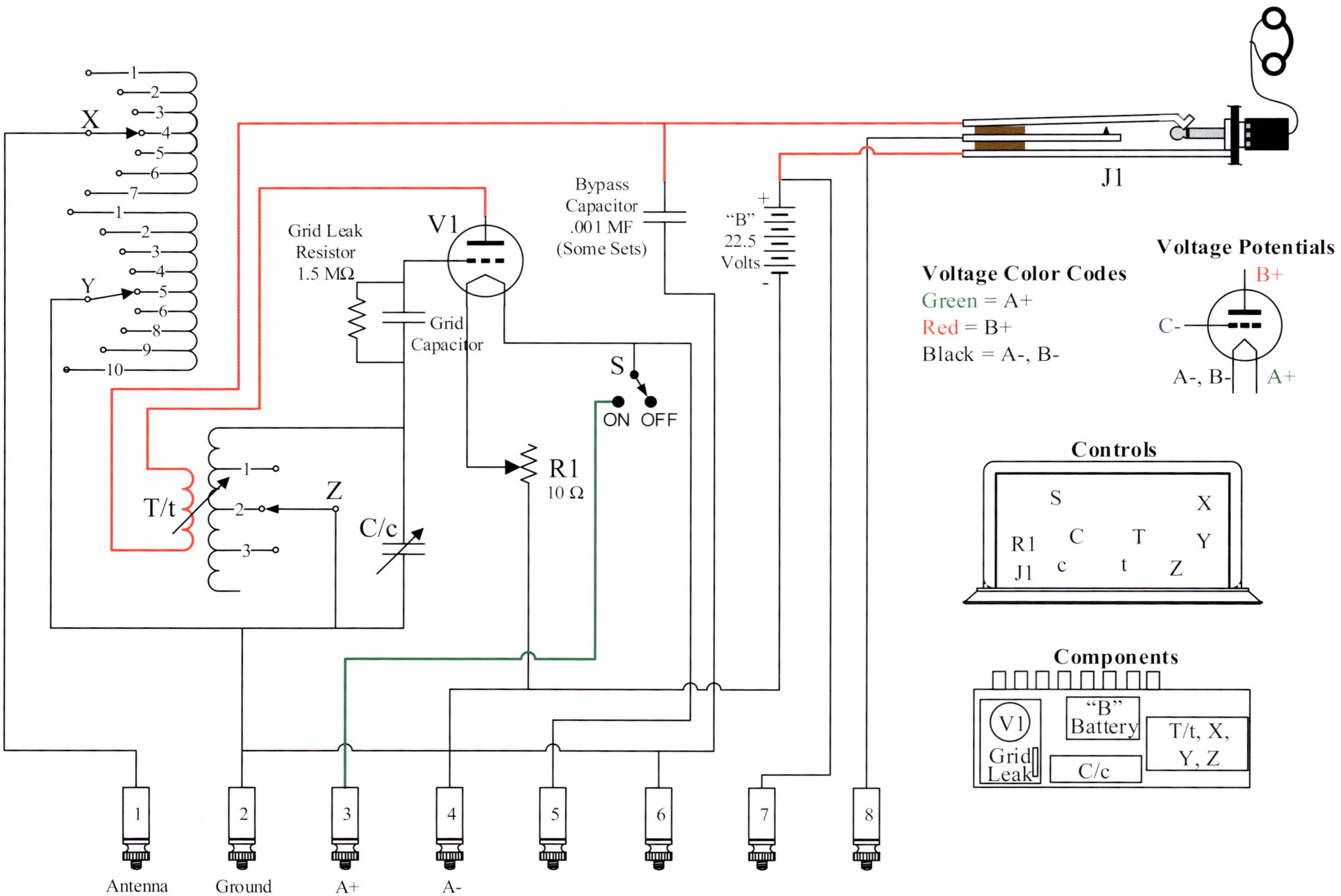

1-R Description of Operation—Switch S On

When switch S is moved to the ON position, the filament circuit is closed and current begins to flow. The radio is now operational and ready for adjustment. The variable capacitor, knob C, is rotated until the desired signal is found. That signal from the antenna (post 1) is inductively coupled to the grid circuit of V1. A portion of the signal from the plate of V1 is fed back (regeneration) to the tube's grid circuit through the tickler coil. The amount of regeneration is controlled by rotating the tickler coil, knob T, located inside the secondary of the main coil until the signal is clear. Tube V1 removes the RF signal (detection) and amplifies the remaining AF signal. The output signal from the plate of V1 is coupled directly to the headphones which convert the electrical impulses into sound waves.

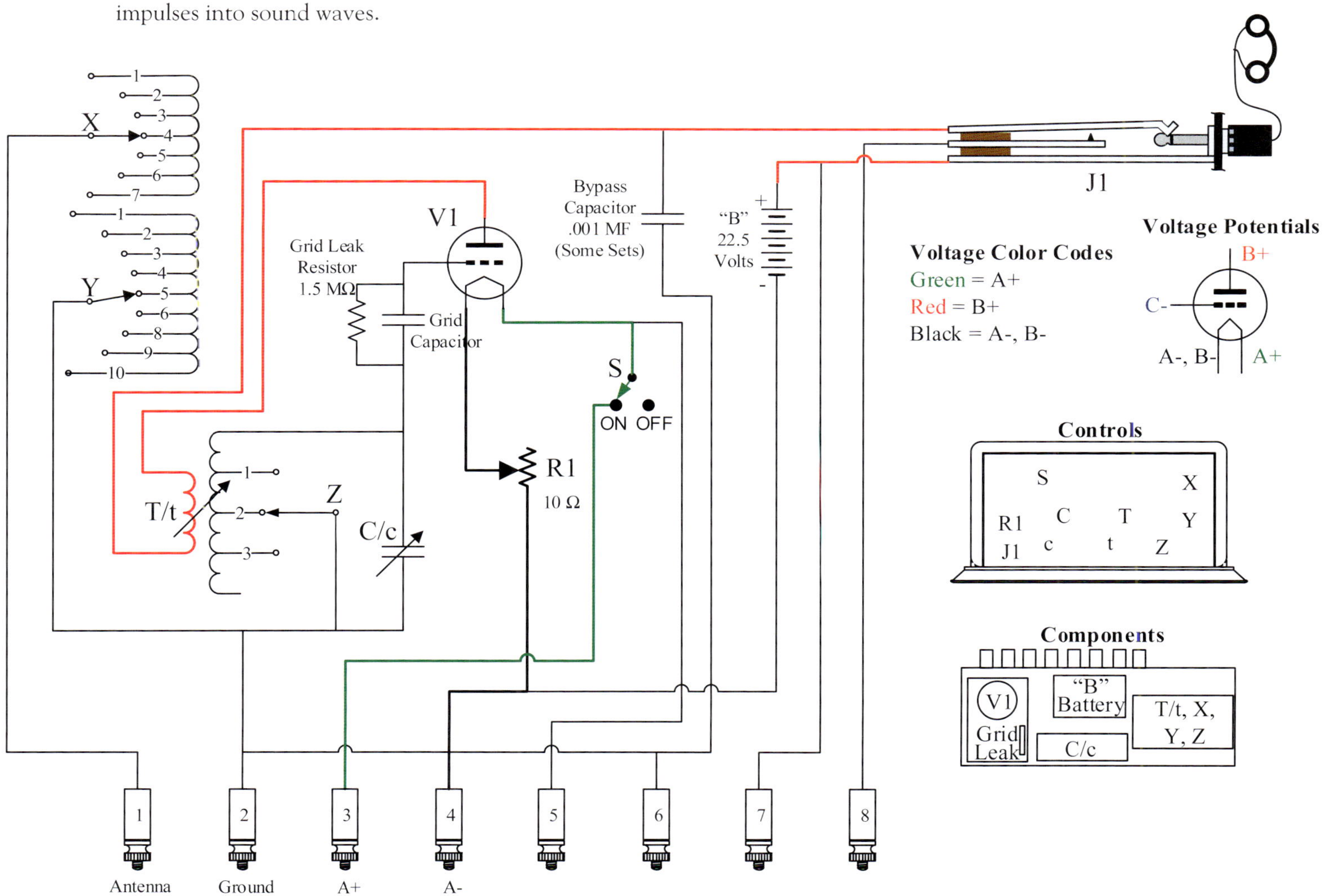

1-R Description of Operation—Switch S On (With AF Amplifier Connected)

When a 2-M or 3-M AF amplifier is connected to the 1-R receiver, remove the phone plug from jack J1, which will apply B+ to binding posts 7 and 8 of the radio. These posts are connected to binding posts 1 and 2 of the AF amplifier, respectively. Binding post 8 on the radio also provides the audio signal to the amplifier. In a like manner, the filament voltage for the amplifier is provided from binding posts 5 and 6 of the radio to terminals 3 and 4 on the AF amplifier. The 1-R now operates as described on the previous page. See Chapter 2—Zenith 2-M, page 31, or Chapter 3—Zenith 3-M, page 46, for descriptions of the operation of these AF amplifiers.

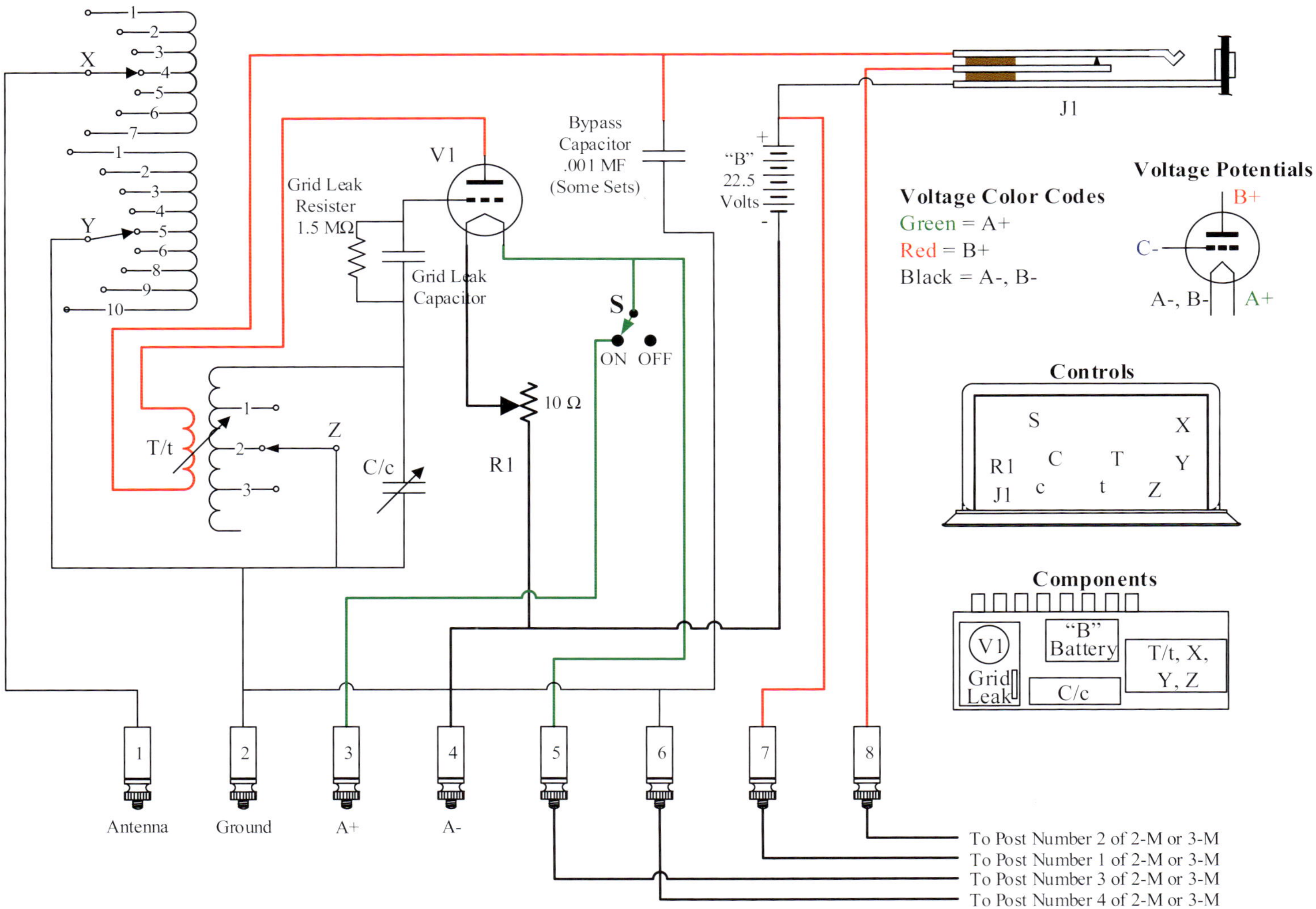

1-R Preparations for Operation

Install the tube and the "B" battery inside the radio. Connect the antenna to Post 1 and the ground to Post 2. Use the battery connection diagram, inside the back door of the radio, to make sure the power will be connected properly and agrees with the 1-R Connections diagram below. If the power connections are not correct or otherwise unavailable, utilize the physical and functional descriptions to assist in determining the correct power connections. Hook up the power to the radio. Proceed to 1-R Operation. See Common Radio Attachments (page 8) for additional information related to antennas, loudspeakers and power options.

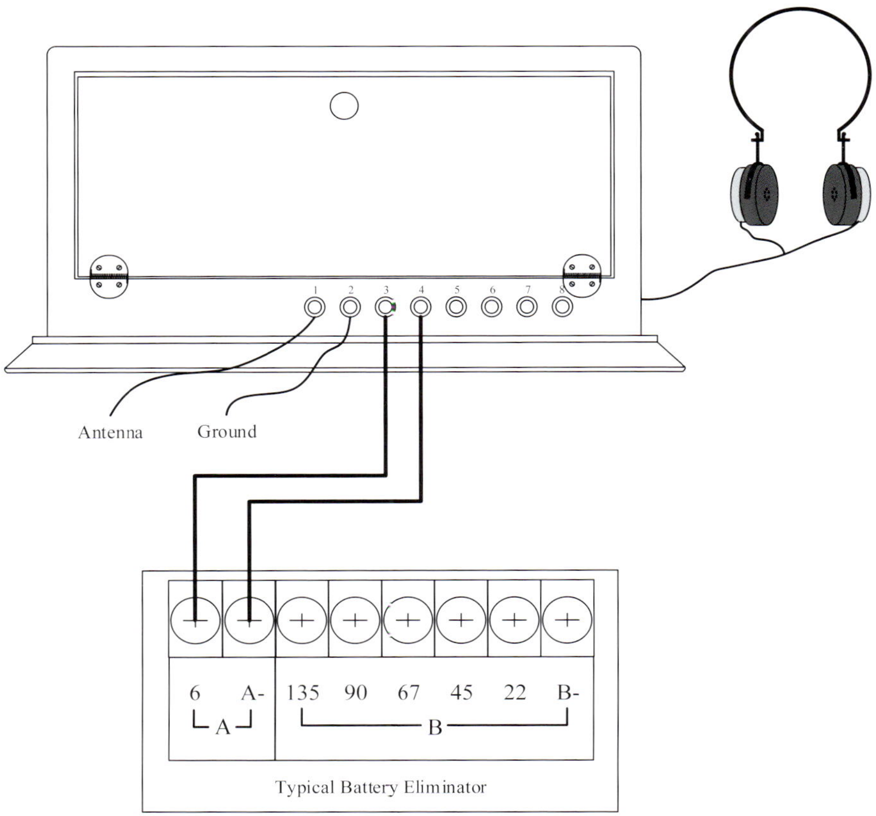

1-R Connections

1-R Operation

Follow the 1-R Preparations for Operation instructions. Set the pointers to the positions shown in the Initial Dial Settings diagram below.

The tuning capacitor C selects the stations and the T tickler provides the feedback (regeneration) that increases the volume of the selected signal. Practice using knobs C and T simultaneously. Utilize the small c and t knobs to make fine adjustments to the C and T mechanisms.

Insert the headphones plug into the J1 jack. Turn the "S" switch to the ON position.

Turn "C" knob to the right until the desired signal is heard. Now gradually turn "T" knob to the right until the volume improves. Make small adjustments to C and T knobs until the signal is heard the loudest and clearest. If T knob is turned too far to the right, the signal will be lost and an unpleasant howl will be heard. To correct this condition, slowly rotate knob "T" to the left until the signal is clear.

Knobs X and Y tune the antenna circuit. A little manipulation of these switches will enable the operator to obtain improved results.

Knob Z changes the bandwidth of the receiver and is generally left at position 2. Position 1 decreases the bandwidth and position 3 increases the bandwidth.

Record the control settings for each station in the Station Log (next page). Then, to return to a particular station, set the controls to the settings that were previously recorded in the log for that station.

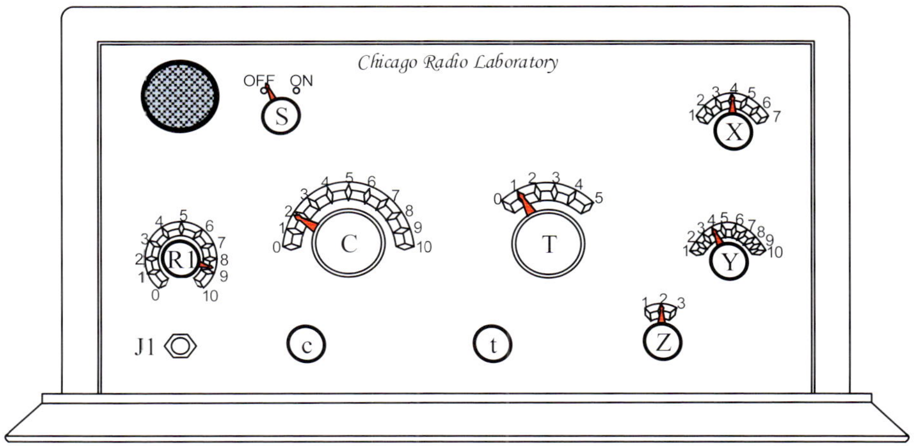

Initial Dial Settings (pointers shown in red for clarity)

1-R Station Log

Name: ______________________________

Date	Time	Station	Frequency	C	T	X	Y	Z

Chapter 2: **Zenith 2-M**

2-M General Specifications

Company:	Zenith Radio Corporation
Model:	2-M
Year Manufactured:	1922
Manufacture Location:	48th and Kedzie, Chicago IL (QRS Factory)
Serial Number:	3845
Price:	$50
Tube:	UV-201-A (2)
Circuit Type:	AF amplifier

2-M General Information

Zenith offered the 2-M and 3-M as optional AF amplifiers for the 1-R radio. The 2-M was usually bought with the 1-R, as a matched set. Most 1-R radios that have been found over the years were found with the matching 2-M. The 2-M presented here and the 1-R in chapter 1 were found together and are believed to be a matching set as they show very similar wear and fading. According to the serial numbers (4138, 3845) of these two sets, the 1-R and 2-M were being sold at a fairly even rate of over 92%. The remaining 8% were probably sold as some combination of individual 1-R radios or 1-R and 3-M combinations.

Since a schematic is not available, one was created utilizing this 2-M (3845). This 2-M schematic may not match exactly what the manufacturer would have drawn, but it should be electrically and functionally equivalent.

Restoration details for the set that was used as the example in this chapter. *The only parts on the 2-M that are made of pot metal are the pointers. The pointers on this set were only slightly affected and were not replaced. The AF transformers were good. The set was thoroughly cleaned.*

Zenith evidently used more than one type of "B" battery bracket for the 2-M. The "B" battery bracket that was in this set is believed to be original. Another 2-M set was found that has a different type of "B" battery bracket, but it could not be completely identified because it was partially missing.

2-M Physical Description Quick Reference Guide

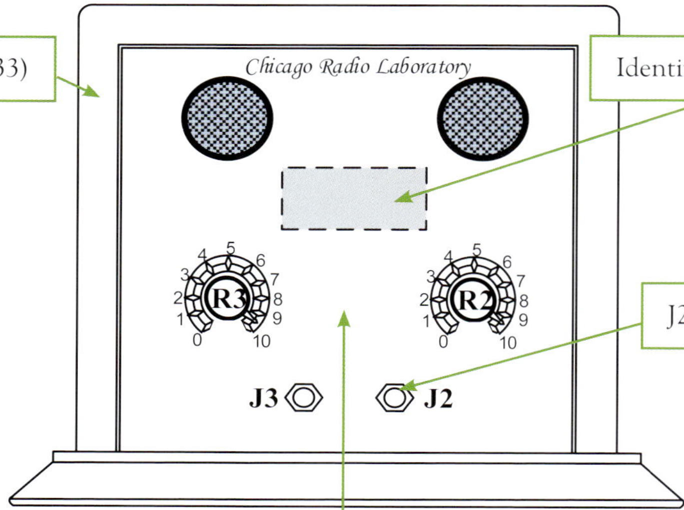

Note: Shaded item is located inside set.

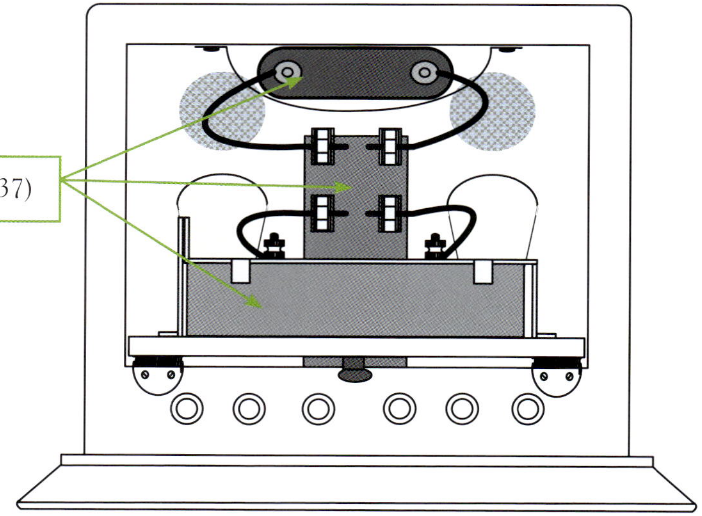

Control	Function
R2	Rheostat (V2 filament voltage control)
R3	Rheostat (V3 filament voltage control)
J2	Headphone Jack (one AF amplifier stage)
J3	Headphone Jack (two AF amplifier stages)

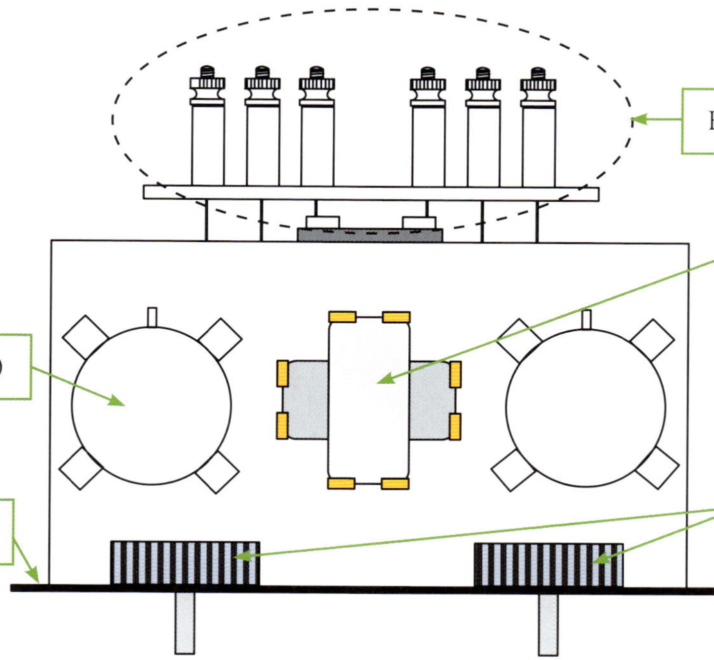

2-M Exterior

The cabinet is made of mahogany and measures 12" x 11" x 9.75". The front panel slopes at an angle of approximately 10° and is made of brown Bakelite with faux wood grain. There are two peep holes for viewing the brightness of the tubes.

Front View

Chicago Radio Laboratory Logo

Top View. Zenith decal is discolored due to the fading of the lacquer overcoat.

Rear View

2-M Identification Labels

Rear View

Inspection and Test
(This 2-M and the 1-R in chapter 1 were inspected and tested by the same person.)

Serial Number

1-R Connections

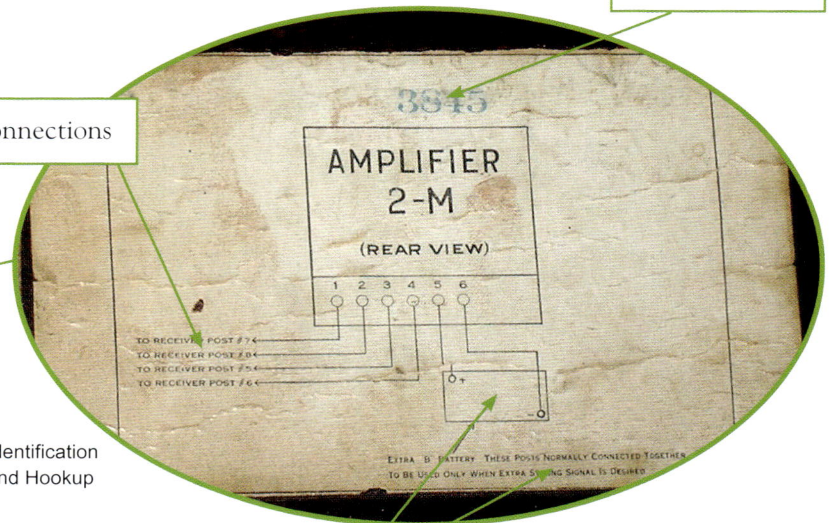

Identification and Hookup

"B" Battery connections and shorting instruction, which states, "EXTRA B BATTERY THESE POSTS NORMALLY CONNECTED TOGETHER. TO BE USED ONLY WHEN EXTRA STRONG SIGNAL IS DESIRED."

2-M Chassis

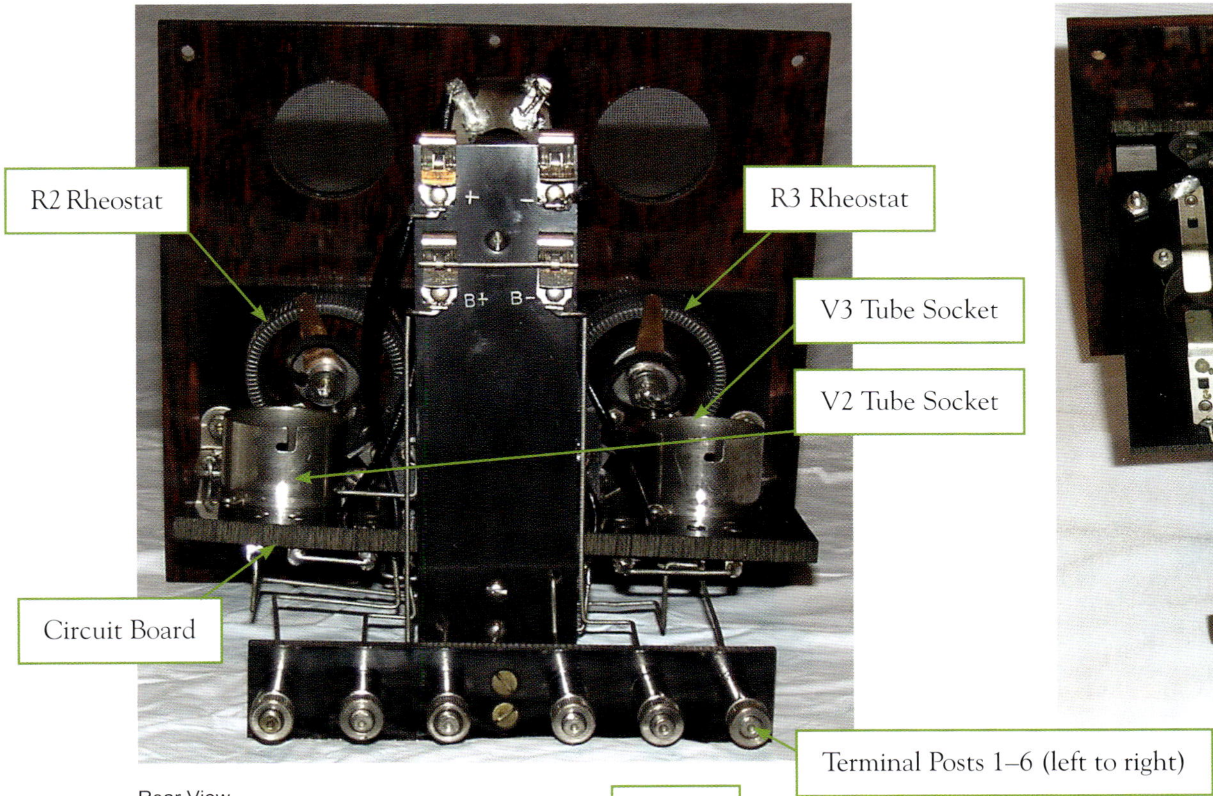

Rear View

Underneath View

Battery View

Front View

Removal of Chassis from Cabinet

1. Inside the rear of the cabinet, disconnect the "B" and "C" batteries. See Battery View.
2. On the front panel, remove the four corner screws with the raised washers and the small screw located at the top. See Front View.
3. The whole chassis slides out the front of the cabinet. Raise the front panel above the bottom cabinet lip and gently push the chassis forward from the rear.

To reinstall the chassis into the cabinet, reverse the removal process.

2-M Chassis (continued)

Top View

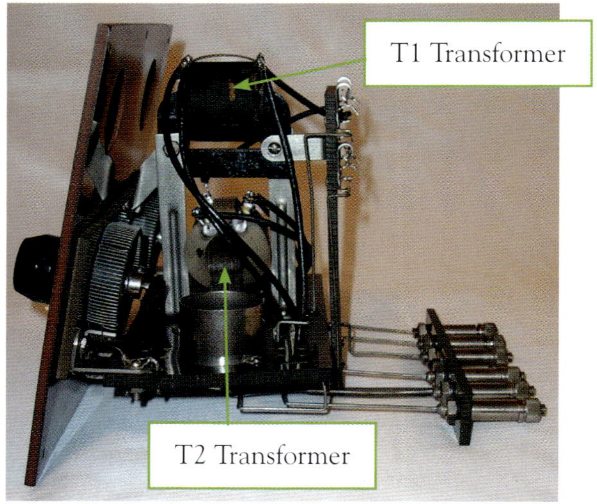

Right Side View

Left Side View

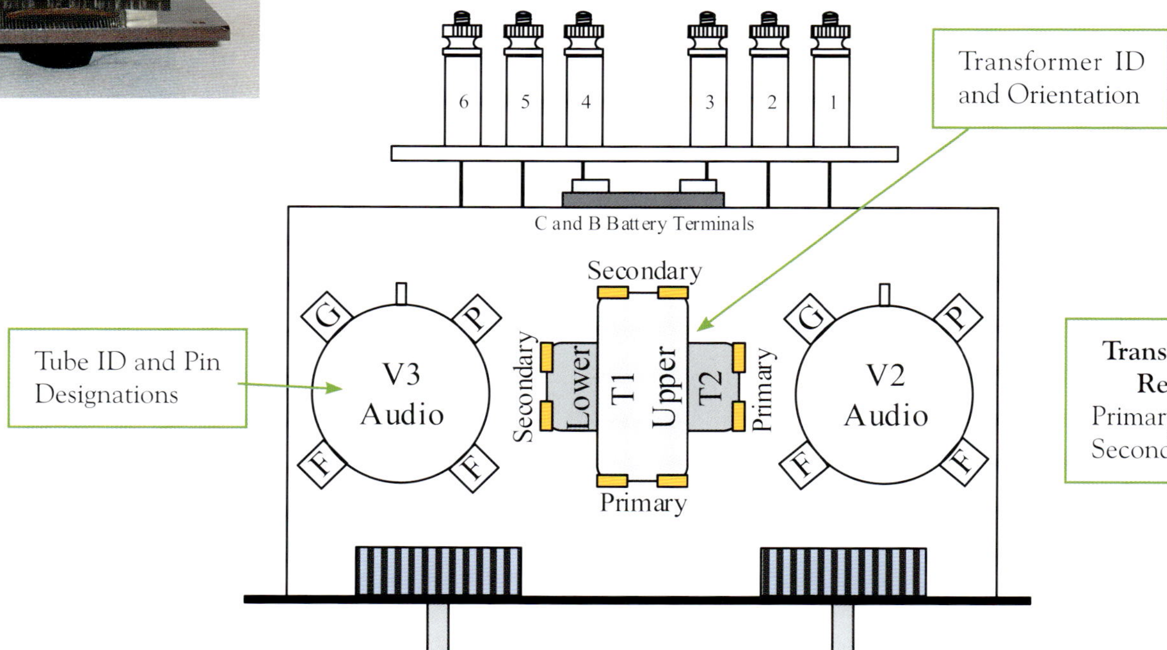

Transformer DC Resistance
Primary ≈ 1.4 KΩ
Secondary ≈ 7 KΩ

Component Layout

2-M Internal Power

The 2-M uses two different internal batteries. The "B" battery provides the plate voltage to each tube and the "C" battery provides the negative bias to the grid of each tube; see Internal Batteries below. There are separate terminals for connecting each battery, as shown in Internal Battery Connections below.

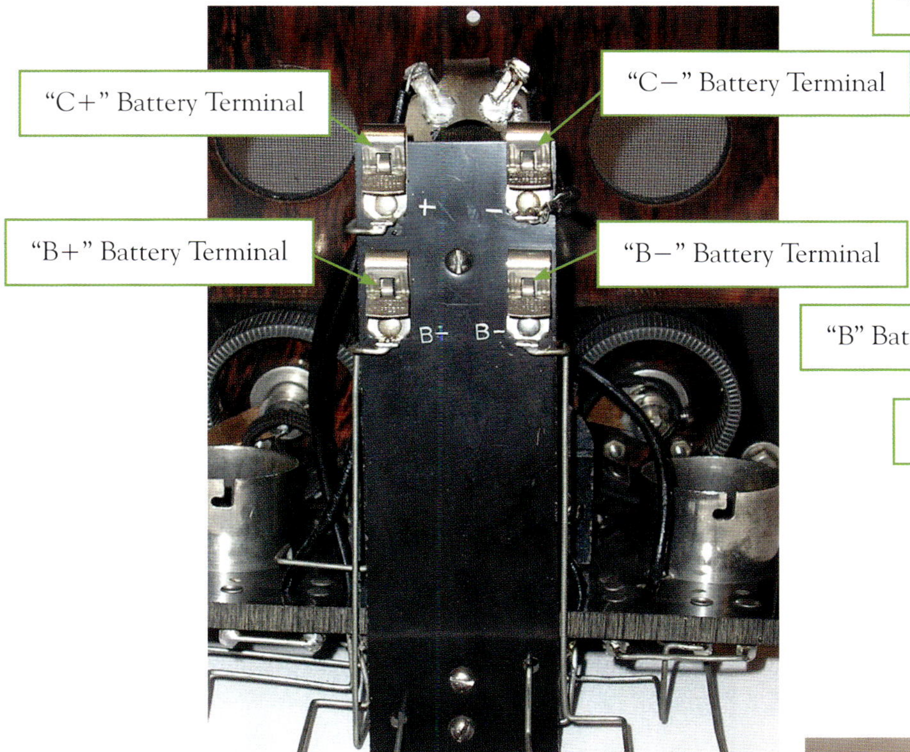

Internal Battery Connections

Internal Batteries

Reproduced "B" and "C" Batteries

2-M External Interfaces

The 1-R and the 2-M are electrically tied together with four wires that are connected to the terminal posts located at the rear of the sets. The 2-M gets its "A" voltage and AF signal from the 1-R. The 2-M also has external terminals (posts) for connecting additional "B" batteries. The purpose of the extra or external "B" battery is to provide additional plate voltage to the tubes, in order to get a stronger signal or for operating a loudspeaker.

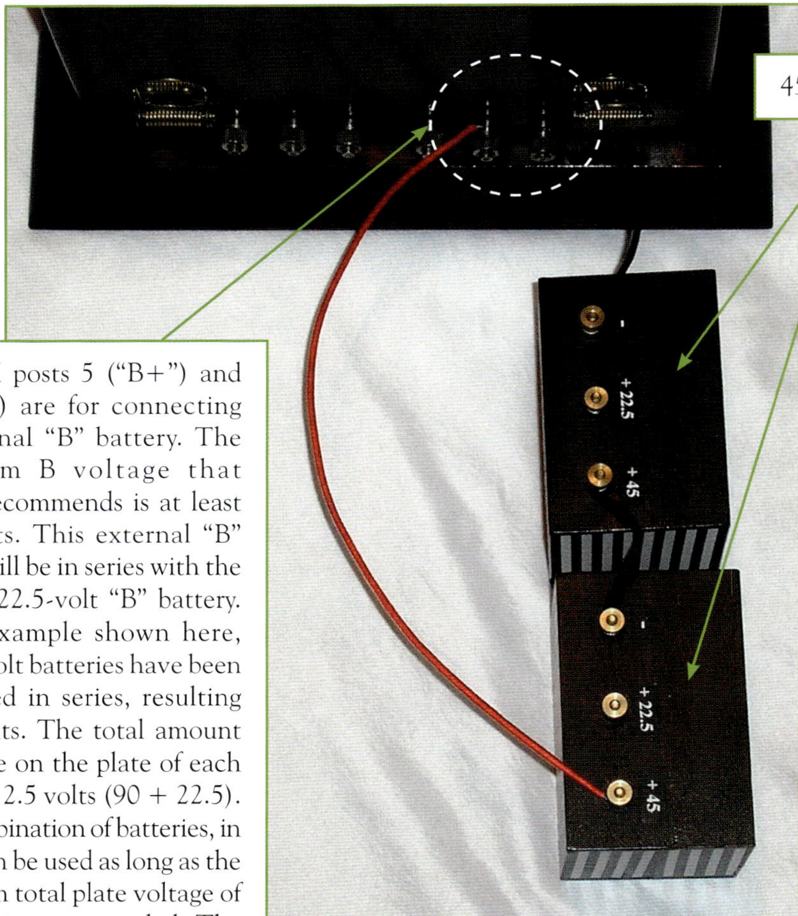

45-volt "B" Batteries

The 2-M posts 5 ("B+") and 6 ("B−") are for connecting an external "B" battery. The minimum B voltage that Zenith recommends is at least 67.5 volts. This external "B" battery will be in series with the internal 22.5-volt "B" battery. In the example shown here, two 45-volt batteries have been connected in series, resulting in 90 volts. The total amount of voltage on the plate of each tube is 112.5 volts (90 + 22.5). Any combination of batteries, in series, can be used as long as the maximum total plate voltage of the tube is not exceeded. The maximum total plate voltage for the UV-201-A is 135 volts.[1]

External "B" Battery Connections

2-M post 5—external battery "B+"
2-M post 6—external battery "B−"
See External "B" Battery Connections.
Note: If no external "B" battery is used, connect posts 5 and 6 together (short), as shown.

Rear View

2-M post 1—connects to 1-R post 7 (audio)
2-M post 2—connects to 1-R post 8 (audio)
2-M post 3—connects to 1-R post 5 ("A+")
2-M post 4—connects to 1-R post 6 ("A−")

2-M Functional Description Quick Reference Guide

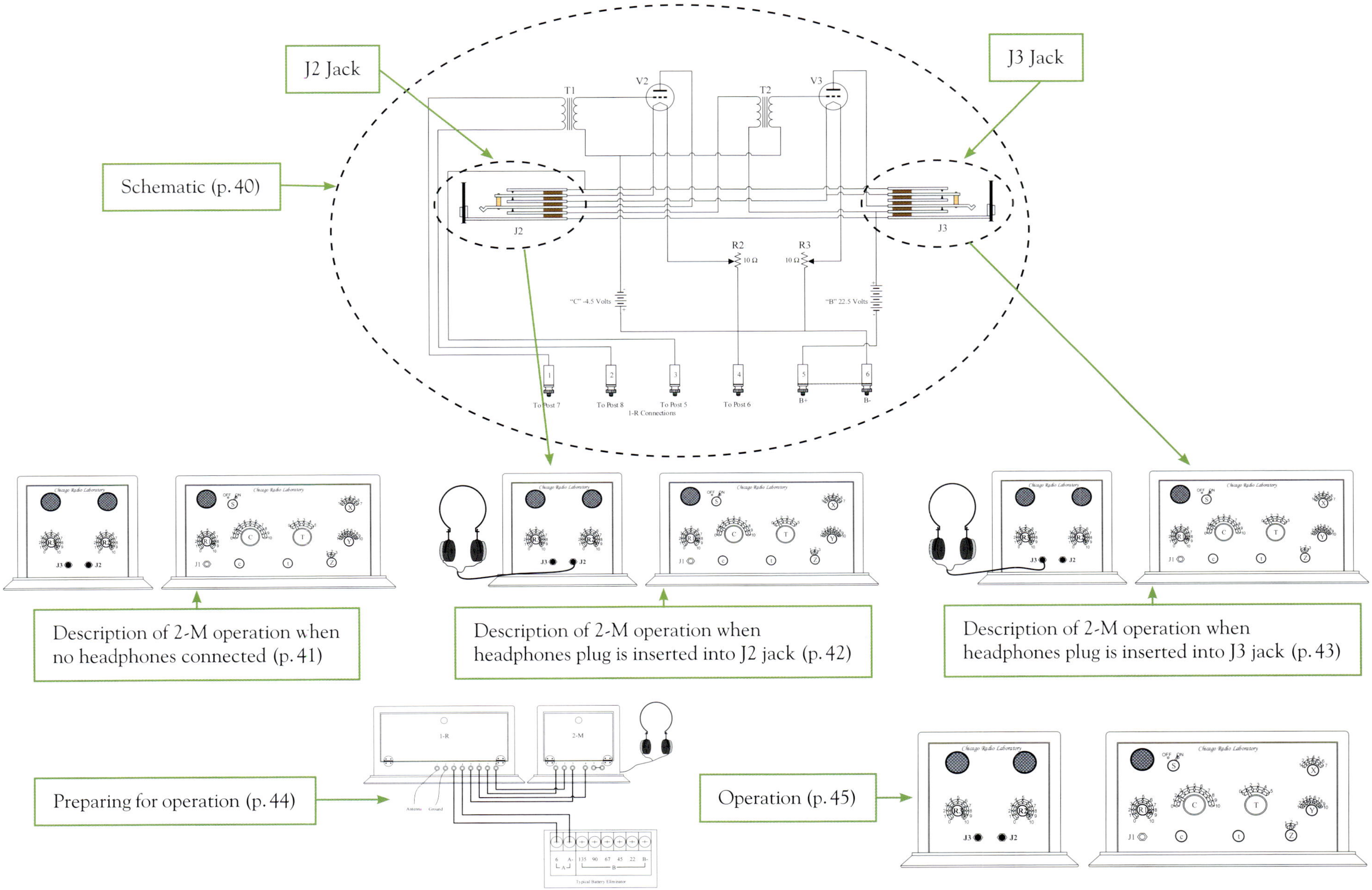

2-M Schematic

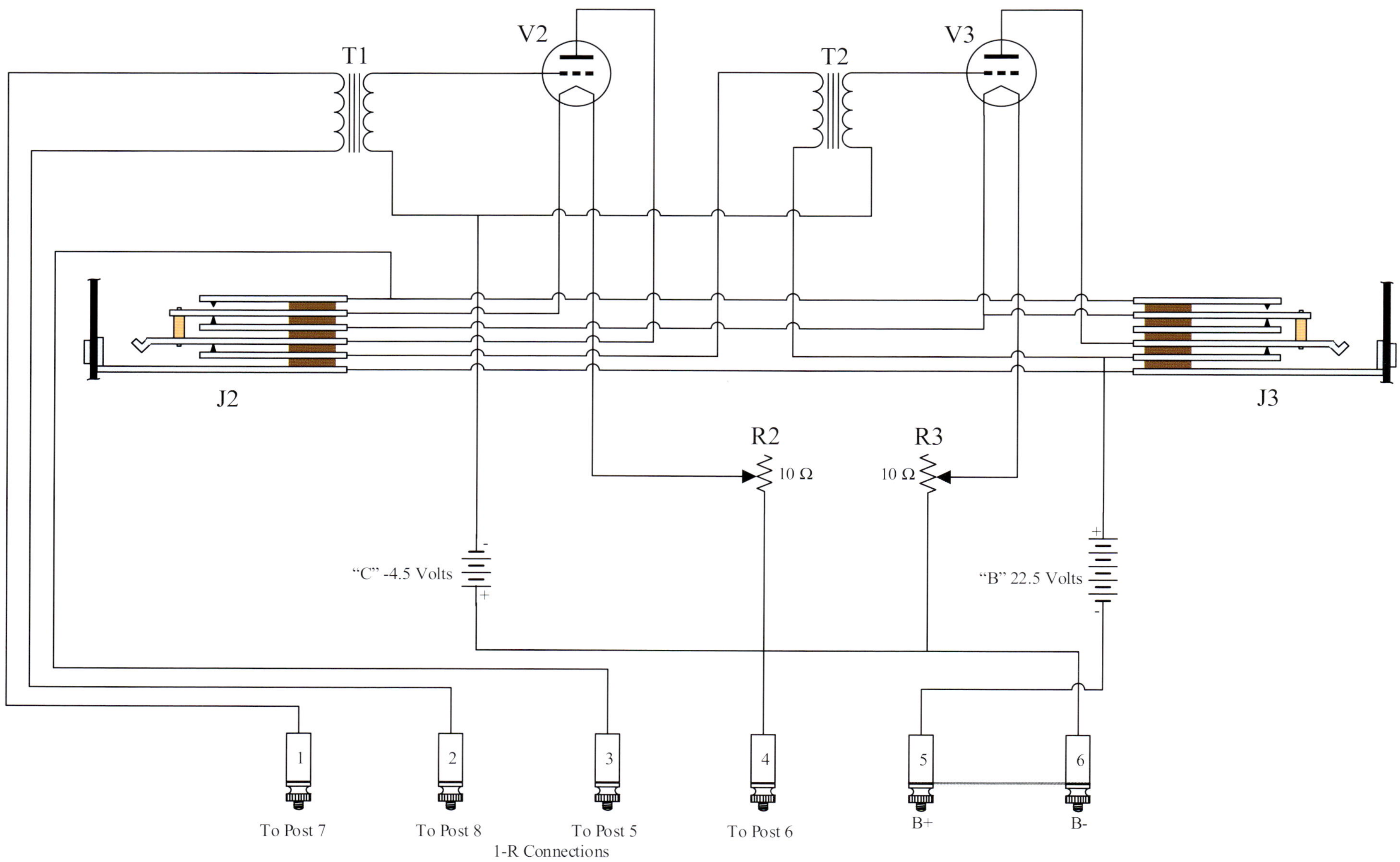

2-M Description of Operation—With No Headphones Connected

The 2-M is connected to the 1-R and the S switch of the 1-R is ON. No headphones are connected. The filament circuits are open and the set is inoperable. Plate voltage is applied to tubes V2 and V3. A negative bias voltage is applied to the grid of tubes V2 and V3. This is considered the initial state of the AF amplifier prior to the insertion of the phone jack into J2 or J3.

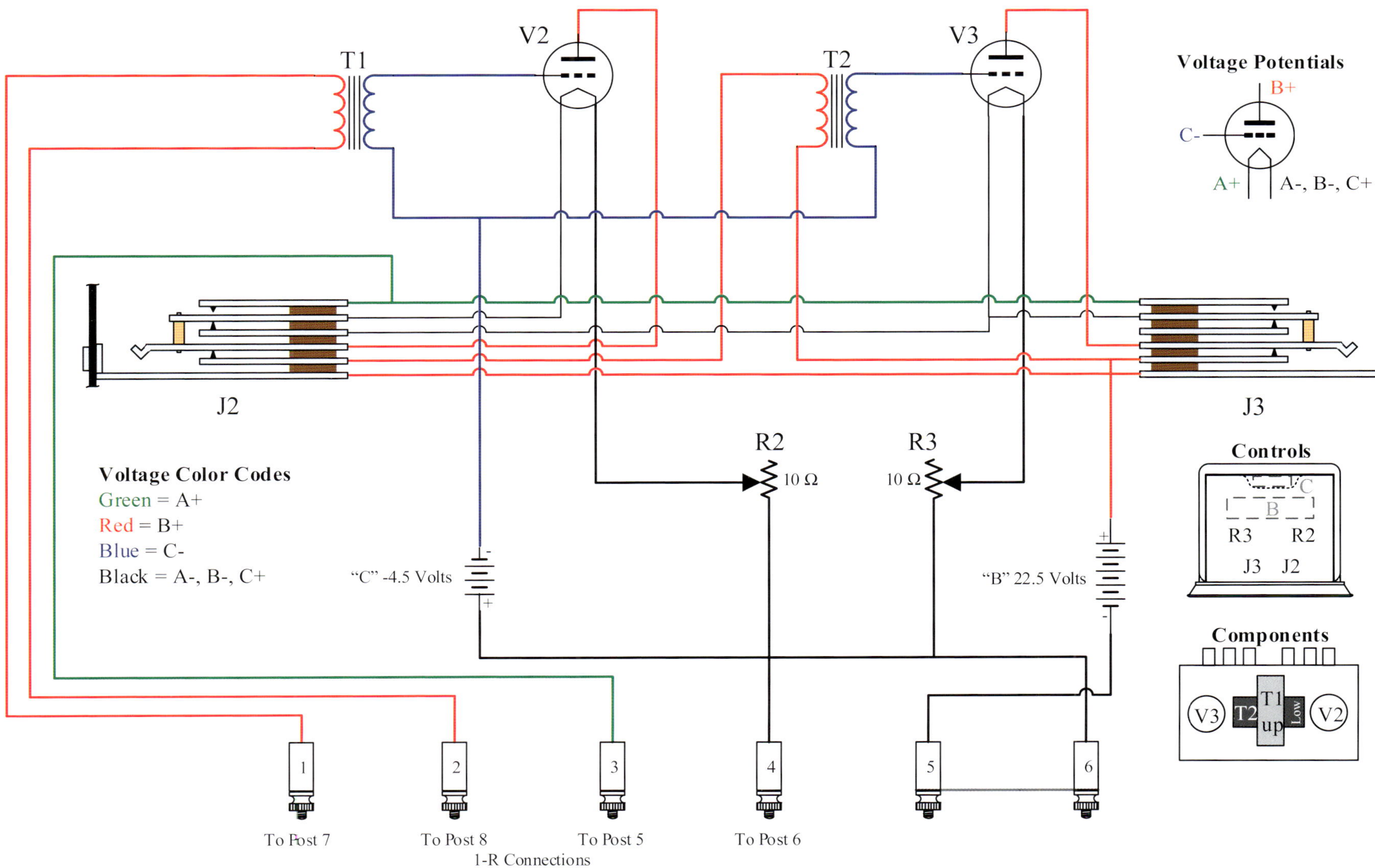

2-M Description of Operation—With Headphones In J2 (One AF Amplifier Stage)

When the headphones plug is inserted into the jack J2 the contact switch action closes energizing the filament circuit of V2. At the same time the contact switch action of J2 disconnects the plate of V2 from primary of transformer T2 and connects the plate of V2 to the headphones circuit. The radio is now playing with one stage (V2) of AF amplification.

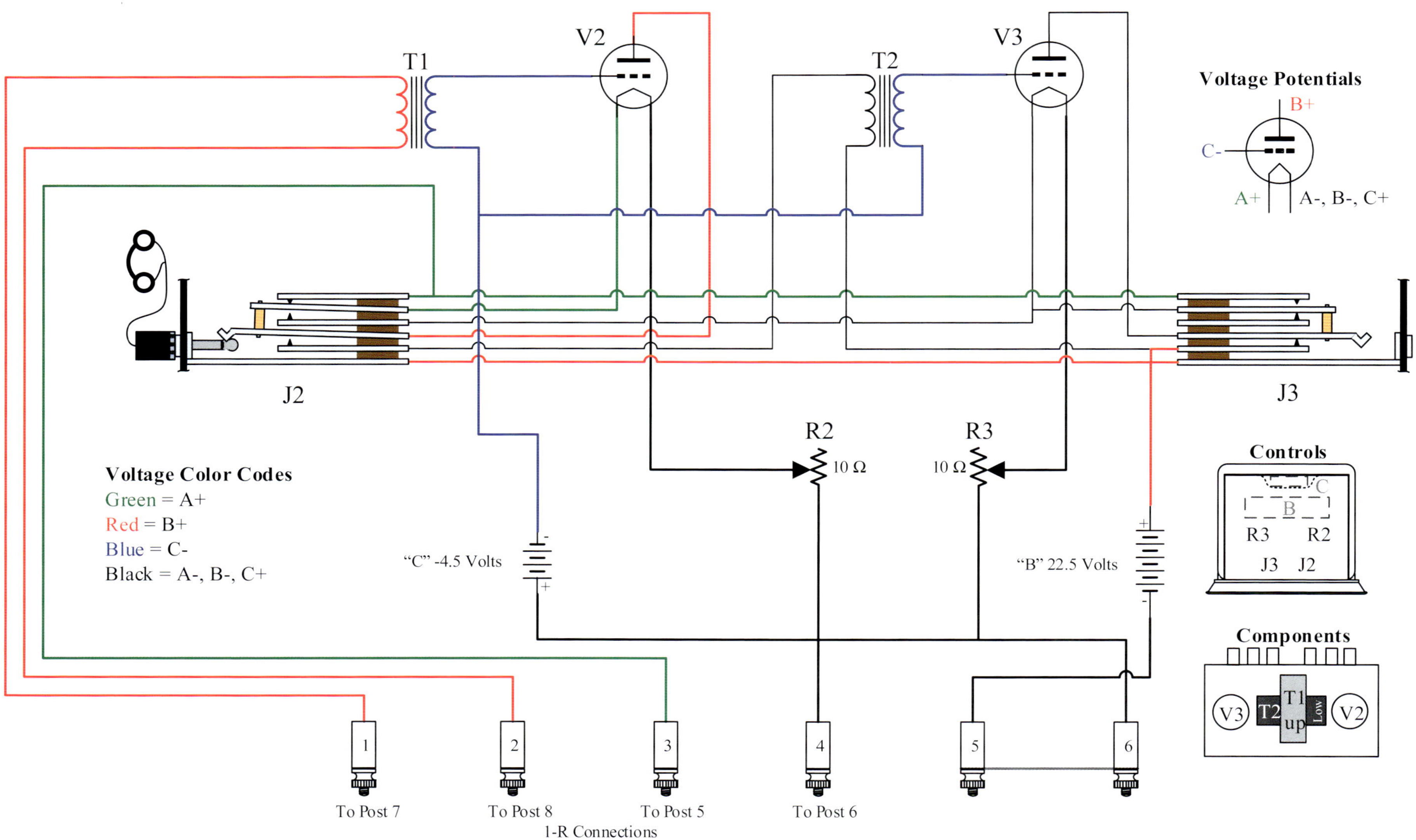

2-M Description of Operation—With Headphones In J3 (Two AF Amplifier Stages)

When the headphones plug is inserted into the jack J3 the contact switch action closes energizing the filament circuits of V2 and V3. At the same time the contact switch action of J3 opens disconnecting the plate of V3 from primary of transformer T2 and connects the plate of V3 to the headphones circuit. The radio is now playing with two stages (V2, V3) of AF amplification.

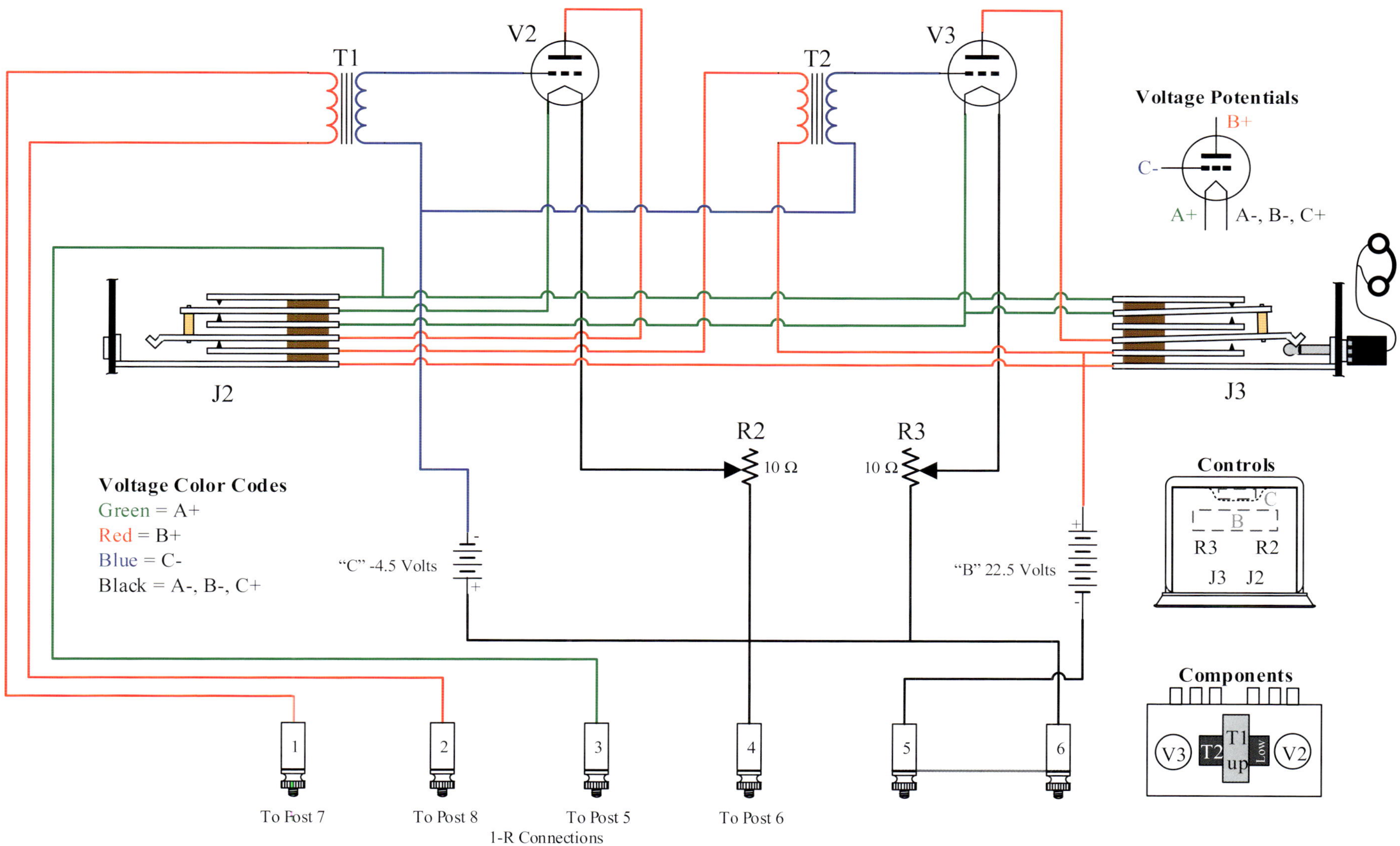

1-R and 2-M Preparations for Operation

Install the tubes and appropriate "B" and "C" batteries inside both sets. Follow the 1-R Preparations for Operation on page 29, to connect the power, antenna and ground to the 1-R radio. Use the appropriate connection diagrams, located inside the back door of each set, to make sure that they can be connected correctly and that they agree with the 1-R and 2-M Connections diagram below. If the connections are not correct or otherwise unavailable, utilize the physical and functional descriptions, for each set, to assist in determining the correct connections between the two sets. Hook up the connections between the sets. Proceed to 1-R and 2-M Operation on the next page. See Common Radio Attachments (page 8) for additional information related to loudspeakers.

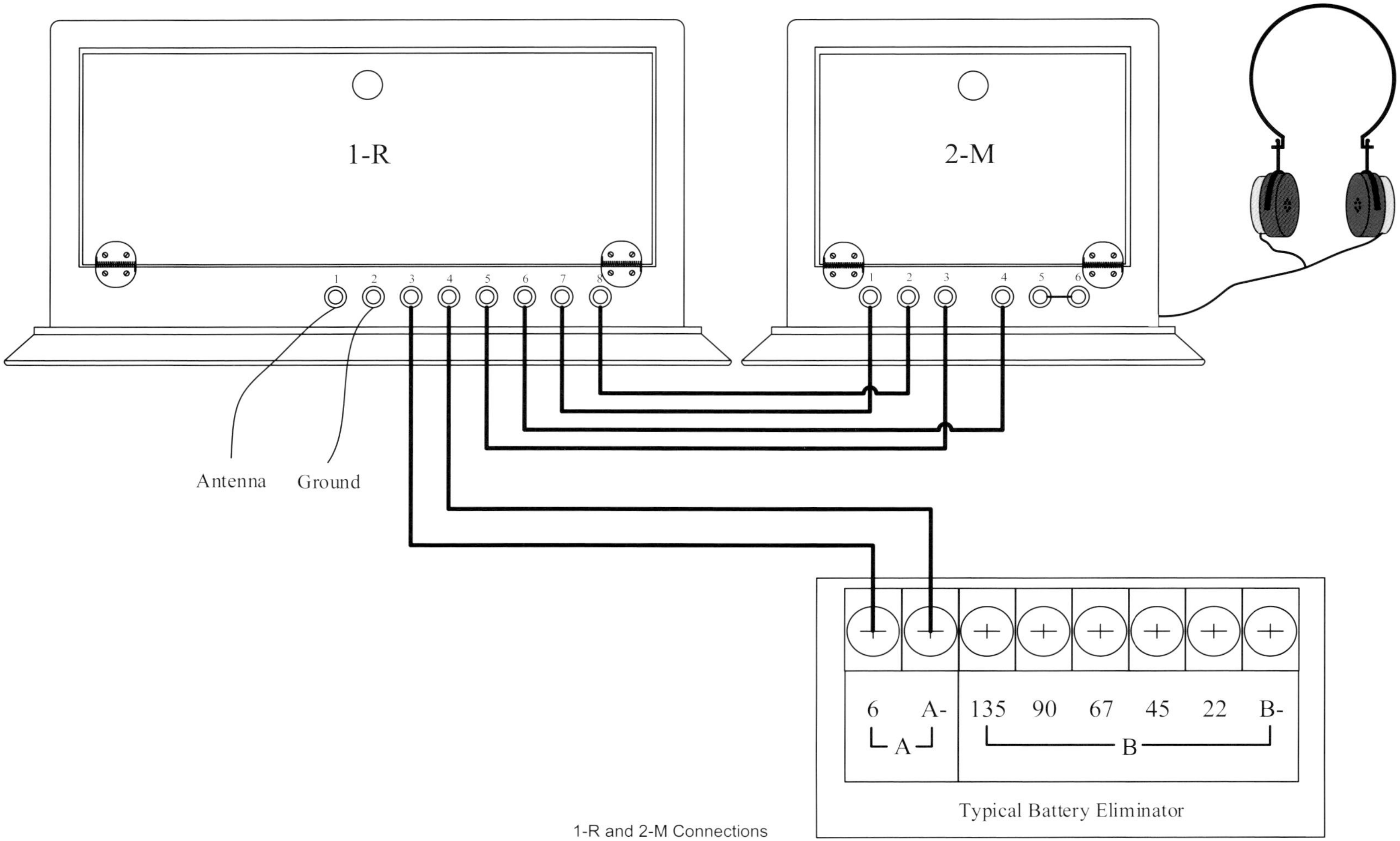

1-R and 2-M Connections

1-R and 2-M Operation

Follow the 1-R and 2-M Preparations for Operation instructions above. Set the pointers to the positions shown in the Initial Dial Settings diagram below.

The tuning capacitor C selects the stations and the T tickler provides the feedback (regeneration) that increases the volume of the selected signal. Practice using knobs C and T simultaneously. Utilize the small c and t knobs to make fine adjustments to the C and T mechanisms.

Insert the headphones plug into J1 jack. Turn the “S” switch to the ON position.

Turn “C” knob to the right until the desired signal is heard. Now gradually turn “T” knob to the right until the volume improves. Make small adjustments to knobs C and T until the signal is heard the loudest and clearest. If T knob is turned too far to the right, the signal will be lost and an unpleasant howl will be heard. This howl (oscillations) is an indication that the T knob is turned too far to the right and should be eliminated by turning it back to the left until the howl stops.

Adjust the audio level (volume) by inserting the headphones plug into jack J1, J2, or J3, as appropriate.

Switches X and Y tune the antenna circuit. A little manipulation of these switches will enable the operator to obtain improved results.

Knob Z changes the bandwidth of the receiver and is generally left in the position 2. Position 1 decreases the bandwidth and position 3 increases the bandwidth.

Record the control settings for each station in the Station Log (see 1-R Station Log, page 30). Then, to return to a particular station, set the controls to the settings that were previously recorded in the log for that station.

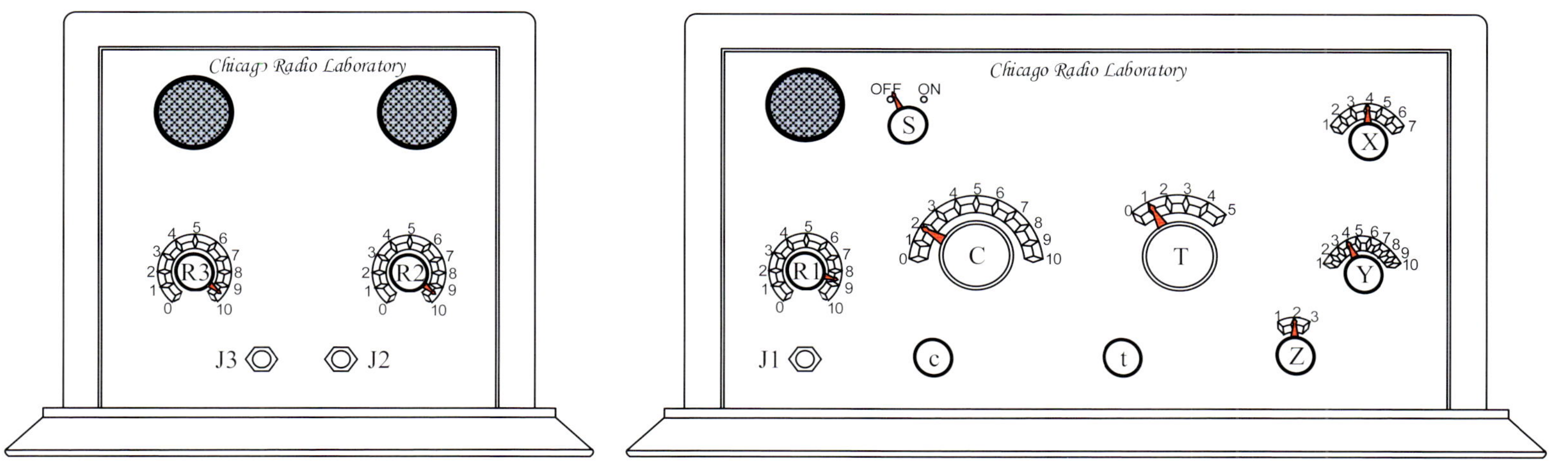

Initial Dial Settings (pointers shown in red for clarity)

Chapter 3: **Zenith 3-M**

3-M General Specifications

Company:	Zenith Radio Corporation
Model:	3-M
Year Manufactured:	1922
Manufacture Location:	48th and Kedzie, Chicago IL (QRS Factory)
Serial Number:	299
Price:	$75
Tube:	UV-201-A (3)
Circuit Type:	AF amplifier

3-M General Information

As described in Chapter 3—Zenith 2-M, most of the 1-R and 2-M or 3-M combinations that were sold, included the more popular 2-M amplifier. The 3-M was evidently produced in limited quantities and is a fairly rare set.

Since no schematic diagram is available, one was created utilizing this 3-M (SN 299). This 3-M schematic may not match exactly what the manufacturer would have drawn, but it should be electrically and functionally equivalent.

Restoration details for the set that was used as the example in this chapter. *When this 3-M was acquired, it was in rough shape. The cabinet, which had been painted with brown paint, was stripped, stained and refinished. A reproduced decal was added to the top. The "B" battery brackets were missing, but the mounting holes were there, so battery brackets were made to match those holes. These brackets are similar to the "B" battery bracket in the 1-R. The engravings around the three rheostat knobs have been worn down significantly and a suitable fix has not been found. The three pointers were badly distorted and were replaced with handmade aluminum pointers.*

The internal electronics had been extensively reworked at some point in its life. The original AF transformers had been removed and replaced with Thordarson transformers. These Thordarson transformers were replaced with reproduced transformers that are believed to be the original style that would have been initially installed in the 3-M. The original letter B, that was etched into the two Bakelite "B" battery terminal strips, had been scratched off and the letter C was scratched in by hand. The circuitry was reverted back to the original configuration as closely as possible, and then verified. The Bakelite terminal strips were sanded smooth and then polished with Novus scratch remover.

When this restored 3-M amplifier was hooked up to the 1-R, it played great on all three stages.

3-M Physical Description Quick Reference Guide

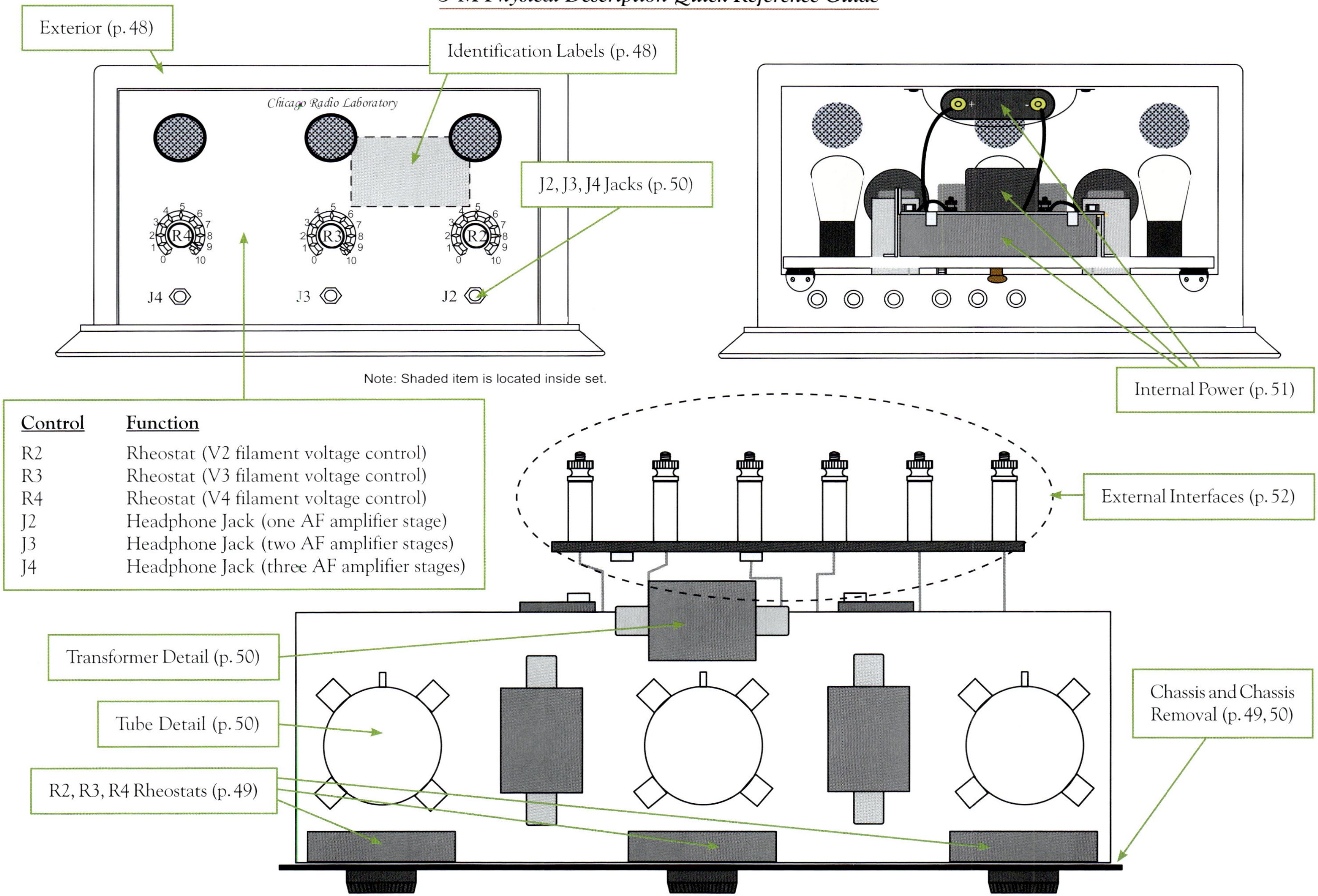

Control	Function
R2	Rheostat (V2 filament voltage control)
R3	Rheostat (V3 filament voltage control)
R4	Rheostat (V4 filament voltage control)
J2	Headphone Jack (one AF amplifier stage)
J3	Headphone Jack (two AF amplifier stages)
J4	Headphone Jack (three AF amplifier stages)

3-M Exterior

The cabinet is made of mahogany and measures 16" by 11" by 9.75". The front panel slopes at an angle of approximately 10° and is made of brown Bakelite with faux wood grain. Peep holes are provided for viewing the brightness of each tube.

Chicago Radio Laboratory Logo

Front View

Top View. Reproduced decal.

Rear View

3-M Identification Labels

Rear View (battery removed for clarity)

Serial Number

1-R Connections

Identification and Hookup

"B" Battery connections and shorting instruction, which states, "EXTRA B BATTERY THESE POSTS NORMALLY CONNECTED TOGETHER. TO BE USED ONLY WHEN EXTRA STRONG SIGNAL IS DESIRED."

3-M Chassis

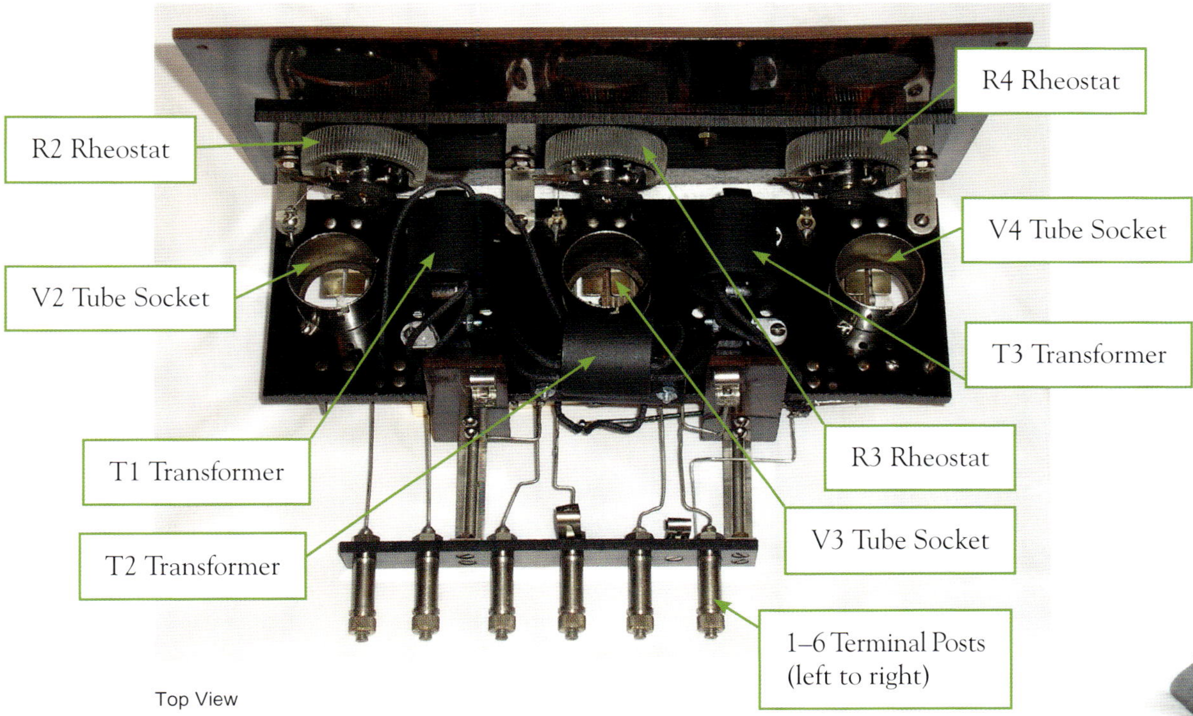

Top View

Front View

Rear View

Removal of Chassis from Cabinet

1. Inside the rear of the cabinet, disconnect the "B" and "C" batteries. See Rear View.
2. On the front panel, remove the four corner screws with the raised washers and the two small screws located at the top. See Front View.
3. The whole chassis slides out the front of the cabinet. Raise the front panel above the bottom cabinet lip and gently push the chassis forward from the rear.

To reinstall the chassis into the cabinet, reverse the removal process.

3-M Chassis (continued)

Rear View

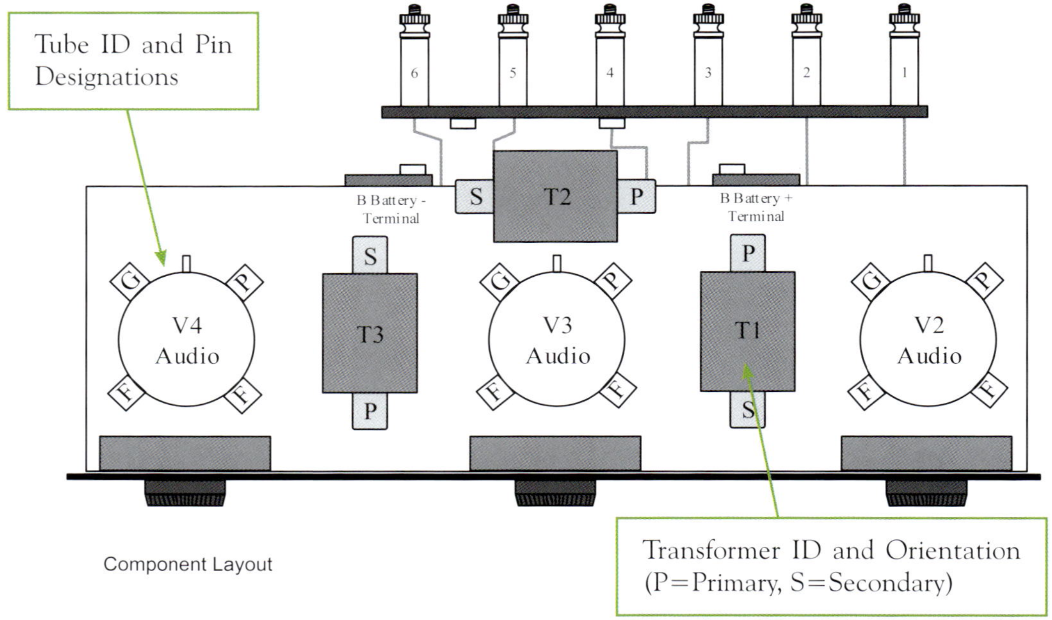

Component Layout

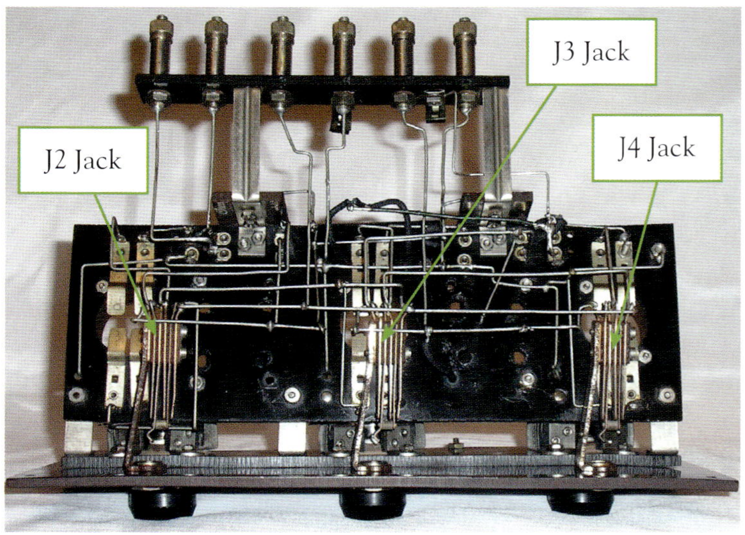

Underneath View

Right Side View

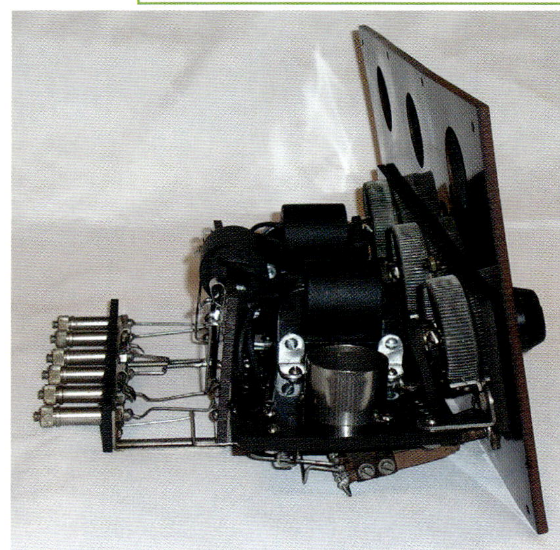

Left Side View

3-M Internal Power

Internally the 3-M uses two different types of batteries. The "B" battery provides the voltage to the plate of each tube and the "C" battery provides the negative bias to the grid of each tube; see Internal Batteries below. There are separate terminals for connecting each battery; see Internal Battery Connections below.

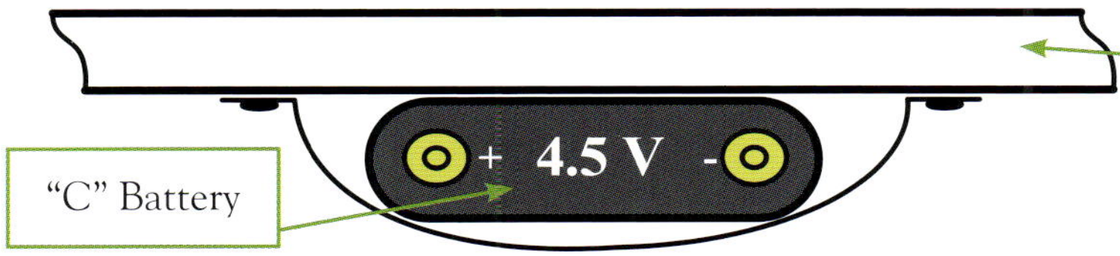

"C" Battery and Holder Detail

Internal Batteries

Internal Battery Connections

Reproduced "B" and "C" Batteries

There is an extra "B" battery on the back door. The intention behind this second "B" battery is unclear. It was included because the holes were there for the bracket. The wear under the battery, on the right, is not as severe as under the battery on the left, which indicates that it may not have been used. It could be connected in series with the other "B" battery, which is not covered in the Zenith manual, or it could just be an extra "B" battery. The extra "B" battery is not connected to the terminals.

3-M External Interfaces

The 1-R and the 3-M are electrically tied together with four wires that are connected to the terminal posts located at the rear of the sets. The 3-M gets its "A" voltage and AF signal from the 1-R. It also has external terminals (posts) for connecting additional "B" batteries. The purpose the extra or external "B" battery is to provide additional plate voltage to the three tubes in order to get a stronger signal or for operating a loudspeaker.

3-M post 1—connects to 1-R post 7 (audio)
3-M post 2—connects to 1-R post 8 (audio)
3-M post 3—connects to 1-R post 5 ("A+")
3-M post 4—connects to 1-R post 6 ("A−")

3-M post 5—external battery "B+"
3-M post 6—external battery "B−"
See External "B" Battery Connections below.
Note: If no external "B" battery is used, connect posts 5 and 6 together (short), as shown.

Rear View

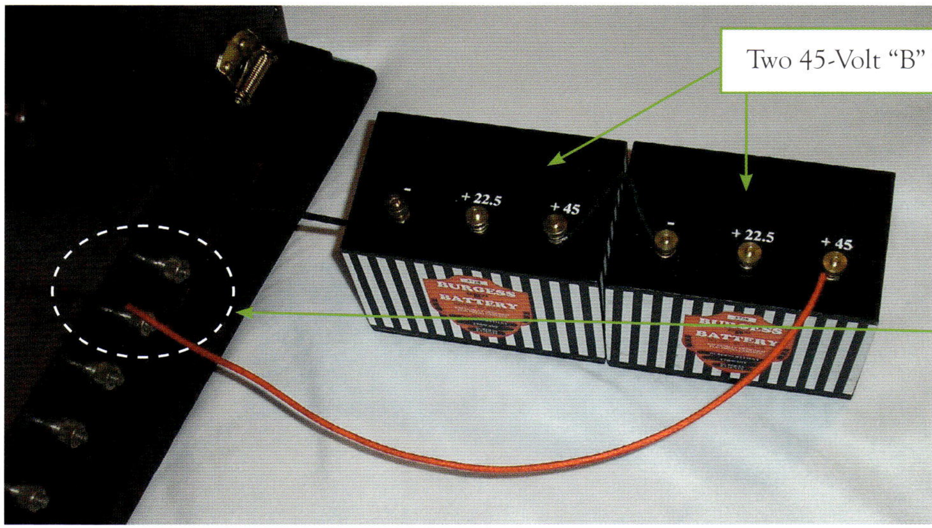

Two 45-Volt "B" Batteries

Posts 5 ("B+") and 6 ("B−") are for connecting an external "B" battery. The minimum "B" voltage that Zenith recommends is at least 67.5 volts. This external "B" battery will be in series with the internal 22.5-volt "B" battery. In the example here, two 45-volt batteries have been connected in series, resulting in 90 volts. The total amount of voltage on the plates of each tube is 112.5 volts (90 + 22.5). Any combination of batteries in series can be used as long as the maximum total plate voltage is not exceeded. The maximum total plate voltage for the UV-201-A is 135 volts.[1]

External "B" Battery Connections

3-M Functional Description Quick Reference Guide

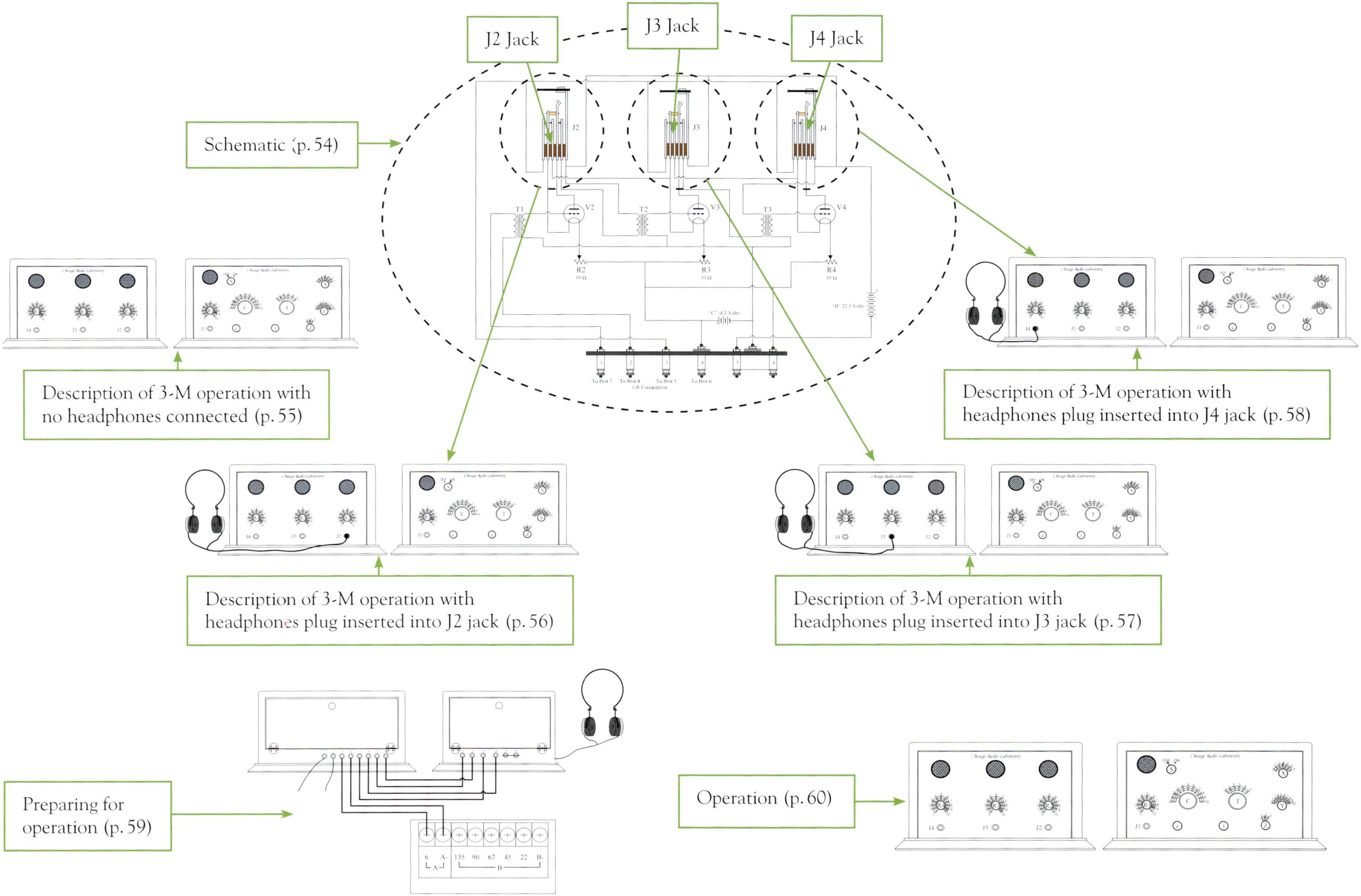

3-M Schematic

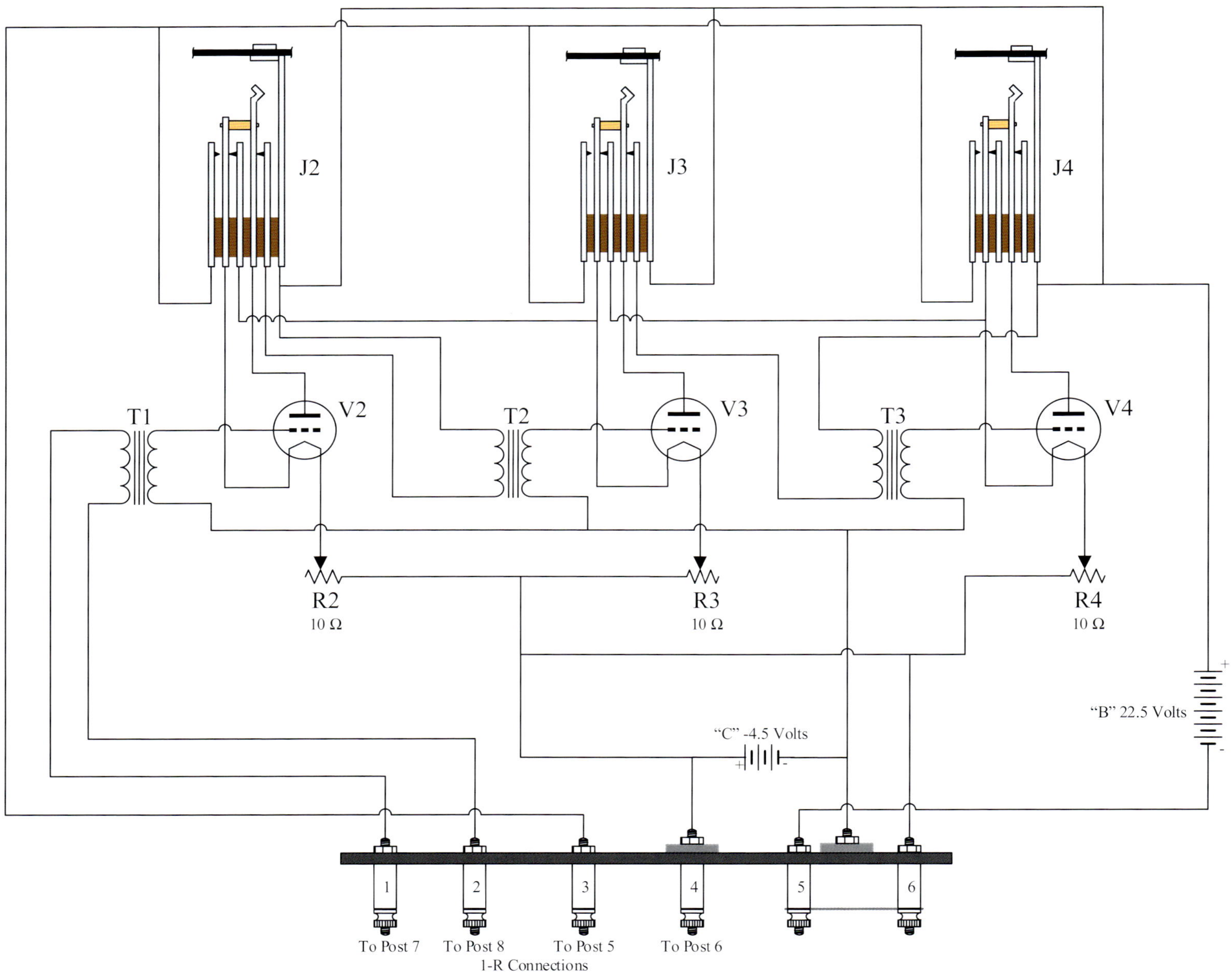

3-M Description of Operation—With No Headphones Connected

The 3-M is connected to the 1-R and the switch S of the 1-R is ON. No headphones are connected. The filament circuits are open and the set is inoperable. A negative bias voltage is applied to the grid of tubes V2, V3 and V4. This is considered the initial state of the AF amplifier prior to the insertion of the phone jack into J2 or J3 or J4.

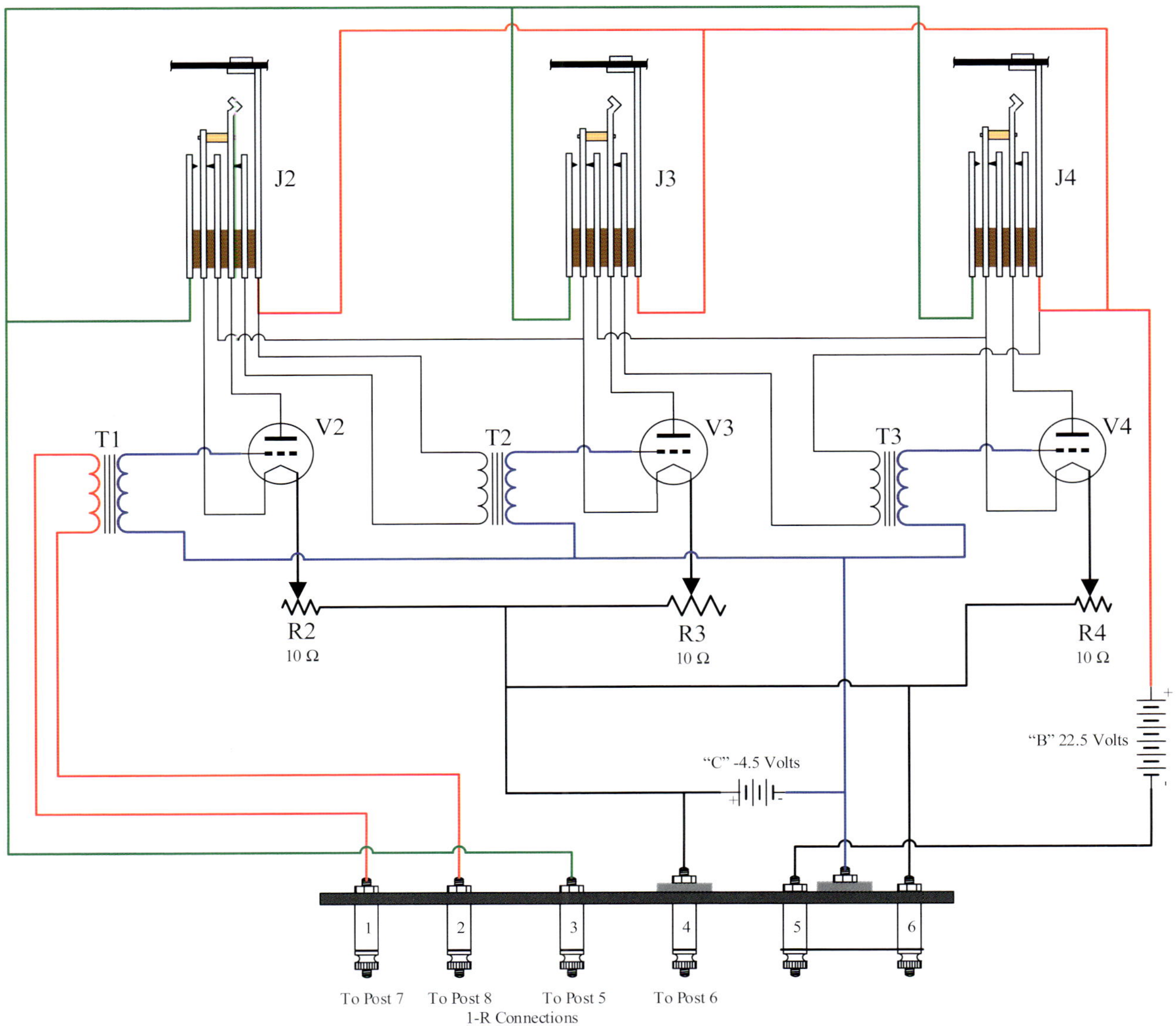

Voltage Color Codes
Green = A+
Red = B+
Blue = C-
Black = A-, B-, C+

Voltage Potentials

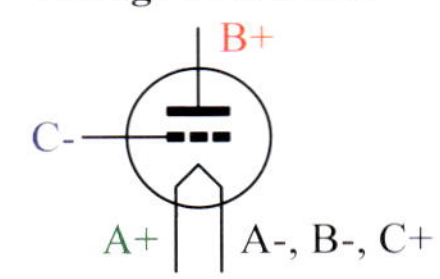

Controls

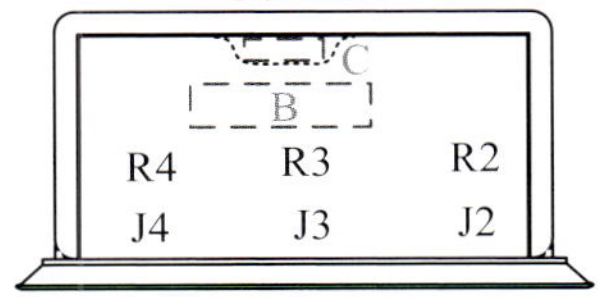

Components

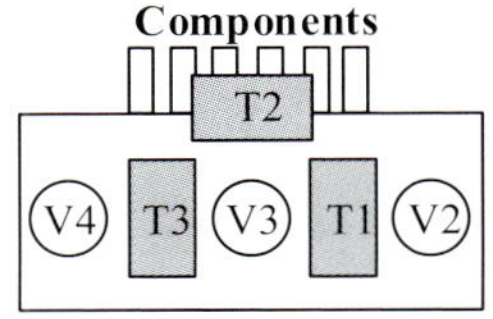

3-M Description of Operation—With Headphones in J2 (One AF Amplifier Stage)

When the headphones plug is inserted into the jack J2 the contact switch action closes energizing the filament circuit of V2. At the same time the contact switch action of J2 disconnects the plate of V2 from primary of transformer T2 and connects the plate of V2 to the headphones circuit. The radio is now playing with one stage (V2) of AF amplification.

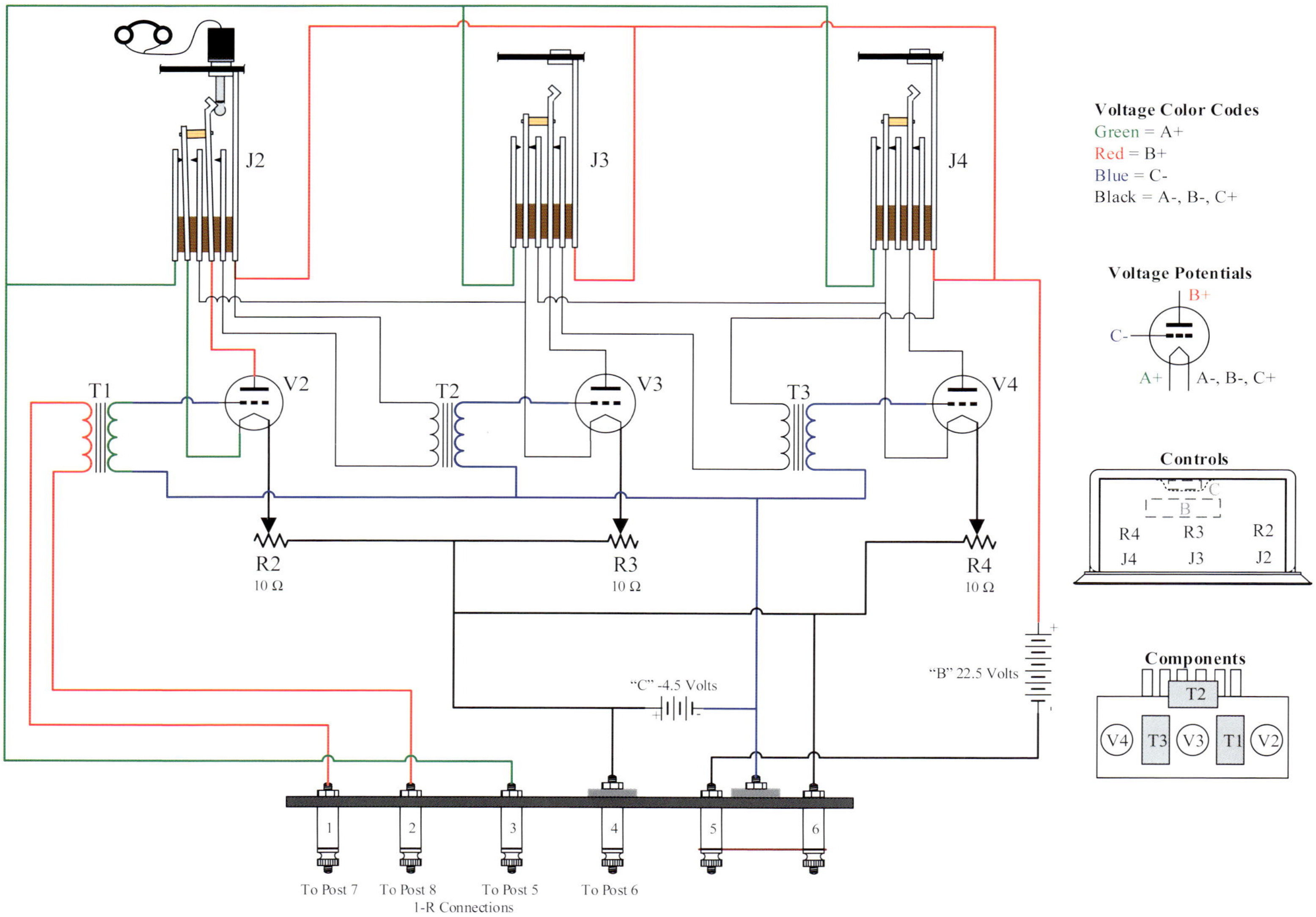

3-M Description of Operation—With Headphones in J3 (Two AF Amplifier Stages)

When the headphones plug is inserted into the jack J3 the contact switch action closes energizing the filament circuits of V2 and V3. At the same time the contact switch action of J3 disconnects the plate of V3 from primary of transformer T3 and connects the plate of V3 to the headphones circuit. The radio is now playing with two stages (V2, V3) of AF amplification.

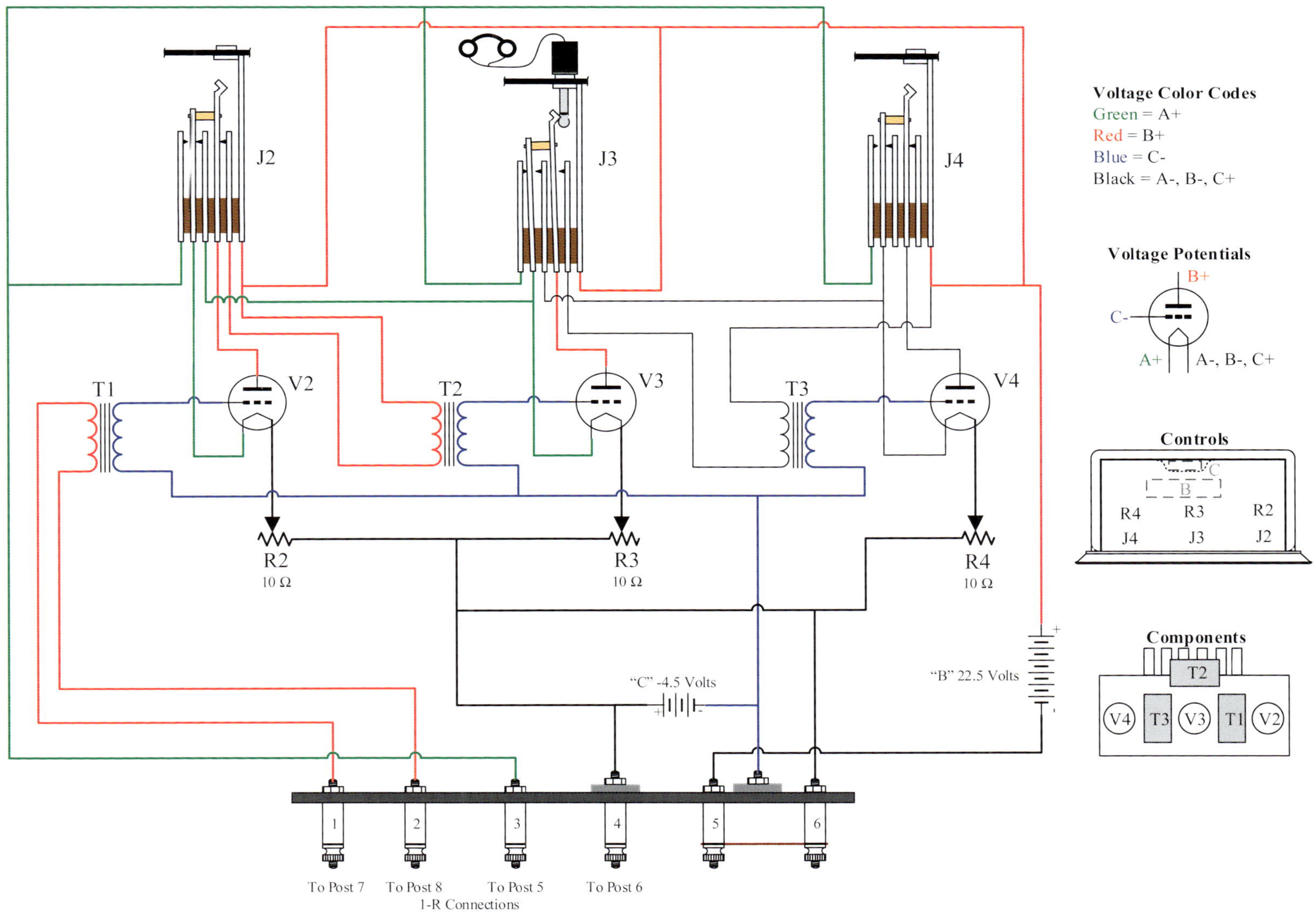

3-M Description of Operation—With Headphones in J4 (Three AF Amplifier Stages)

When the headphones plug is inserted into the jack J4 the contact switch action closes energizing the filament circuits of V2, V3 and V4. At the same time the contact switch action of J4 connects the plate of V4 to the headphones circuit. The radio is now playing with three stages (V2, V3, and V4) of AF amplification.

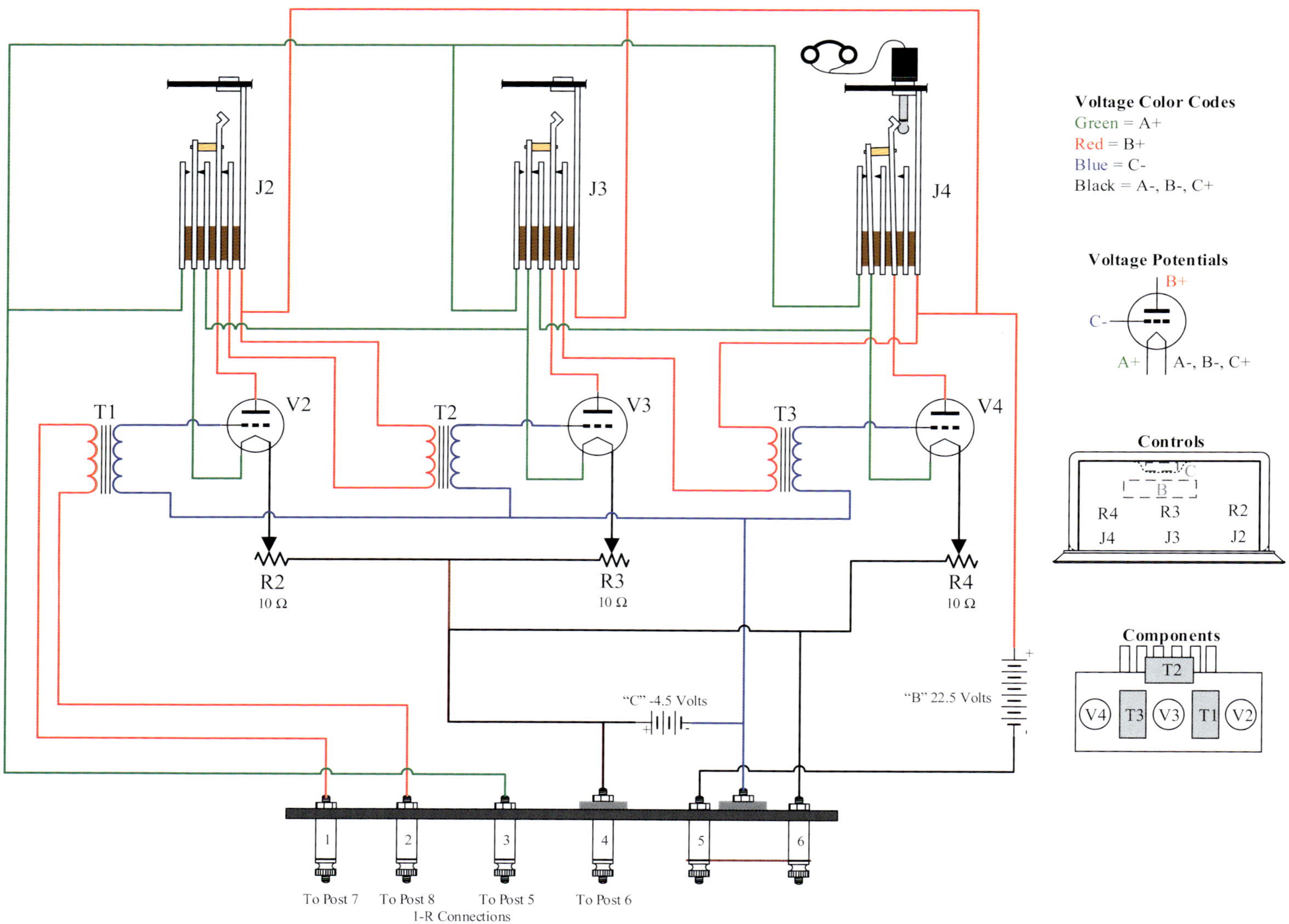

1-R and 3-M Preparations for Operation

Install the tubes and appropriate "B" and "C" batteries inside both sets. Follow the 1-R Preparations for Operation, page 29, to connect the power, antenna and ground to the 1-R radio. Use the appropriate connection diagrams, located inside the back door of each set, to make sure that they can be connected correctly and that they agree with the 1-R and 3-M Connections diagram below. If they are not correct or otherwise unavailable, utilize the Physical and Functional descriptions, for each set, to assist in determining the correct connections between the two sets. Hook up the connections between the sets. Proceed to 1-R and 3-M Operation on the next page. See Common Radio Attachments, (page 8), for additional information related to loudspeakers.

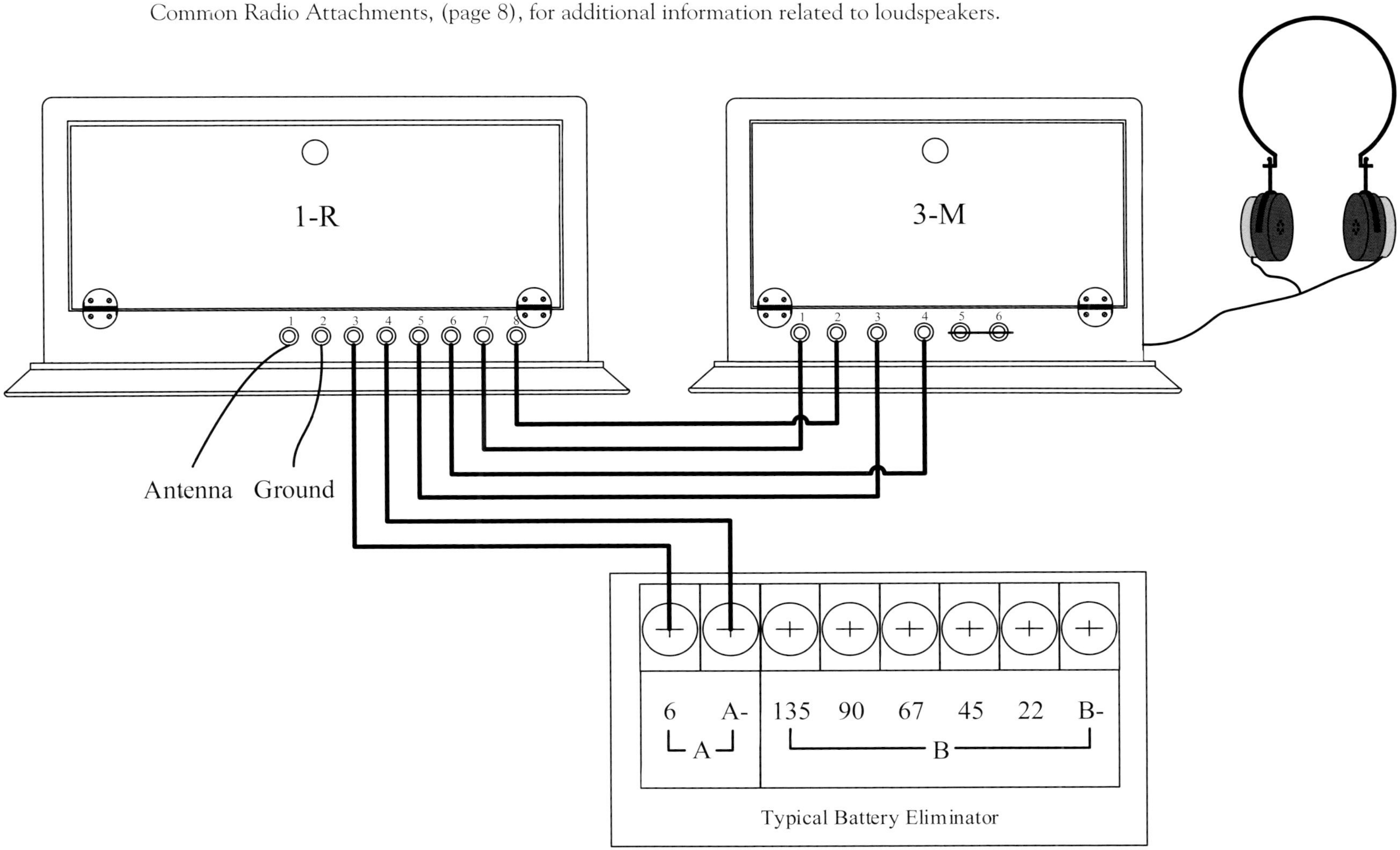

1-R and 3-M Connections

1-R and 3-M Operation

Follow the 1-R and 3-M Preparations for Operation instructions on the previous page. Set the pointers to the positions shown in the Initial Dial Settings diagram below.

The C tuning capacitor selects the stations and the T tickler provides the feedback (regeneration) that increases the volume of the selected signal. Practice using knobs C and T simultaneously. Utilize the small c and t knobs to make fine adjustments to the C and T mechanisms.

Insert the headphones plug into J1 jack. Turn the “S” switch to the ON position.

Turn “C” knob to the right until the desired signal is heard. Now gradually turn “T” knob to the right until the volume improves. Make small adjustments to knobs C and T until the signal is heard the loudest and clearest. If T knob is turned too far to the right, the signal will be lost and an unpleasant howl will be heard. This howl (oscillations) is an indication that the T knob is turned too far to the right and should be eliminated by turning it back to the left until the howl stops.

Adjust the audio level (volume) by inserting the headphones plug into jack J1, J2, J3 or J4, as appropriate.

Switches X and Y tune the antenna circuit. A little manipulation of these switches will enable the operator to obtain improved results.

Knob Z changes the bandwidth of the receiver and is generally left in the position 2. Position 1 decreases the bandwidth and position 3 increases the bandwidth.

Record the control settings for each station in the Station Log (see 1-R Station Log, page 30). Then, to return to a particular station, set the controls to the settings that were previously recorded in the log for that station.

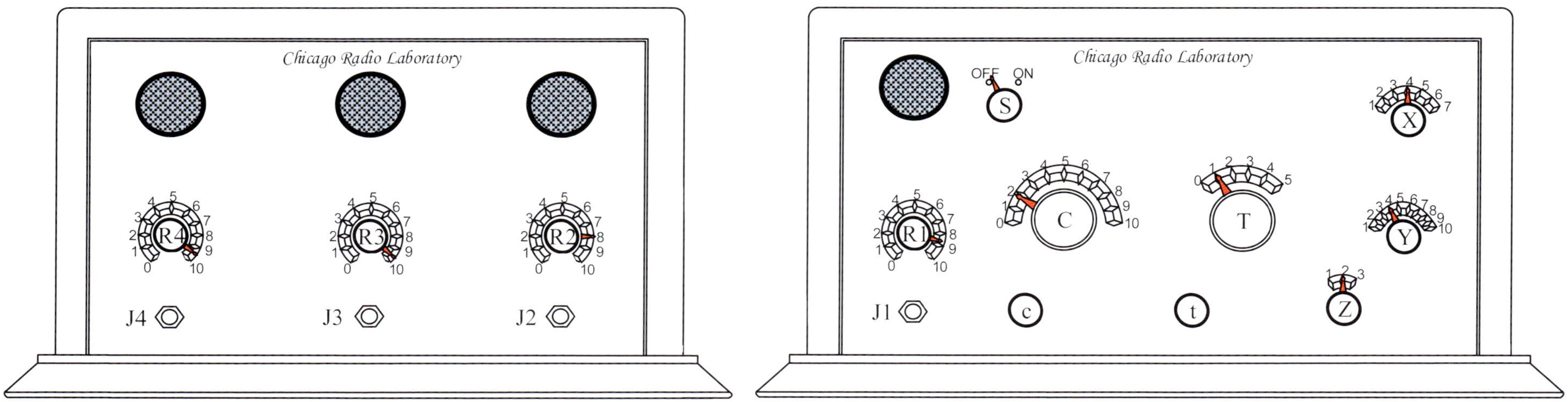

Initial Dial Settings (pointers shown in red for clarity)

Chapter 4: **Zenith 3-R** (Early Version)

3-R(E) General Specifications

Company:	Zenith Radio Corporation
Model:	3-R (Early Version)
Year Manufactured:	1923
Manufacture Location:	48th and Kedzie, Chicago IL (QRS Factory)
Serial Number:	2239
Price:	$160
Tube:	UV-201-A (4), one detector, three AF amplifiers
Circuit Type:	Regenerative

3-R(E) General Information

The 3-R (E) is a direct descendant of the combination 1-R radio and 3-M AF amplifier. It has a detector and three stages of AF amplification, enclosed in a single cabinet. Zenith added a camshaft switch for selecting the desired number of AF amplification stages, which was a significant improvement over the plug and jack method. The AF amplifier circuits were updated with increased multiple "B" and "C" voltages. Also added was a power cable with a keyed plug for connecting the external power. Several features of the 1-R radio were retained: cabinet, front panel layout, knobs, pointers and peep hole. A mahogany trim molding strip was added to support the bottom portion of the front panel.

Restoration details for the set that was used as the example in this chapter. *This particular radio was extensively affected by the pot metal problem. The pointers were replaced with handmade aluminum pointers and the pulleys were replaced with machined aluminum pulleys. Also the two small pulleys for the c and t vernier dials were replaced. The power cable is original. The "C" battery was replaced with a reproduced "C" battery.*

3-R (E) Physical Description Quick Reference Guide

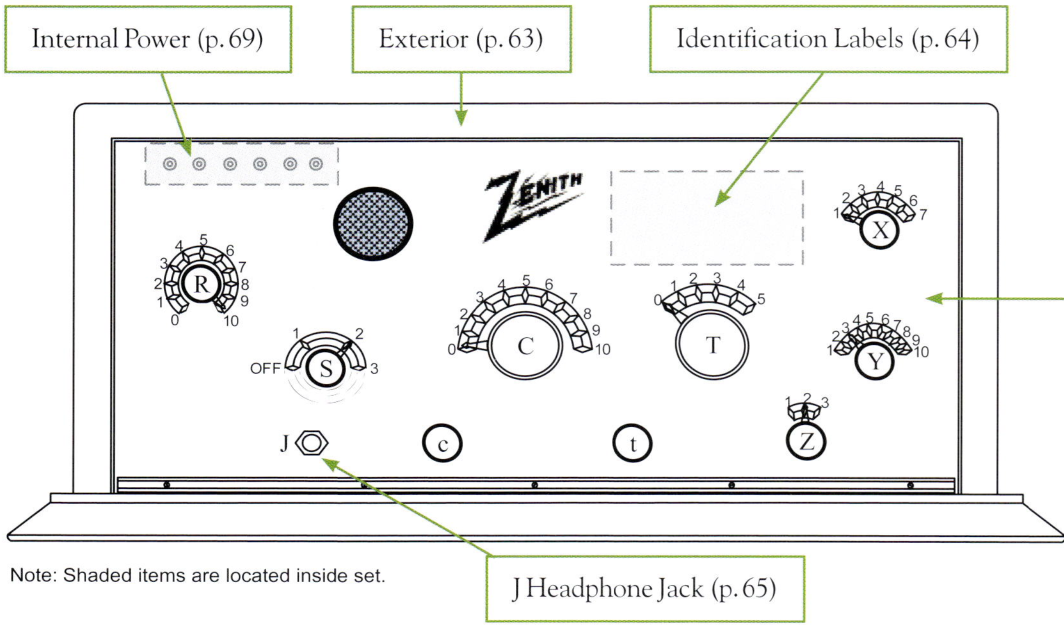

Note: Shaded items are located inside set.

Control	Function
C	Variable Capacitor (tuning)
c	Vernier for Variable Capacitor
J	Headphone Jack
R	Rheostat (filament voltage control)
S	Switch (Off, 1, 2 or 3 AF amp. stages)
T	Tickler (regeneration)
t	Vernier for Tickler
X	Tap Switch (tunes the antenna circuit)
Y	Tap Switch (tunes the antenna circuit)
Z	Tap Switch (selects the bandwidth)

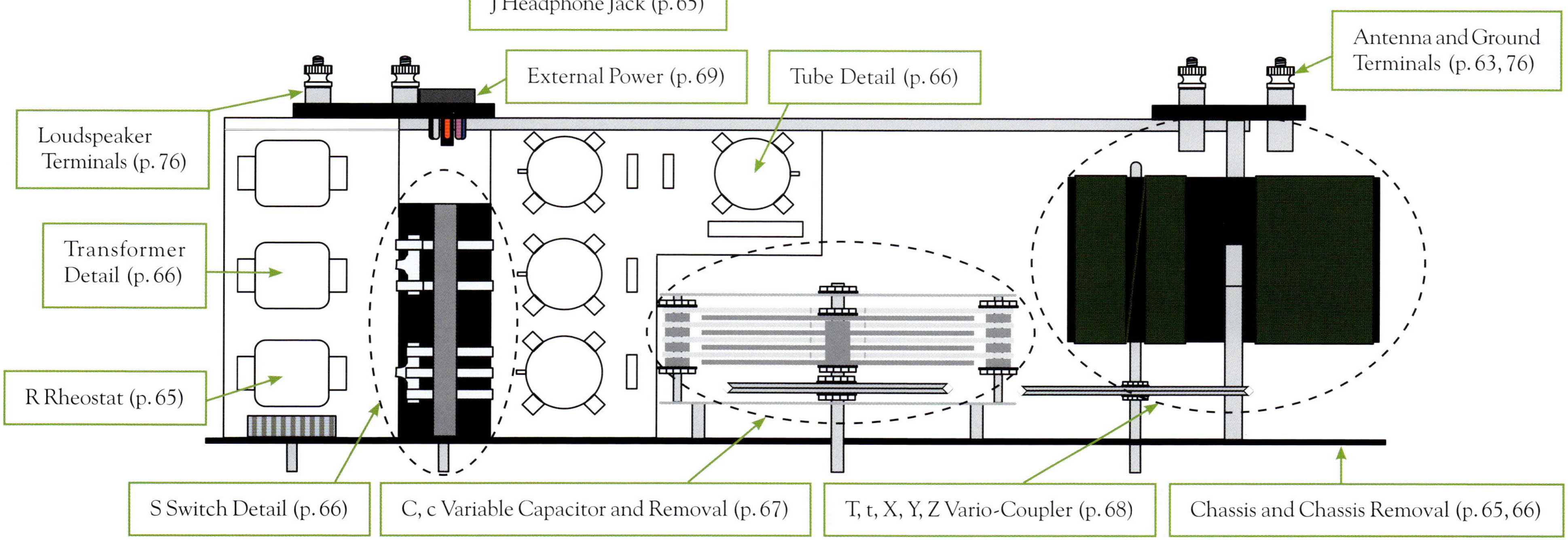

3-R (E) Exterior

The cabinet is made of mahogany and measures 25" by 10.25" by 9.5". The front panel slopes at an angle of approximately 10° and is made of brown Bakelite with faux wood grain. A mahogany molding strip is located on the bottom of the front panel. There is a peep hole for viewing the brightness of the tubes.

Front View

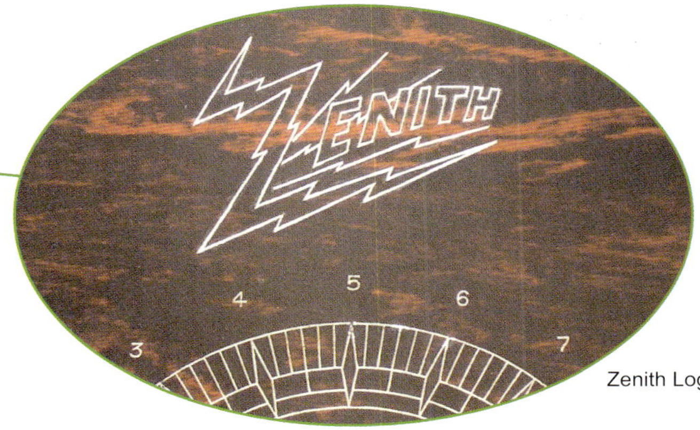

Zenith Logo

Chicago Radio Laboratory logo and Armstrong patent information.

Top View. Original Zenith decal.

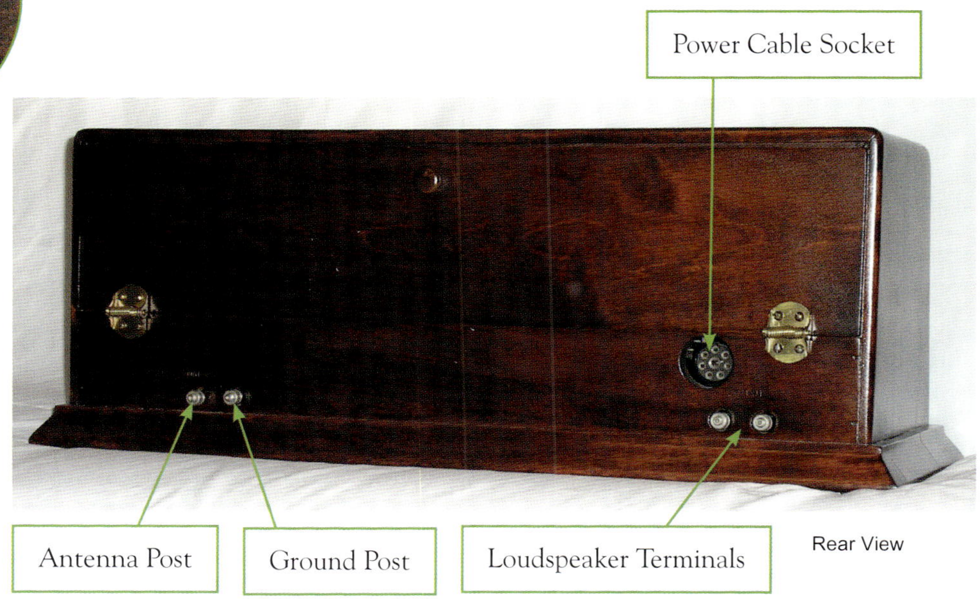

Rear View

3-R (E) Identification Labels

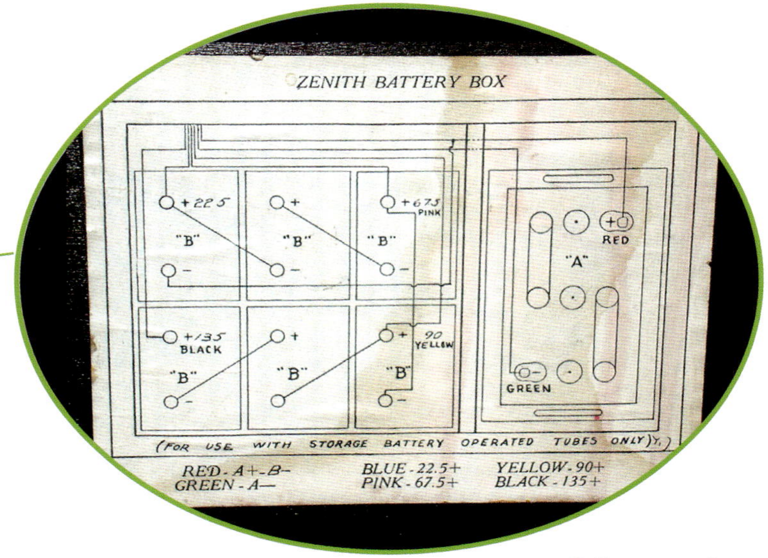

Manufacturer's Labels

Battery connections when using UV-201-A tubes.

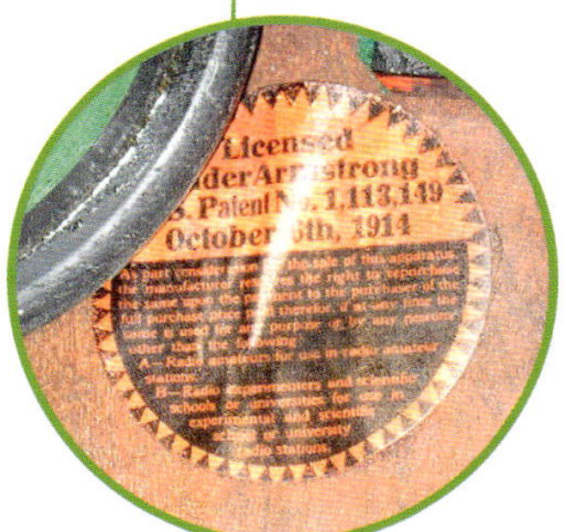

Armstrong
License and Patent Label

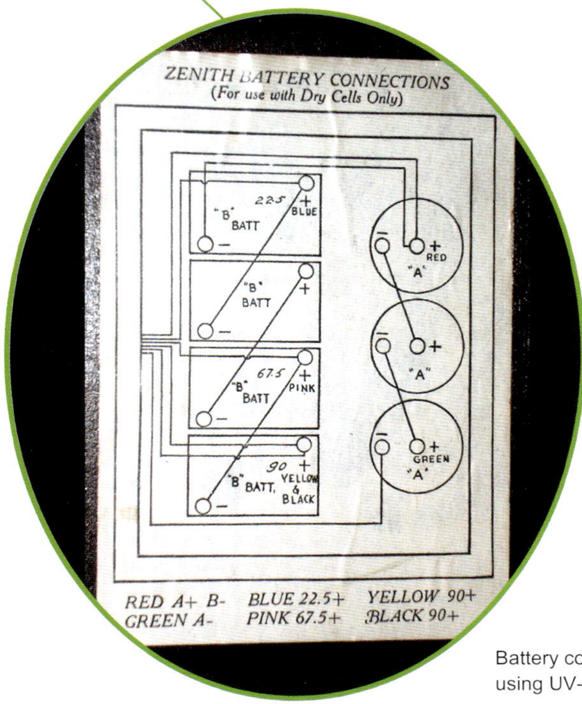

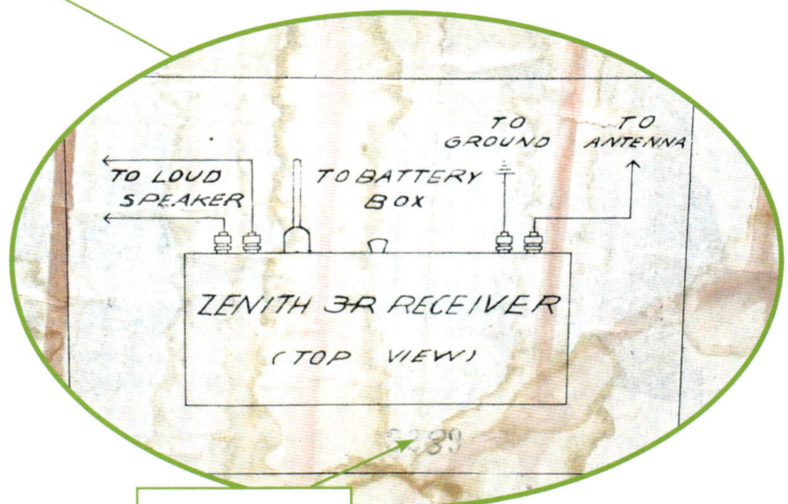

Serial Number

External Connections

Battery connections when using UV-199 tubes (dry cells).

3-R (E) Chassis

Rear View

Underneath View

Top View (looking into cabinet with door open)

Cabinet Rear View

Front View

Removal of Chassis from the Cabinet

1. Inside the rear of the cabinet, remove the two screws holding the chassis to the bottom of the cabinet; see Top View. Disconnect the "C" battery; see Cabinet Rear View.
2. On the front panel, remove trim molding. Remove the four corner screws with the raised washers. Remove the two small screws located at the top of the front panel; see Front View.
3. The whole chassis slides out the front of the cabinet. Gently lift the chassis above the cabinet lip and gently push it from the rear.

To reinstall the chassis into the cabinet, reverse the removal process.

3-R (E) Chassis (continued)

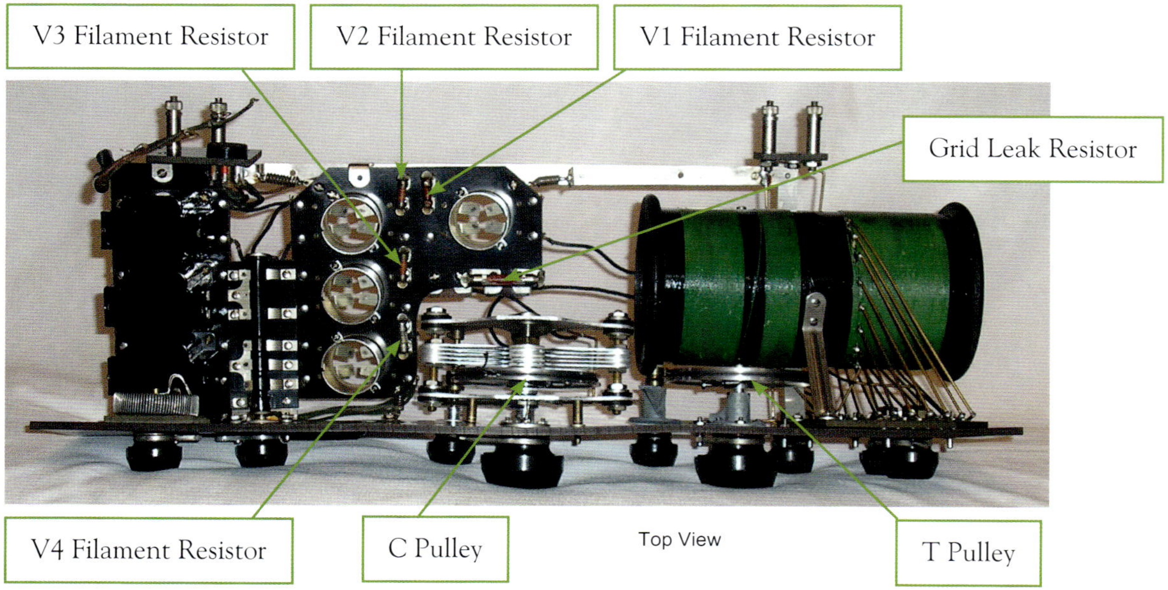

Top View

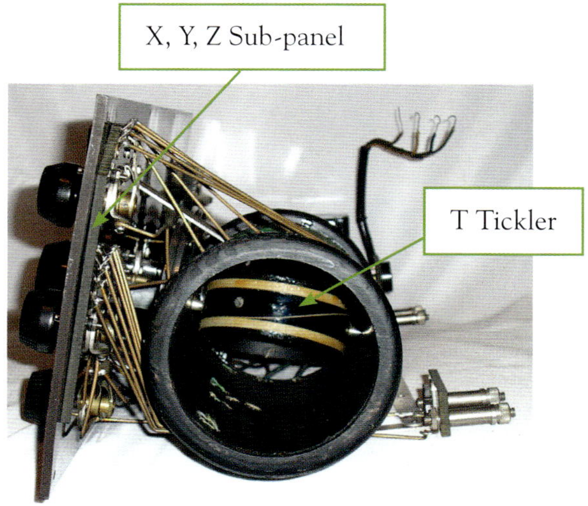

Right Side View

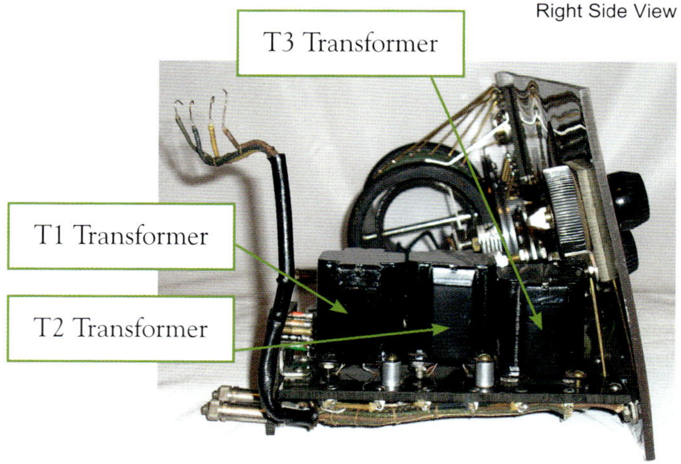

Left Side View

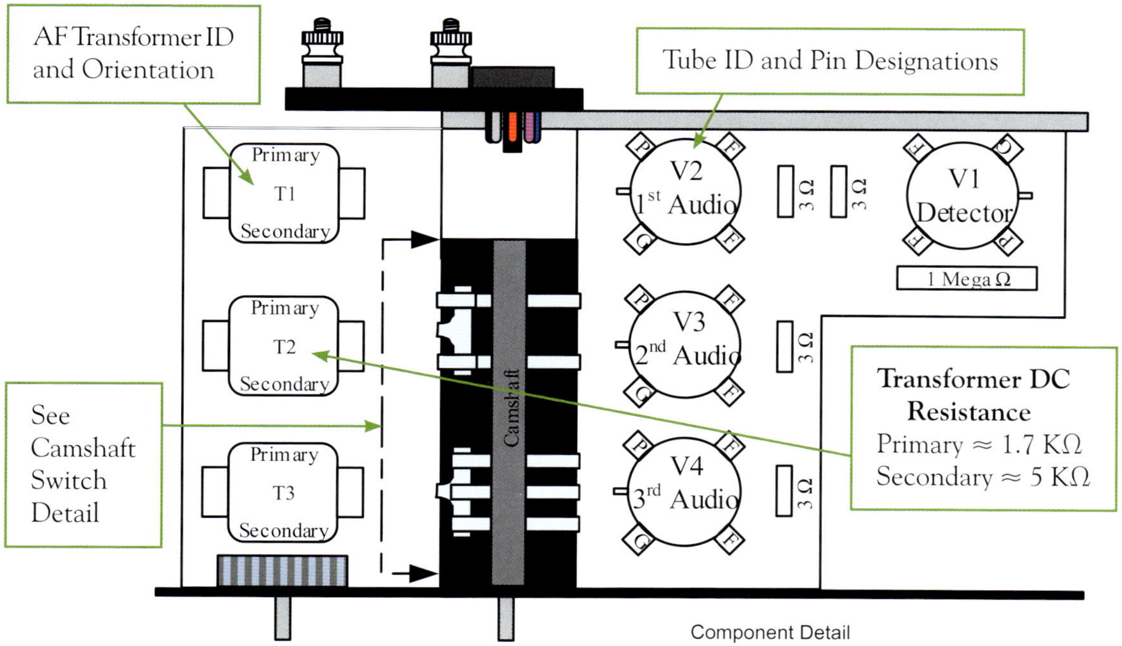

Component Detail

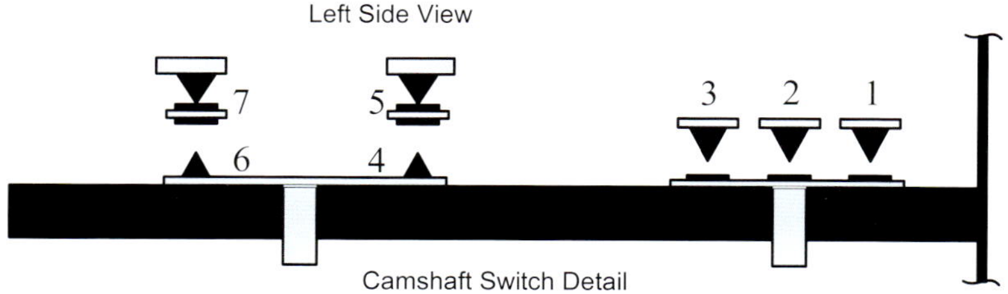

Camshaft Switch Detail

3-R (E) Variable Capacitor

The variable capacitor housing holds a shaft (rotor) that has 5 plates and a large gear C that is driven by the smaller vernier gear c. The shaft rotates the 5 plates, merging them with the 5 fixed plates (stator). The large pulleys were replaced with reproduced pulleys, as shown in Top View and Underside View.

Top View

Underside View

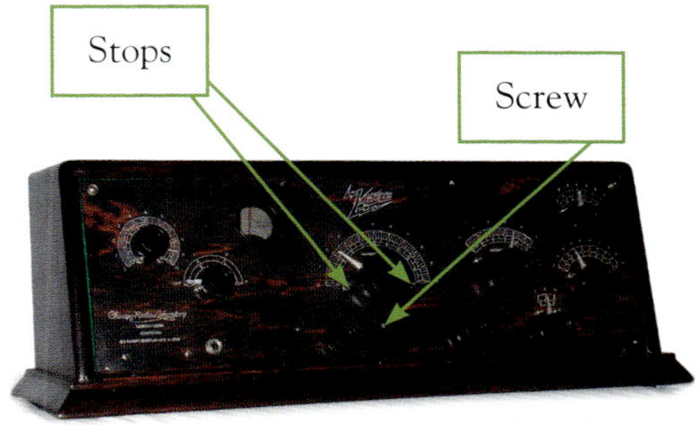

Front Panel

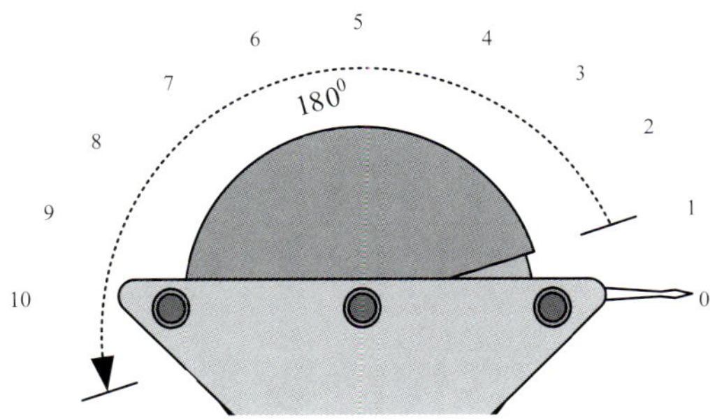

Variable Capacitor Clocking (Viewed from Rear)

Initial starting position of the rotating plates and C pointer in relation to the numbers on the front panel. As the Variable Capacitor rotates 180° right-to-left from the position shown the pointer rotates from 0 to 10, respectively. The pointer and logging numbers are located on front panel, but are shown for clarity.

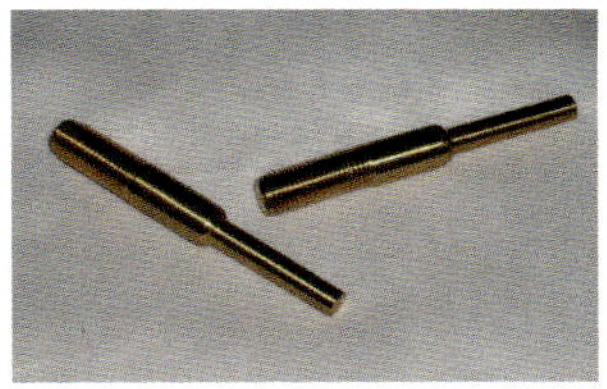
Reproduced Variable Capacitor Shafts

Removal of Variable Capacitor from Chassis

1. Disconnect the bus wires attached to the variable capacitor housing. See Top View and Underside View.
2. On the front panel, remove the two stops and screw directly below the C knob. See Front Panel.
4. Gently remove the variable capacitor assembly from the chassis.

To reinstall the variable capacitor, reverse the removal process.

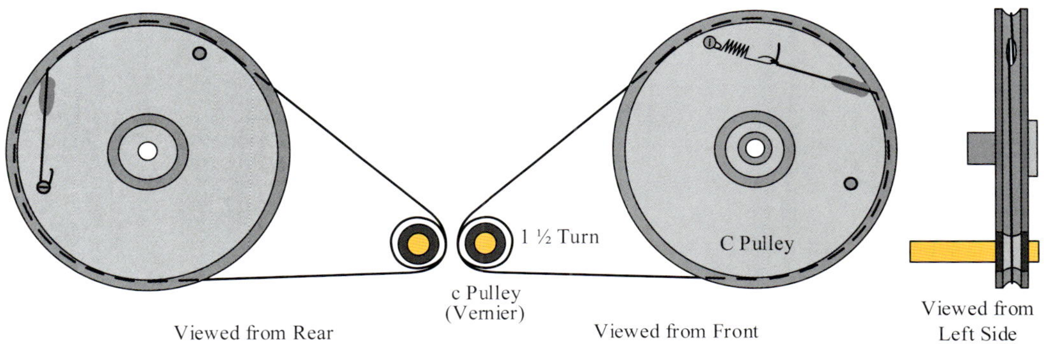

C Pulley Detail

3-R (E) Vario-Coupler

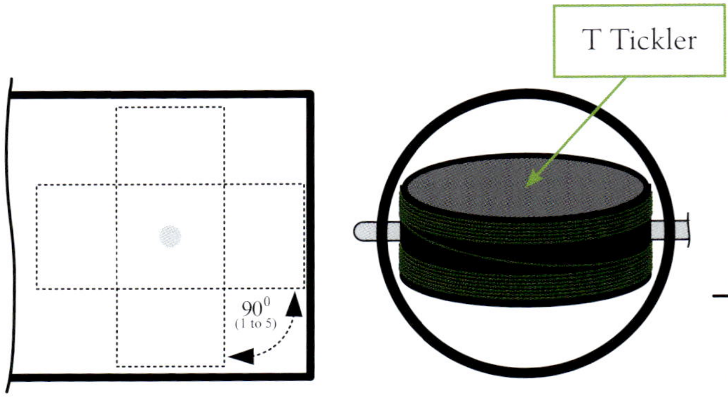

Tickler Clocking

When viewed from the front the tickler rotates from horizontal down to vertical, and T pointer moves from 5 to 0, respectively.

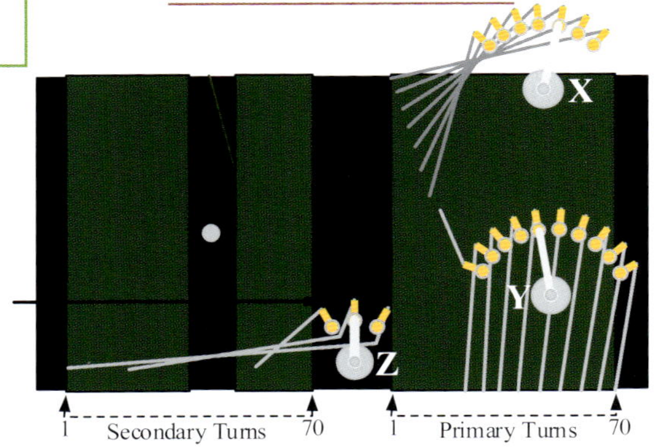

Vario-Coupler Front View

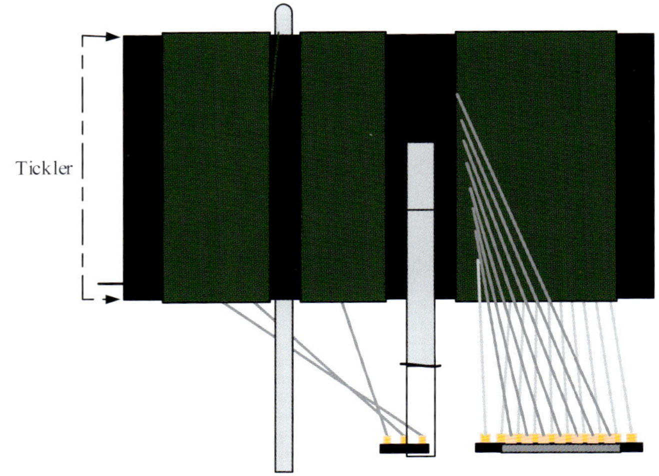

Vario-Coupler Top View

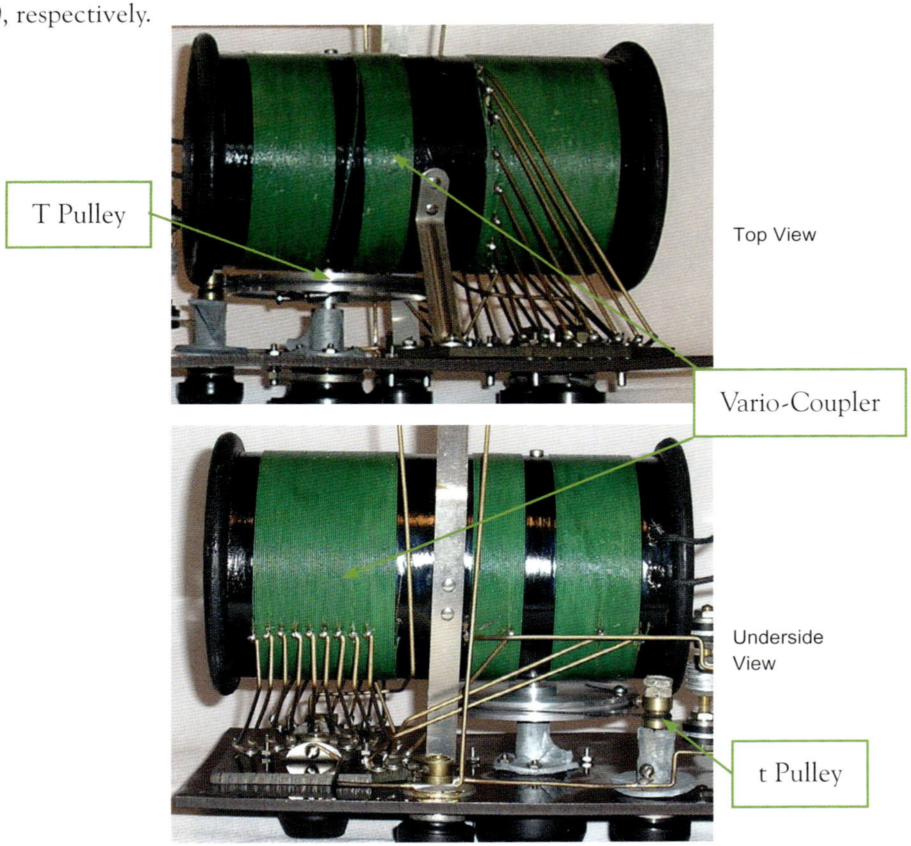

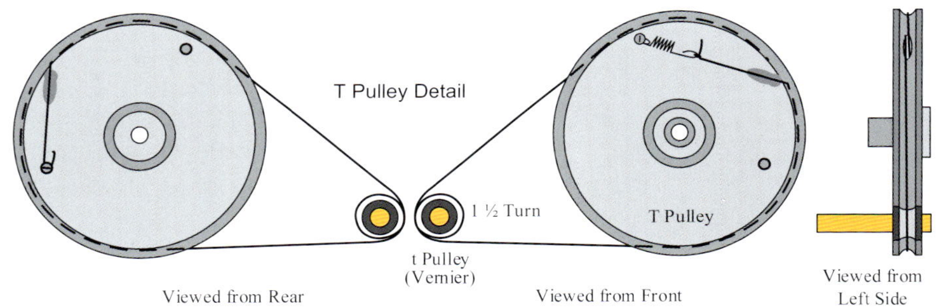

Vario-Coupler Windings and Tap Switch Connections

The main coil of the 3-R (E) consists of a primary and secondary coil each having 70 turns. The secondary coil is physically wound as two coils separated by ½". The coil is constructed in this manner to facilitate the mounting of the small tickler coil inside the main coil. The X-switch tap points 1 through 7 are connected to turns 7 through 1 of the primary coil. In a like manner, Y-switch tap points 1 through 10 are connected to turns 8, 15, 22, 29, 36, 43, 50, 56, 63, and 70 of the primary coil. Additionally, the Y-switch tap point 1 is connected to X-switch tap point 1. The Z-switch taps are connected to the turns of the secondary coil as follows: 1 (tap 3), 20 (tap 2) and 50 (tap 1). The T tickler coil has two connected sets of windings, separated by ½", with each coil section having 13 turns (no taps).

3-R (E) Internal Power

The negative grid bias voltages for the three AF tubes are provided by an internal multi-voltage "C" battery.

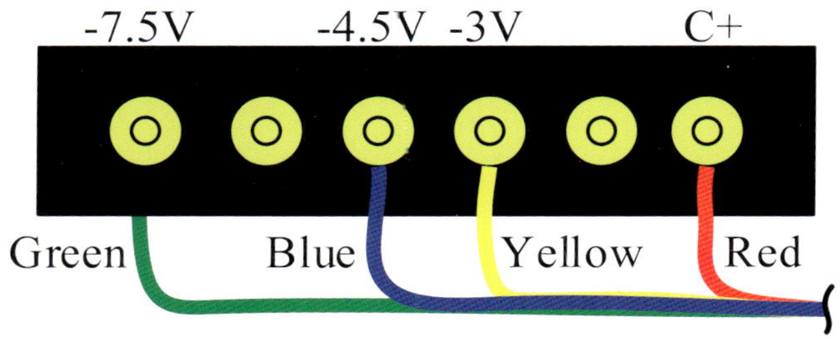

"C' Battery Connections

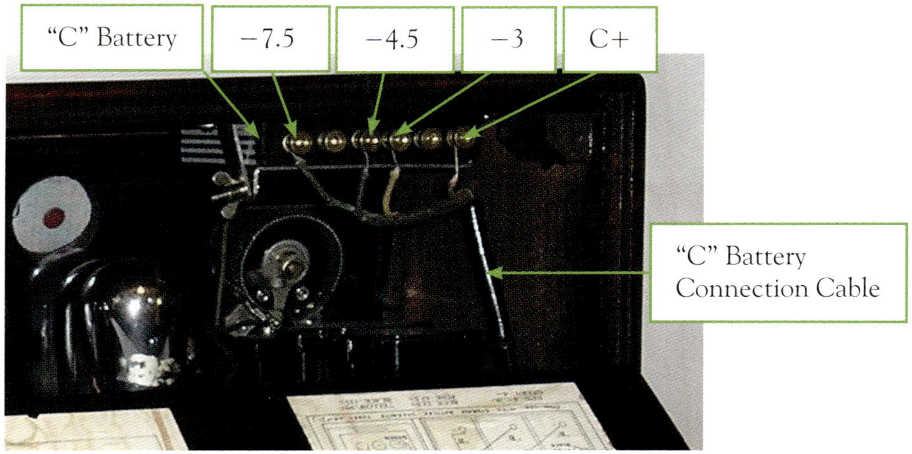

Internal "C" Battery

Reproduced and Original "C" Batteries

3-R (E) External Power

The 3-R (E) uses a 7 conductor power cable to hook up the external batteries. The power cable has a multi-plug on one end that is keyed, so that it only goes into the socket one way. This keying method ensures that the proper voltages are routed to the proper circuits. The loose wires, at the other end, are color coded and should be connected per the manufacturer's instructions, which are located on the inside of the back door of the radio. (Note: Some sets did not have a brown wire.)

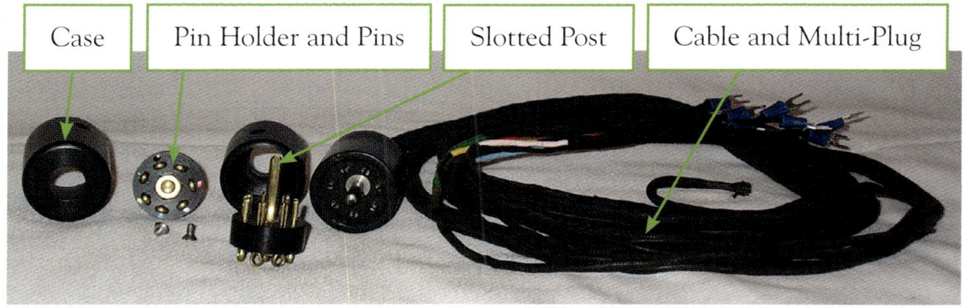

Original Power Cable and Reproduced Multi-Plugs

Power Cable Socket

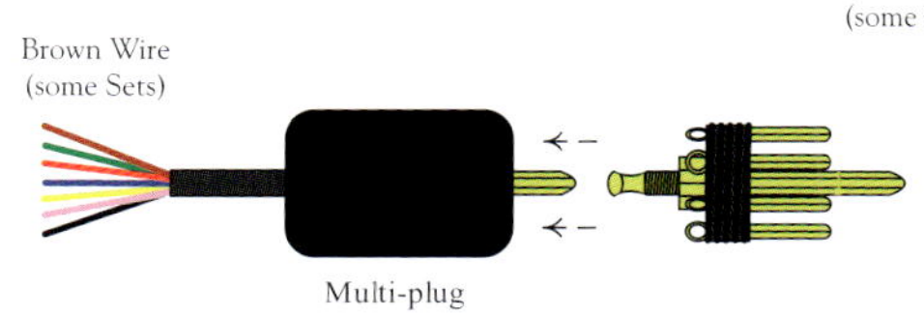

Power Cable and Multi-Plug

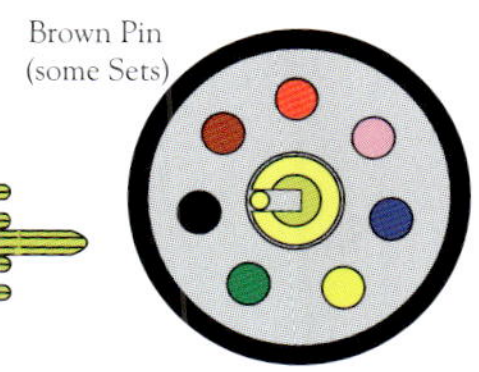

Multi-Plug Color Coding and Clocking

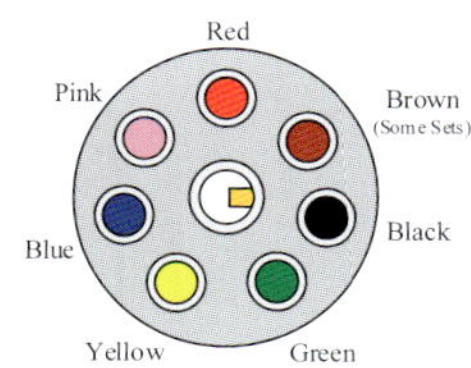

Socket Color Coding and Clocking

3-R (E) Functional Description Quick Reference Guide

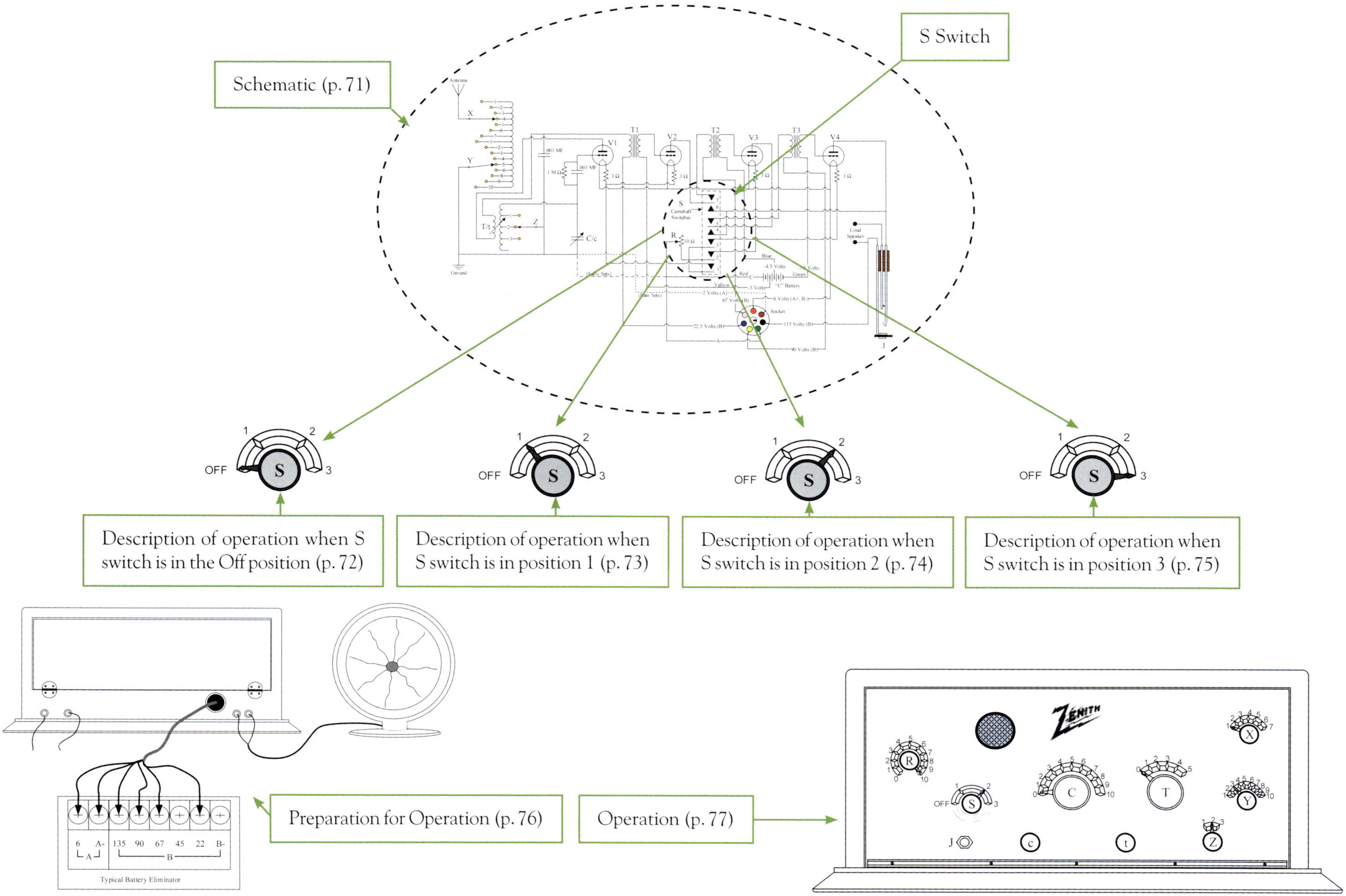

3-R (E) Schematic

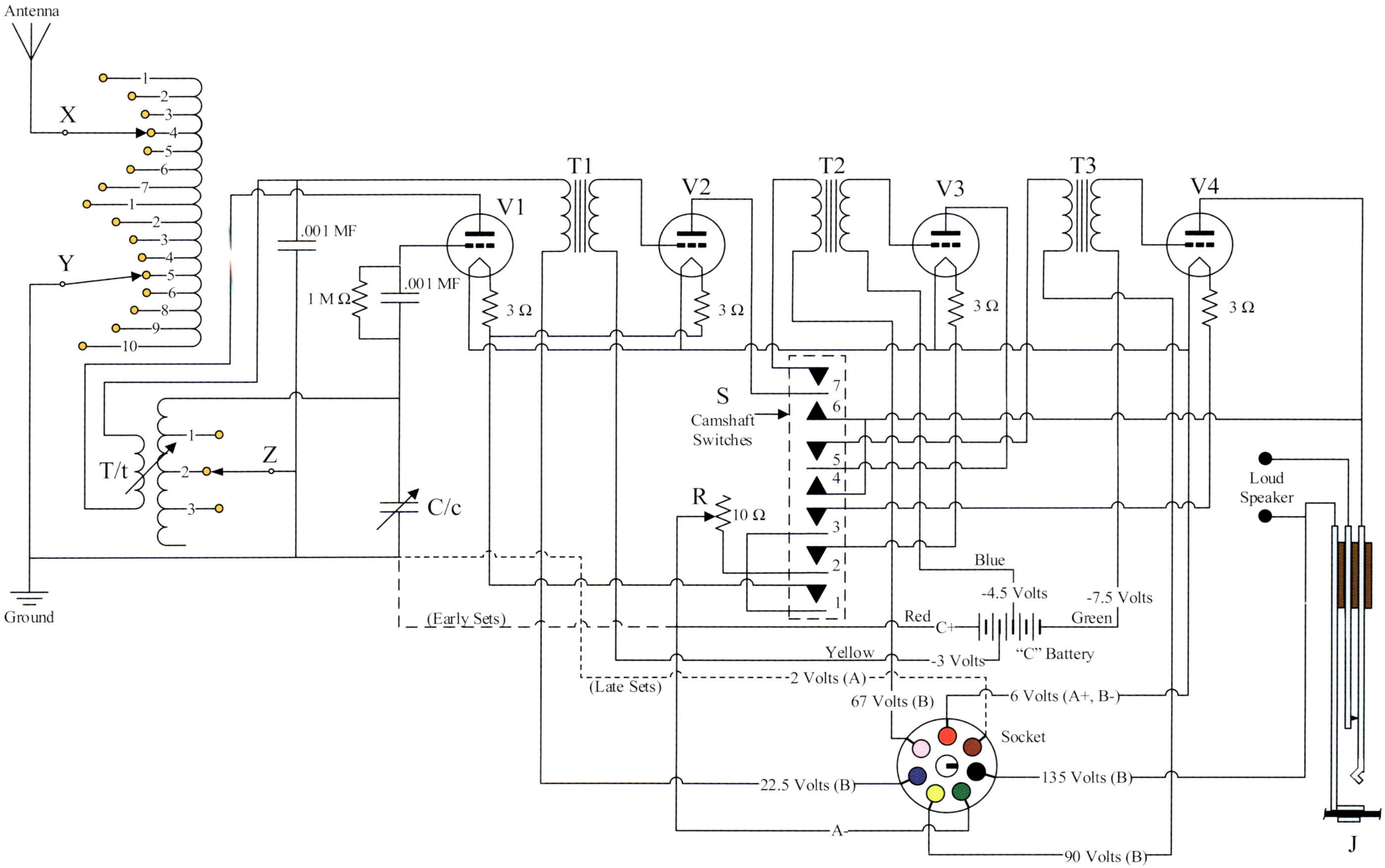

3-R (E) Description of Operation—Switch S in Off position

When switch S is in the OFF position the filament circuits are open and the radio is inoperable. Plate voltage is applied to the plates of V1, V2, V3 and V4. A negative bias voltage is applied to the grids of V2, V3, and V4. This is considered the initial state of the radio for future discussions of circuit operation.

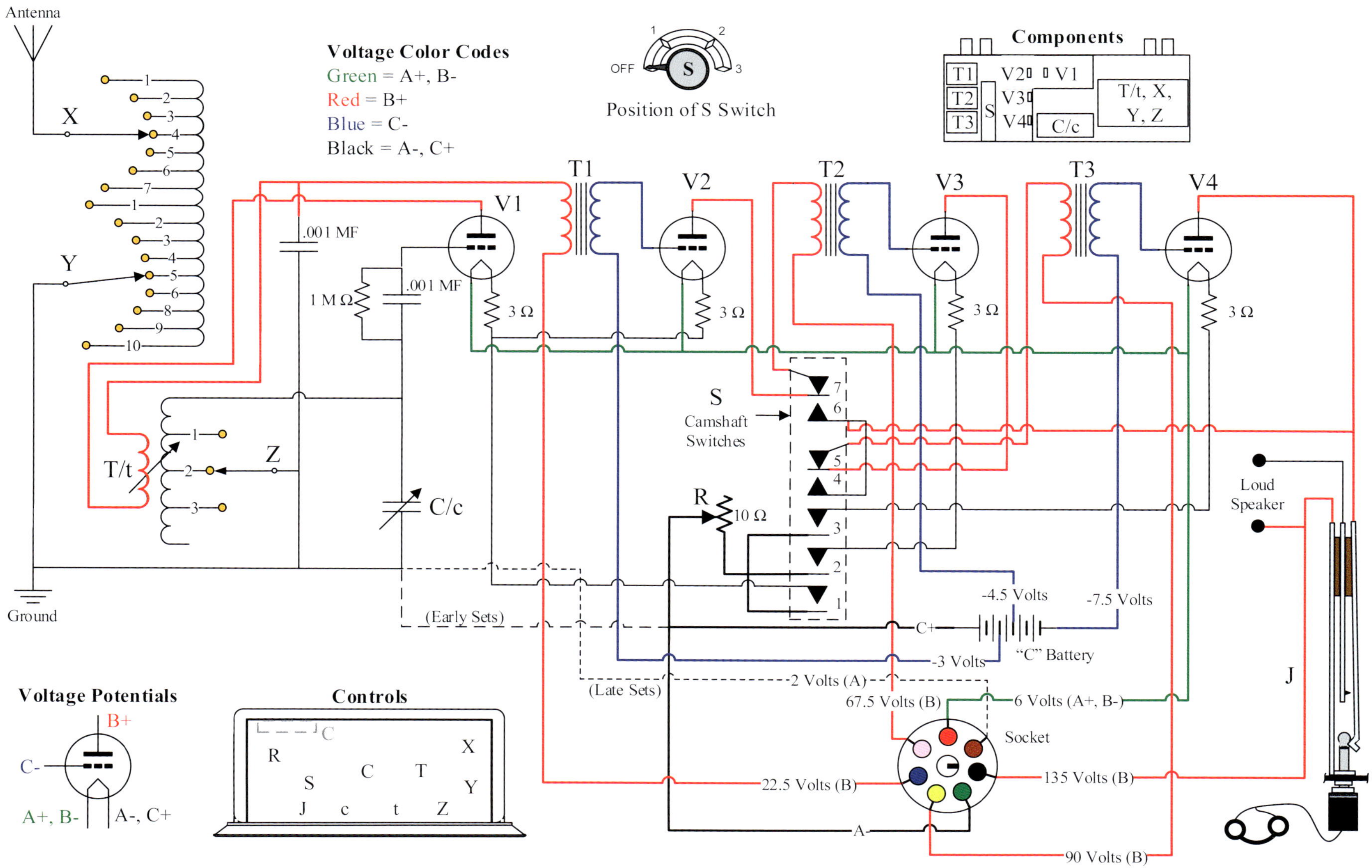

3-R (E) Description of Operation—Switch S in Position 1 (One AF Amplifier Stage)

When switch S is moved to position 1, camshaft switch 1 closes energizing the filament circuits of V1 and V2. At the same time camshaft switch 7 opens disconnecting the plate of V2 from the primary of transformer T2, and camshaft switch 6 closes connecting the plate of V2 to the headphones. The radio is now playing with the detector (V1) and one stage (V2) of AF amplification.

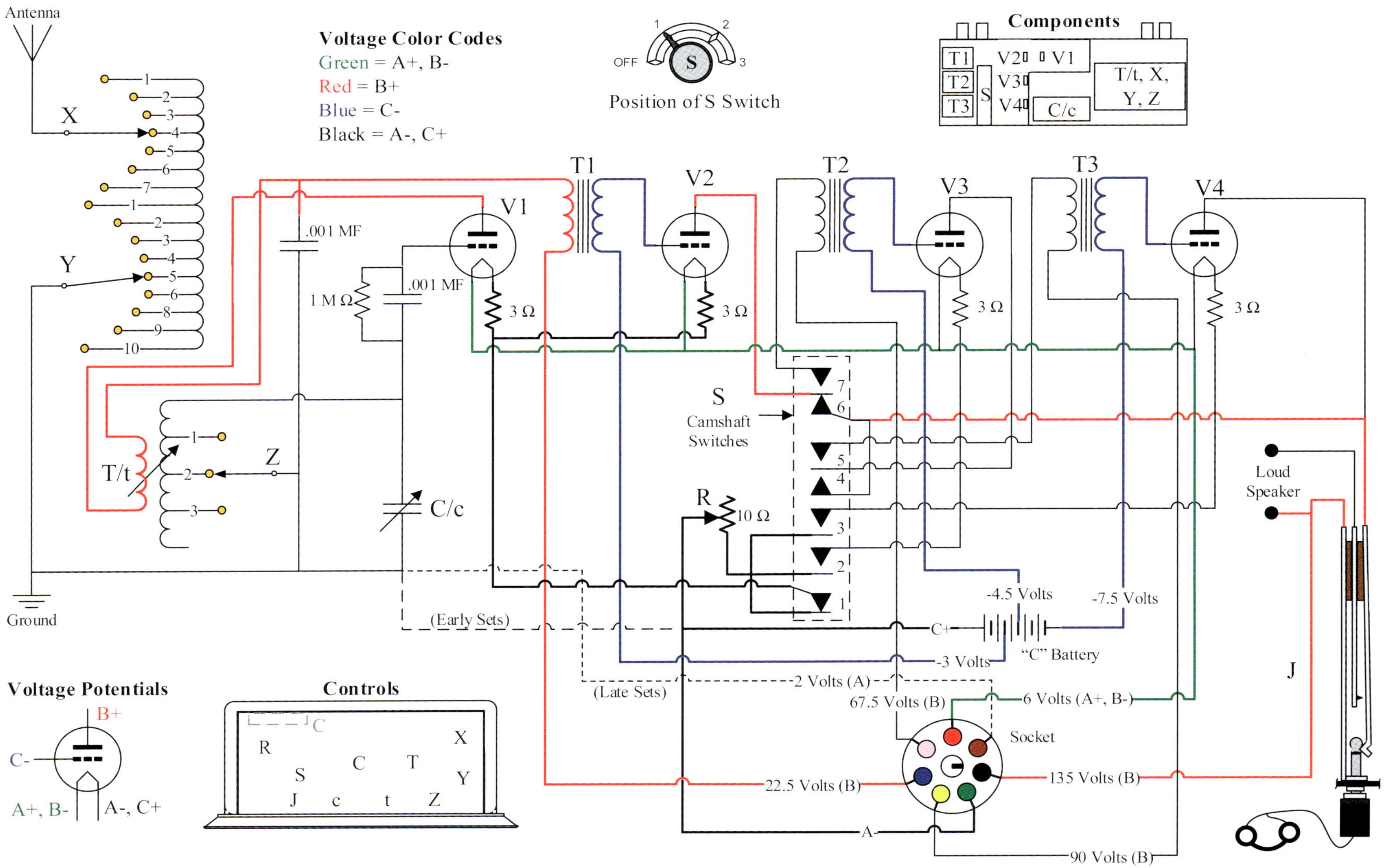

3-R (E) Description of Operation—Switch S in Position 2 (Two AF Amplifier Stages)

When switch S is moved to position 2, camshaft switch 2 closes (camshaft switch 1 remains closed) energizing the filament circuits of V1, V2 and V3. At the same time camshaft switch 6 opens disconnecting the plate of V2 from the headphones circuit, camshaft switch 7 closes connecting the plate of V2 to the primary of transformer T2, and camshaft switch 4 closes connecting the plate of V3 to the headphones circuit. The radio is now playing with the detector (V1) and two stages (V2, V3) of AF amplification.

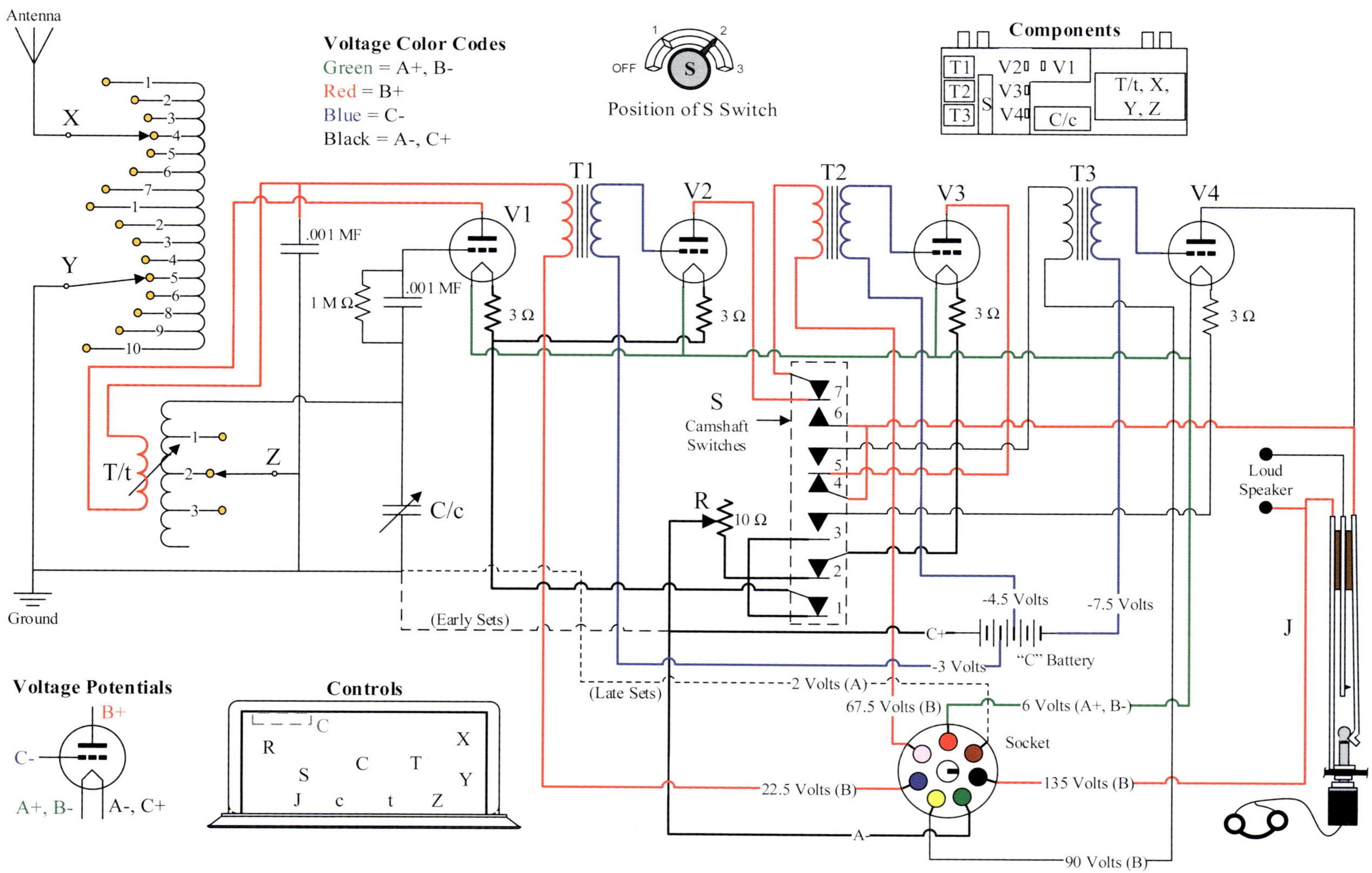

3-R (E) Description of Operation—Switch S in Position 3 (Three AF Amplifier Stages)

When switch S is moved to position 3, camshaft switch 3 closes (camshaft switches 1 and 2 remain closed) energizing the filament circuits of V1, V2, V3 and V4. At the same time camshaft switch 4 opens disconnecting the plate of V3 from the headphones circuit (the plate of V4 is already connected to the headphones circuit), and camshaft switch 5 closes connecting the plate of V3 to the primary of transformer T3. The radio is now playing with the detector (V1) and three stages (V2, V3, V4) of AF amplification.

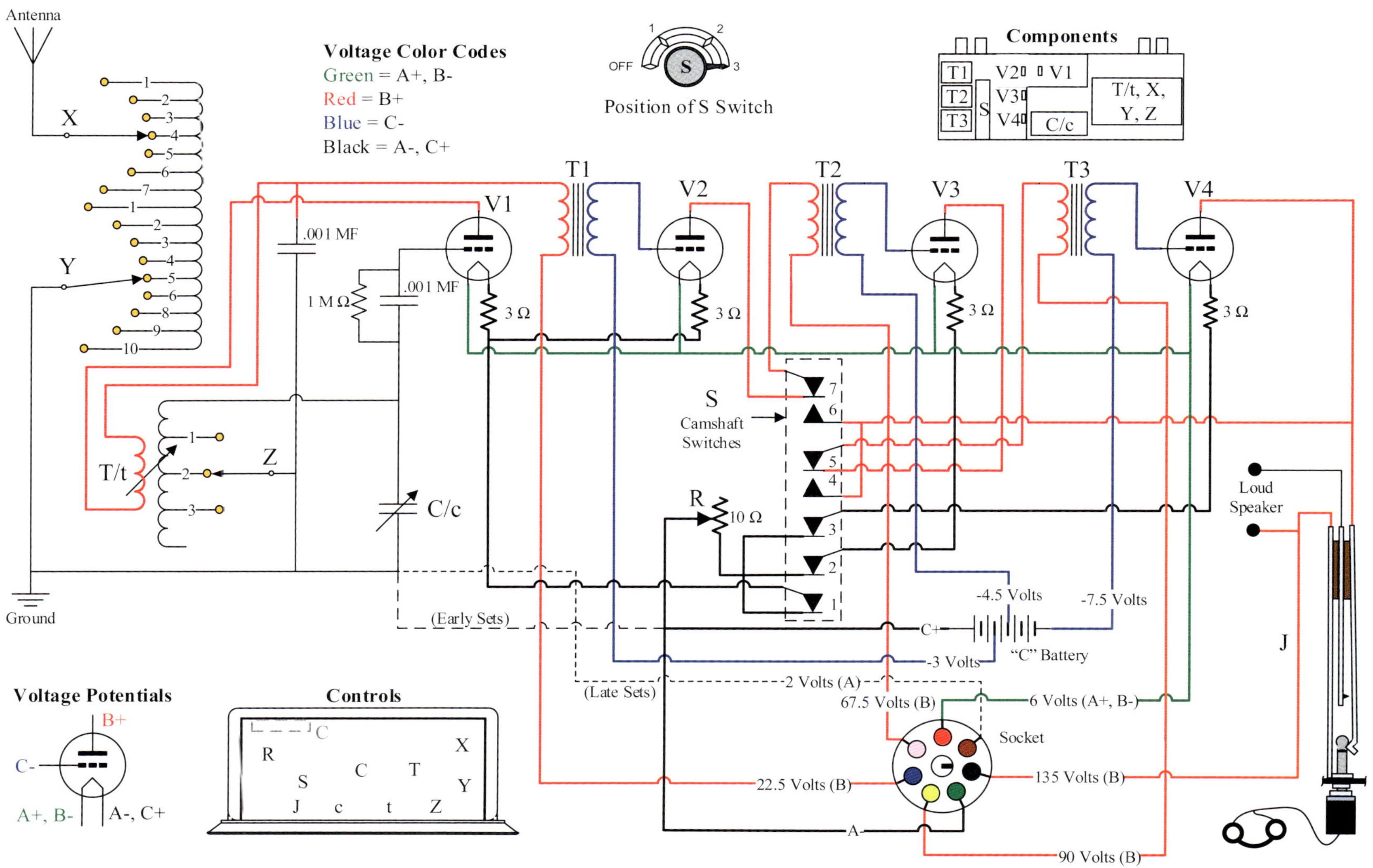

3-R (E) Preparations for Operation

Install the tubes and the "C" battery inside the radio. Connect the antenna and ground to the ANTENNA and GROUND terminals. Use the appropriate battery connection diagram, inside the back door of the radio, to make sure the power can be connected properly and agrees with the 3-R (E) Connections diagram below. If the power connections are not correct or otherwise unavailable, utilize the physical and functional descriptions to assist in determining the correct power connections. Hook up the power to the radio. Proceed to 3-R (E) Operation on the next page. See Common Radio Attachments (page 8) for additional information related to antennas, loudspeakers and power options.

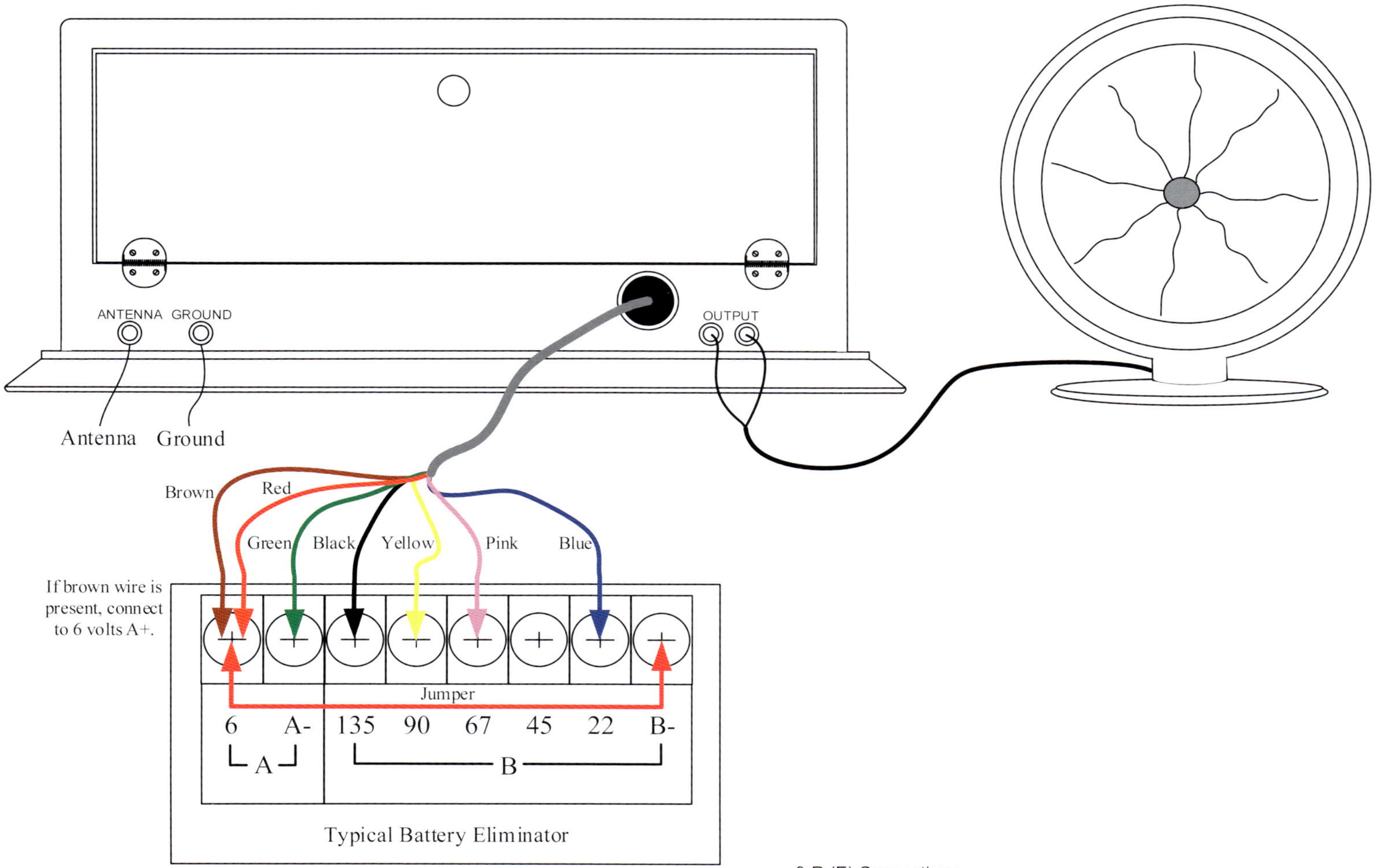

3-R (E) Connections

3-R (E) Operation

Follow the 3-R (E) Preparation for Operation instructions on the previous page. Set the pointers to the positions shown in the Initial Dial Settings diagram below.

The tuning capacitor C selects the stations and the tickler T provides the feedback (regeneration) that increases the volume of the selected signal. Practice using knobs C and T simultaneously. Utilize the small c and t knobs to make fine adjustments to the C and T mechanisms.

Insert the headphone plug into J jack. Turn the "S" switch to position 1 or 2.

Gradually move "T" to the right until a hissing sound is heard and hold the pointer to the beginning of that sound. Gradually move "C" to the right until a bird whistle is heard; first high, then descending to low; then a pause will be heard. As you continue to move "C" the whistle is again low, then ascending to high. The station is between these low points or valleys. Move "T" to the beginning of the whistle, while "C" is being moved. If you have C in the valley of the whistle, slightly move "T" to the left to clear the station. For best results operate C and T simultaneously. If an unpleasant howl (oscillations) is heard and the station is lost, turn "T" back to the left until the howl stops.

Adjust the audio level (volume) by setting switch S to position 1, 2 or 3, as appropriate.

If a speaker is connected to the rear terminals and a headphone plug is inserted into the J jack, the loudspeaker will be automatically turned OFF, and when the plug is pulled out of the J jack, the loudspeaker will be automatically turned ON.

Knobs X and Y tune the antenna circuit. A little manipulation of these switches will enable the operator to obtain improved results.

Knob Z changes the bandwidth of the receiver and is generally left in the position 2. Position 1 decreases the bandwidth and position 3 increases the bandwidth.

Record the control settings for each station in Station Log (next page). Then, to return to a particular station, set the controls to the settings that were previously recorded in the log for that station.

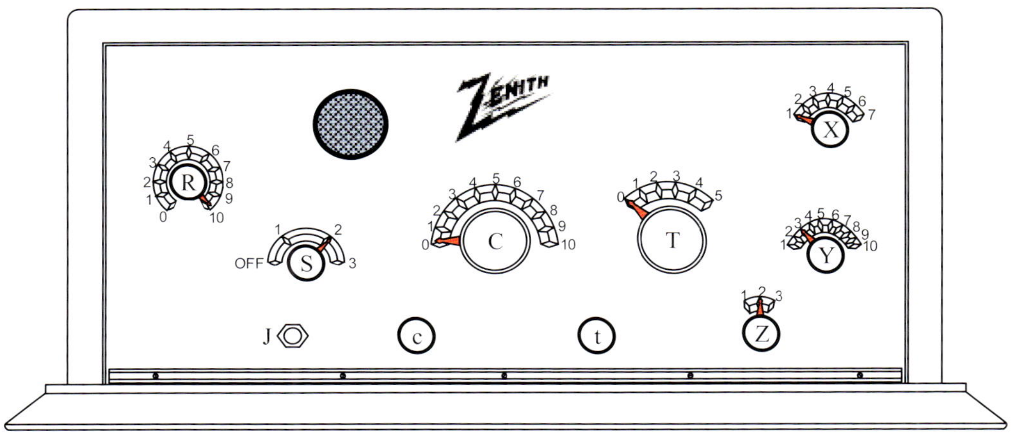

Initial Dial Settings (pointers shown in red for clarity)

3-R (E) Station Log

Name: ______________________________

Date	Time	Station	Frequency	C	T	X	Y	Z

This page may be copied, scanned or reproduced.

From *The Early Zenith Radios: The Battery Powered Table Sets 1922–1927,* © 2014 by Gilbert M. Hedge with Durell M. Roth. ISBN 978-0-7643-4674-3. Schiffer Publishing Ltd. www.schifferbooks.com

Chapter 5: **Zenith 4-R** (Early Version)

4-R (E) General Specifications

Company:	Zenith Radio Corporation
Model:	4-R (Early Version)
Year Manufactured:	1923
Manufacture Location:	48th and Kedzie, Chicago IL (QRS Factory)
Serial Number:	6717
Price:	$95
Tube:	UV-201-A (4), one detector, three AF amplifiers
Circuit Type:	Regenerative

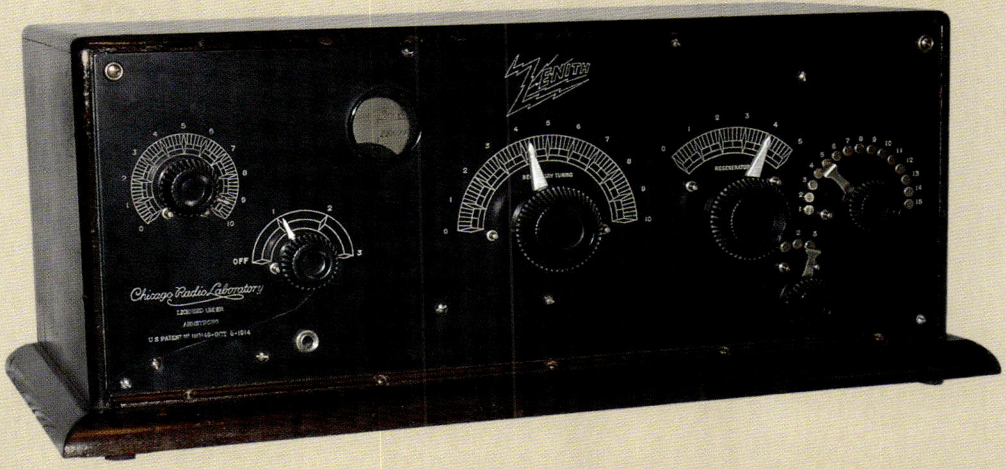

4-R (E) General Information

The 4-R (E) radio was a less expensive version of the 3-R (E) radio with most of the same features. It did not have the vernier dials or the rear posts for attaching a horn speaker. The circuitry was the same except the X tap switch was eliminated and the Y tap switch was expanded from 10 to 15 taps. The cabinet base was very plain, as compared to the 3-R, and the front panel was straight and did not slope like the 3-R. The front panel can be either black or brown. A mahogany trim molding strip was added to support the bottom portion of the front panel. The tap switches Y and Z are exposed on the front panel.

Restoration details for the set that was used as the example in this chapter. *The pointers are the only parts that were affected by the pot metal problem. They had been previously replaced with handmade aluminum pointers. The power cable is original. The "C" battery was replaced with a reproduced battery.*

4-R (E) Physical Description Quick Reference Guide

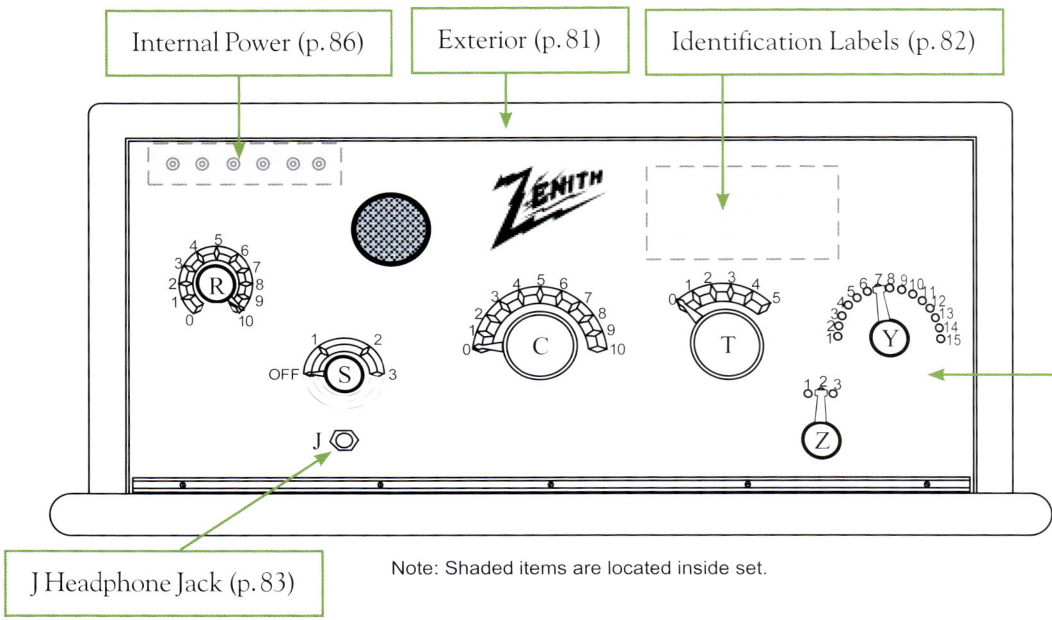

Control	Function
C	Variable Capacitor (tuning)
J	Headphone Jack
R	Rheostat (filament voltage control)
S	Switch (Off, 1, 2 or 3 AF amp. stages)
T	Tickler (regeneration)
Y	Tap Switch (tunes the antenna circuit)
Z	Tap Switch (selects the bandwidth)

Note: Shaded items are located inside set.

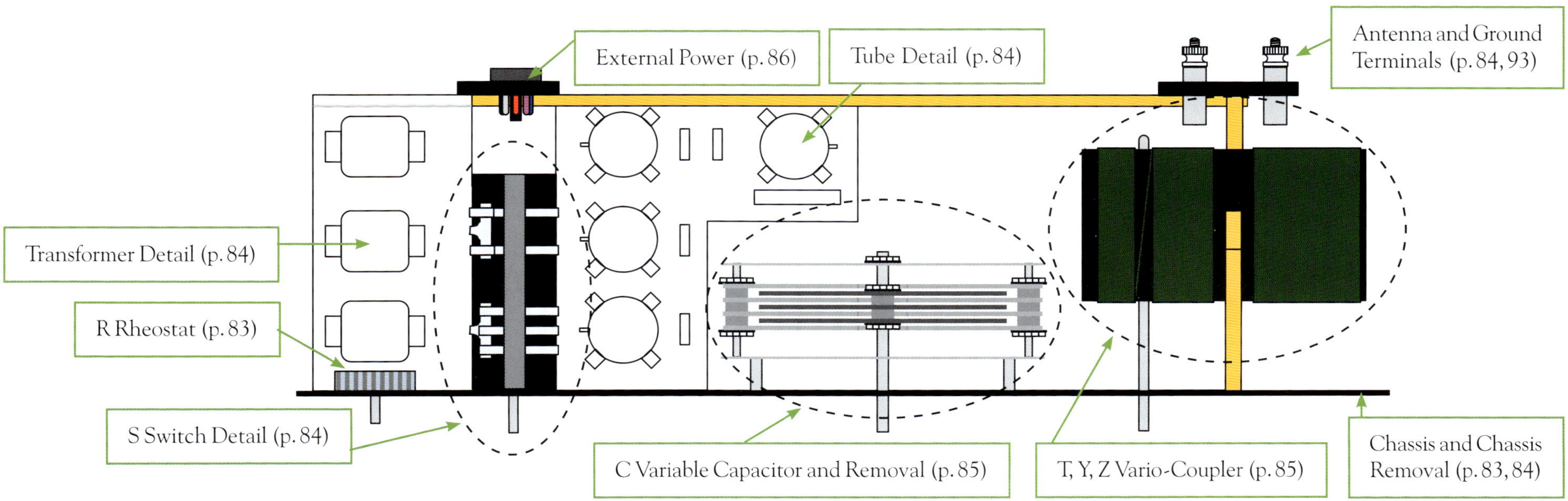

4-R (E) Exterior

The cabinet is made of mahogany and measures 21.75" by 10" by 9". The front panel is made of black Bakelite and a mahogany trim molding strip is located at the bottom. There is a peep hole for viewing the brightness of the tubes.

Front View

Rear View

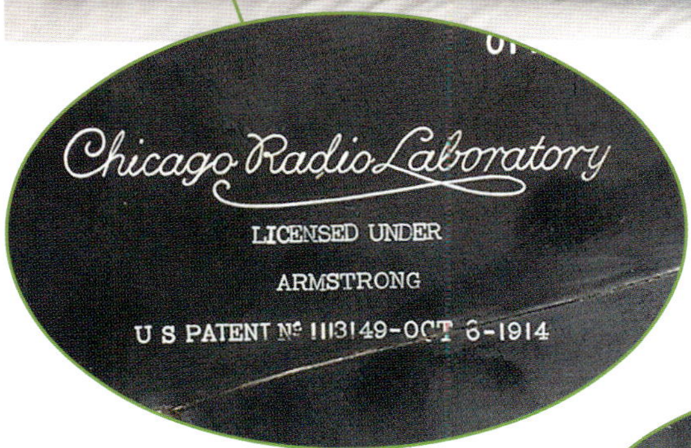

Chicago Radio Laboratory
logo and Armstrong patent information.

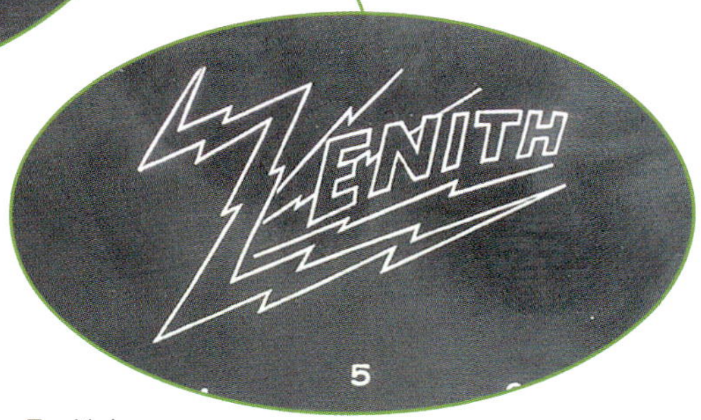

Zenith Logo

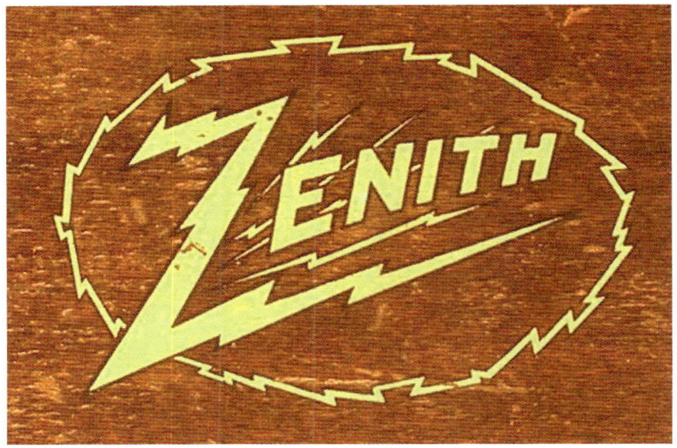

Top View. Original Zenith decal.

4-R (E) Identification Labels

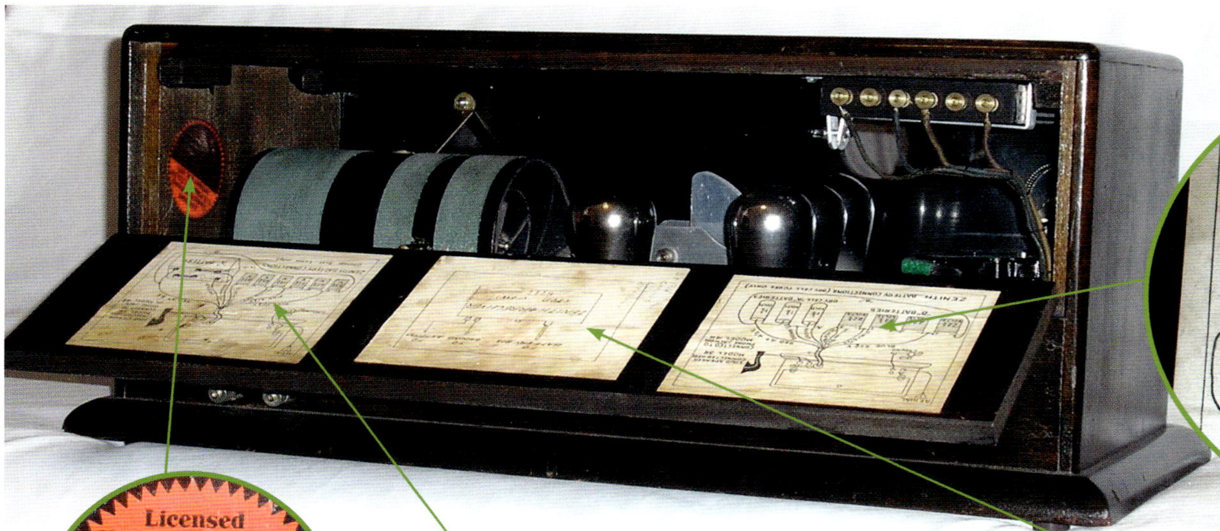

Manufacturer's Labels

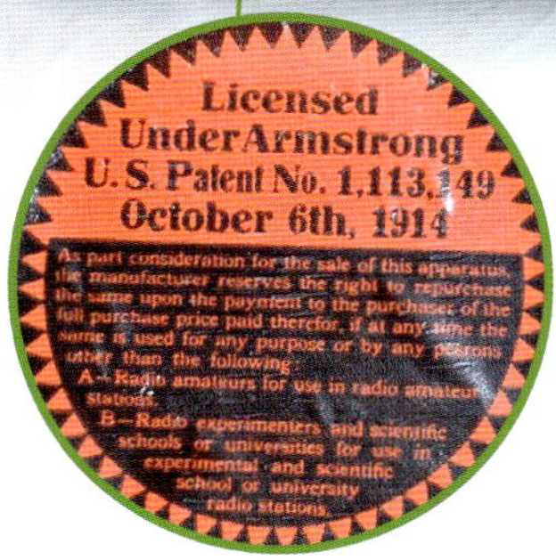

Armstrong License and Patent label

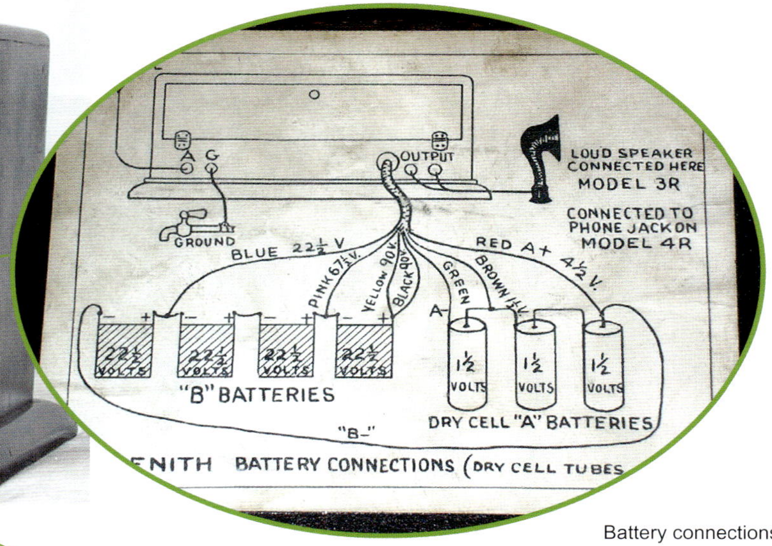

Battery connections when using UV-199 tubes (dry cells).

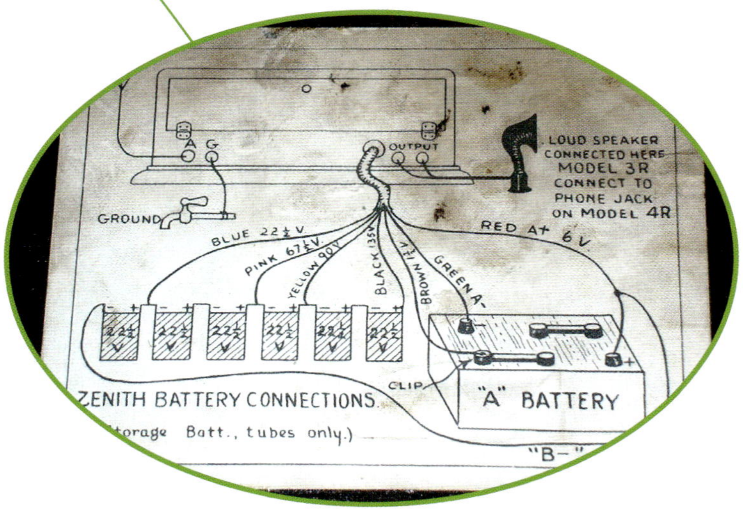

Battery connections when using UV-201-A tubes.

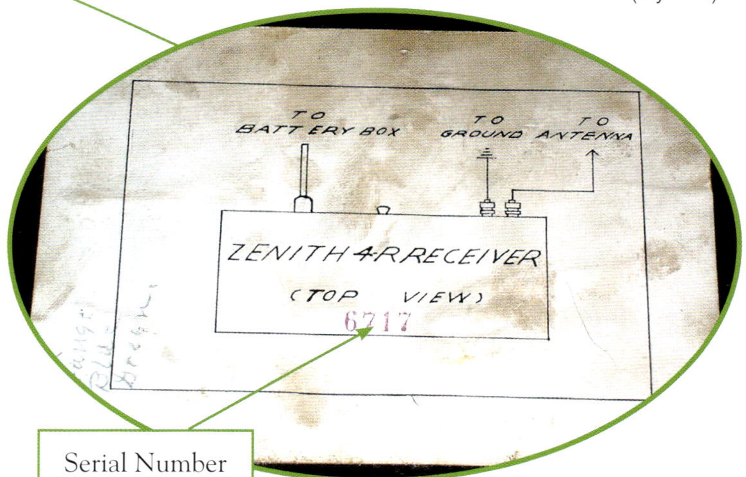

Serial Number

External Connections

4-R (E) Chassis

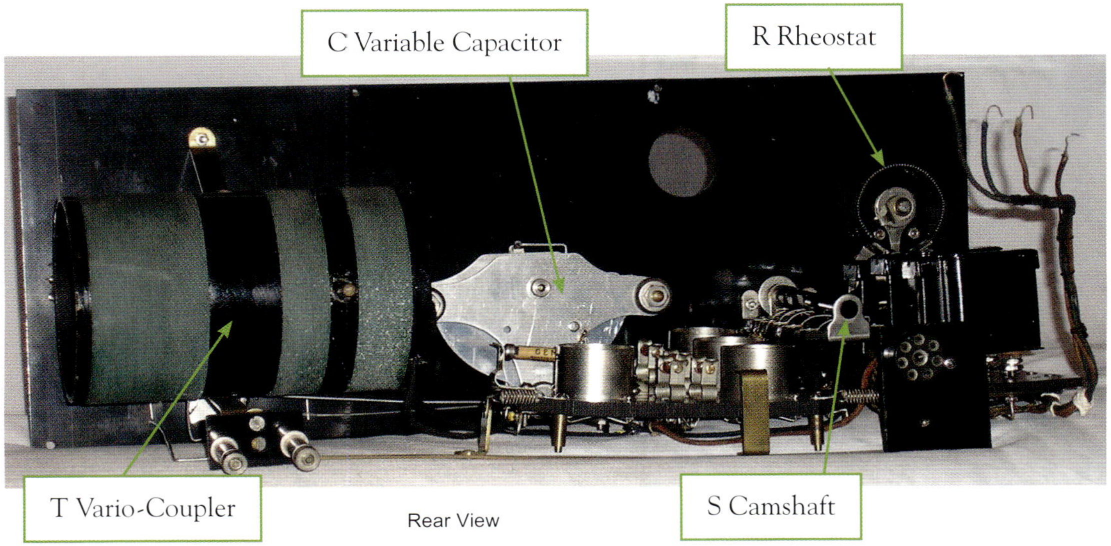

Rear View

Top View (looking into cabinet with door open)

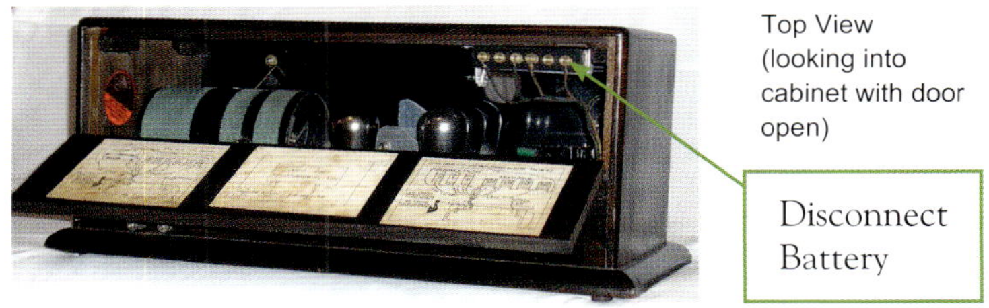

Back View

Front View

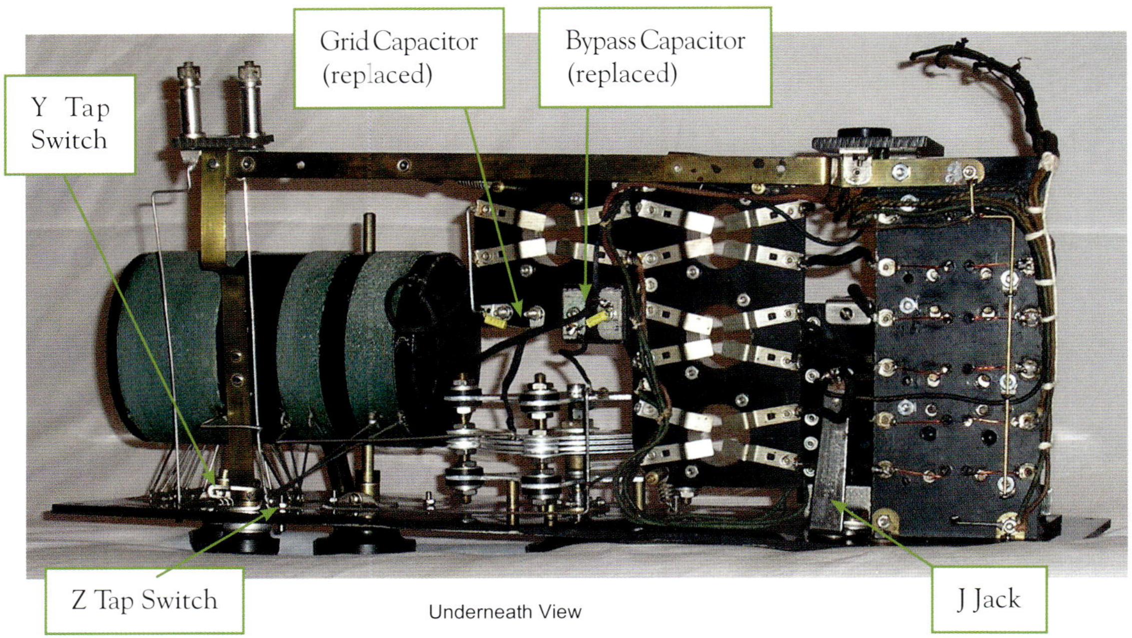

Underneath View

Removal of Chassis from the Cabinet

1. Inside the rear of the cabinet, remove the two screws holding the chassis to the bottom of the cabinet; see Top View. Disconnect the "C" battery, see Back View.
2. On the front panel, remove trim molding. Remove the four corner screws with the raised washers. Remove the two small screws located at the top of the front panel; see Front View.
3. The whole chassis slides out the front of the cabinet. Lift the chassis above the cabinet lip and gently push it from the rear.

To reinstall the chassis into the cabinet, reverse the removal process.

4-R (E) Chassis (continued)

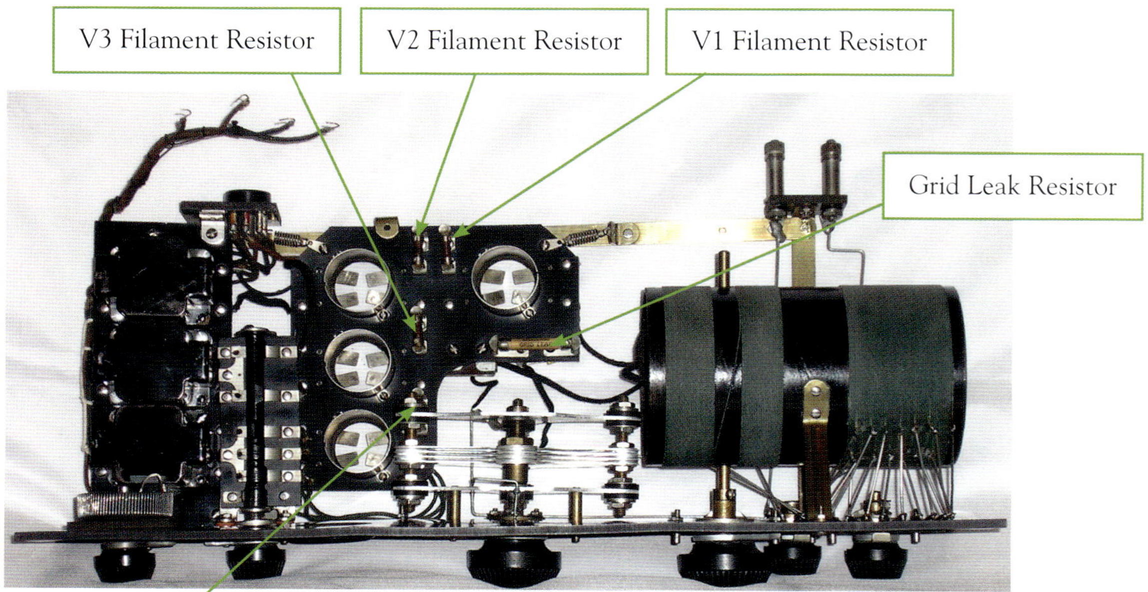

Top View

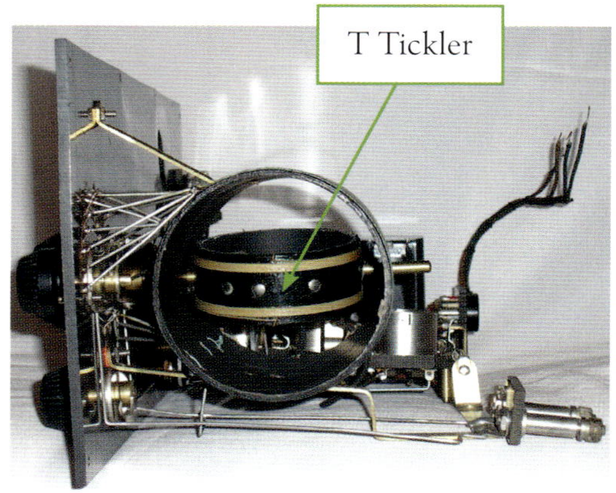

Right Side

Left Side

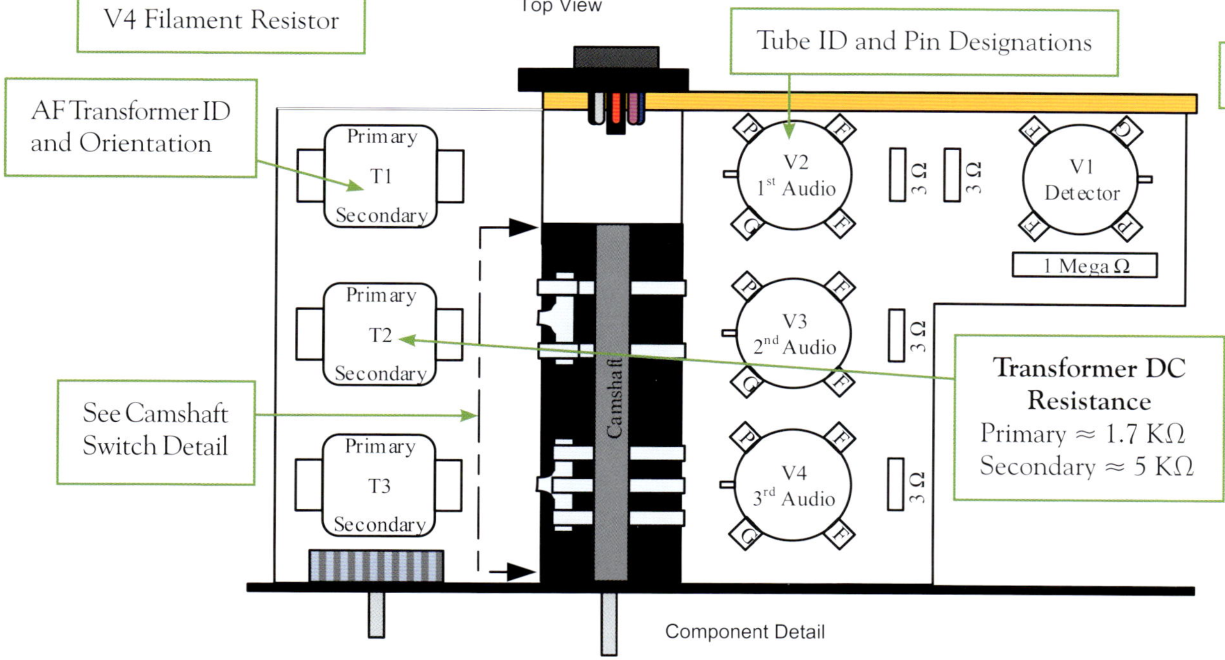

Component Detail

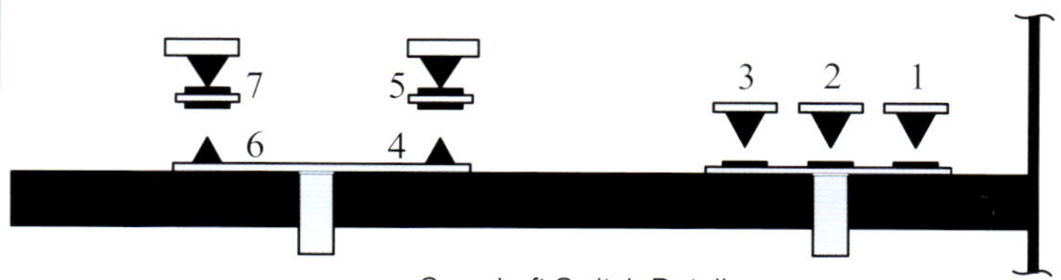

Camshaft Switch Detail

4-R (E) Variable Capacitor

The variable capacitor housing holds a shaft (rotor) that rotates 3 plates, merging them with 4 fixed plates (stator).

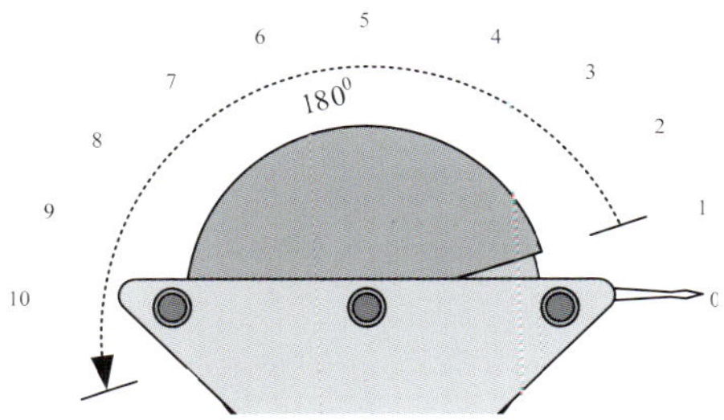

Variable Capacitor Clocking (Viewed from Rear)

As the Variable Capacitor rotates 180° left-to-right (viewed from front) the pointer rotates 1 to 10. Pointer and numbers are shown for clarity and are located on the front panel.

Removal of Variable Capacitor from Chassis

1. Disconnect the bus wires attached to the variable capacitor housing. See Underside View and Top View.
2. On the front panel, remove the two stops for the C pointer and screw directly below the C knob. See Front View.
4. Gently remove the variable capacitor assembly from the chassis.

To reinstall the variable capacitor, reverse the removal process.

Underside View

Top View

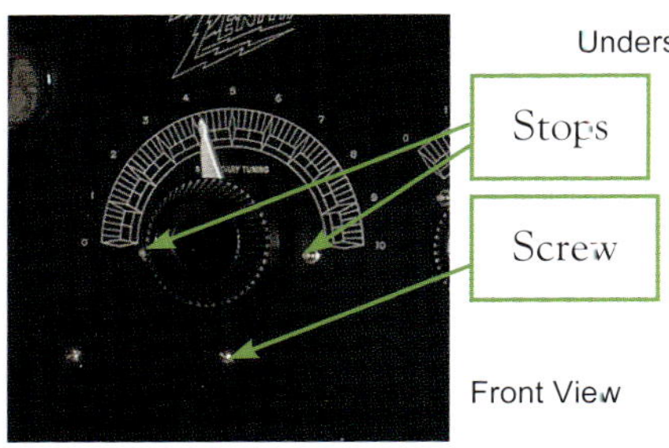

Front View

4-R (E) Vario-Coupler

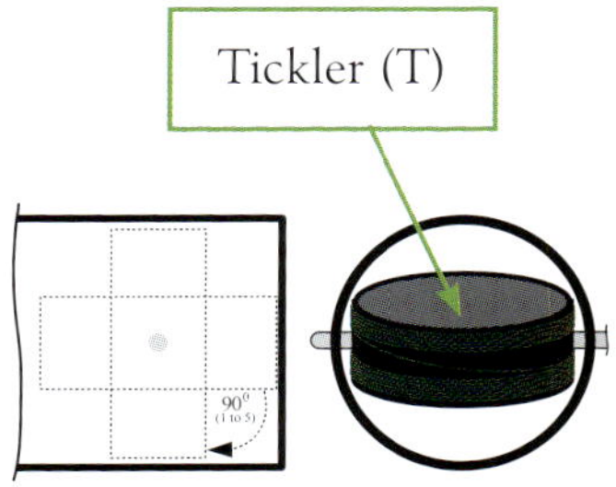

Tickler Clocking

When viewed from front the tickler rotates horizontal, down to vertical, and T pointer moves from 5 to 0, respectively.

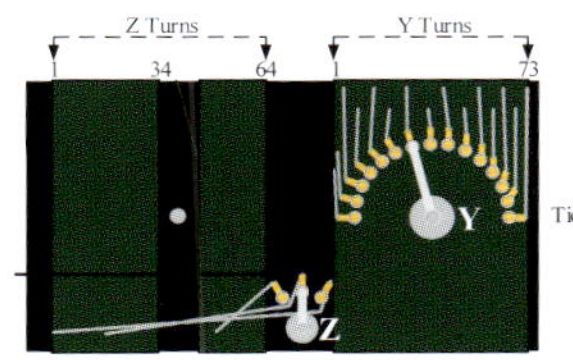

Front View

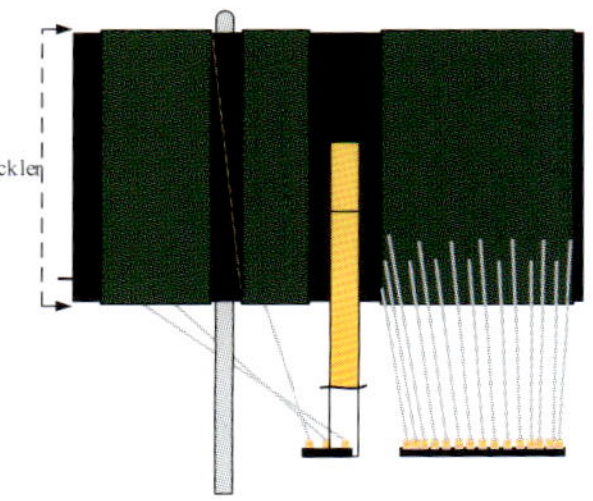

Top View

Vario-Coupler Windings and Tap Switch Connections

The main coil of the 4-R (E) consists of a primary coil with 73 turns and secondary coil with 64 turns. The secondary coil is physically wound as two coils separated by 5/8". The coil is constructed in this manner to facilitate the mounting of the small tickler coil inside the main coil. The Y-switch tap points 1 through 15 are connected to turns 1, 2, 8, 13, 18, 23, 28, 33, 38, 43, 48, 53, 58, 63, and 73 of the primary coil. In a like manner the Z-switch taps are connected to the turns of the secondary coil as follows: 1 (tap 3), 20 (tap 2) and 44 (tap 1). The tickler T coil has two connected sets of windings, separated by ½", with each coil section having 10 turns (no taps).

4-R (E) Internal Power

The negative grid bias voltages for the three tubes (V2, V3, V4) are provided by an internal multi-voltage "C" battery.

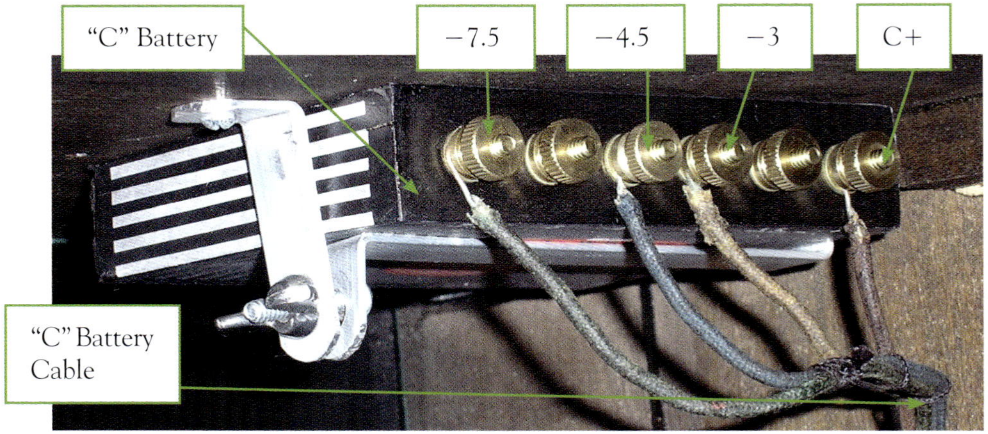

Internal "C" Battery

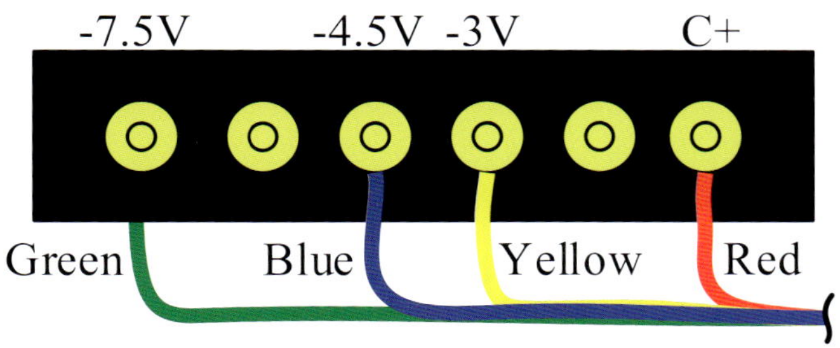

"C" Battery Connections

Reproduced and Original "C" Batteries

4-R (E) External Power

External power for the 4-R (E) is provided by a 7 conductor cable that has a multi-plug on one end. The multi-plug and socket are keyed so that the plug only goes into the socket one way. This keying method ensures that the proper voltages are routed to the proper circuits. The loose wires at the other end of the power cable are color coded and should be connected per the manufacturer's instructions, which are located on the inside of the back door. (Note: Some sets may not have the brown wire.)

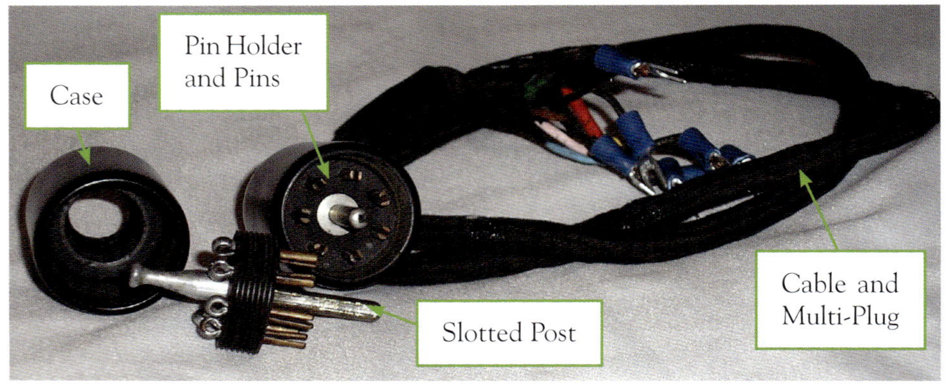

Original Power Cable and Reproduced Multi-Plugs

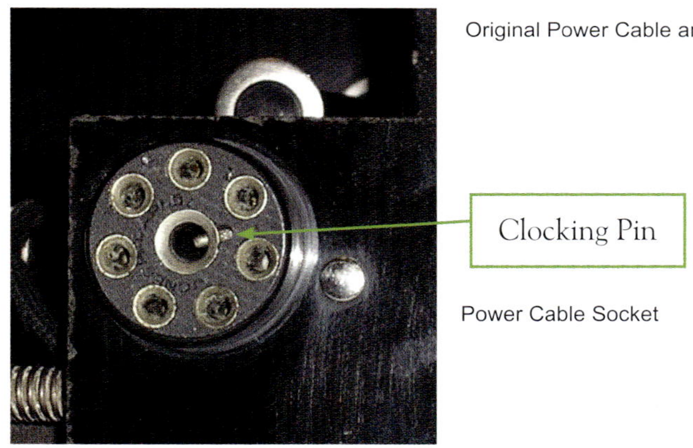

Power Cable Socket

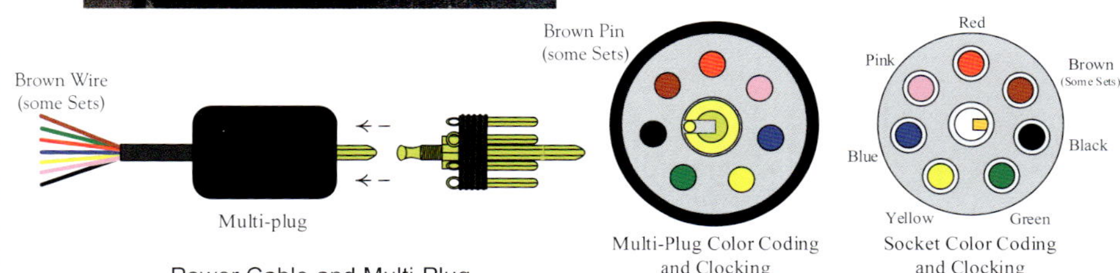

Power Cable and Multi-Plug

4-R (E) Functional Description Quick Reference Guide

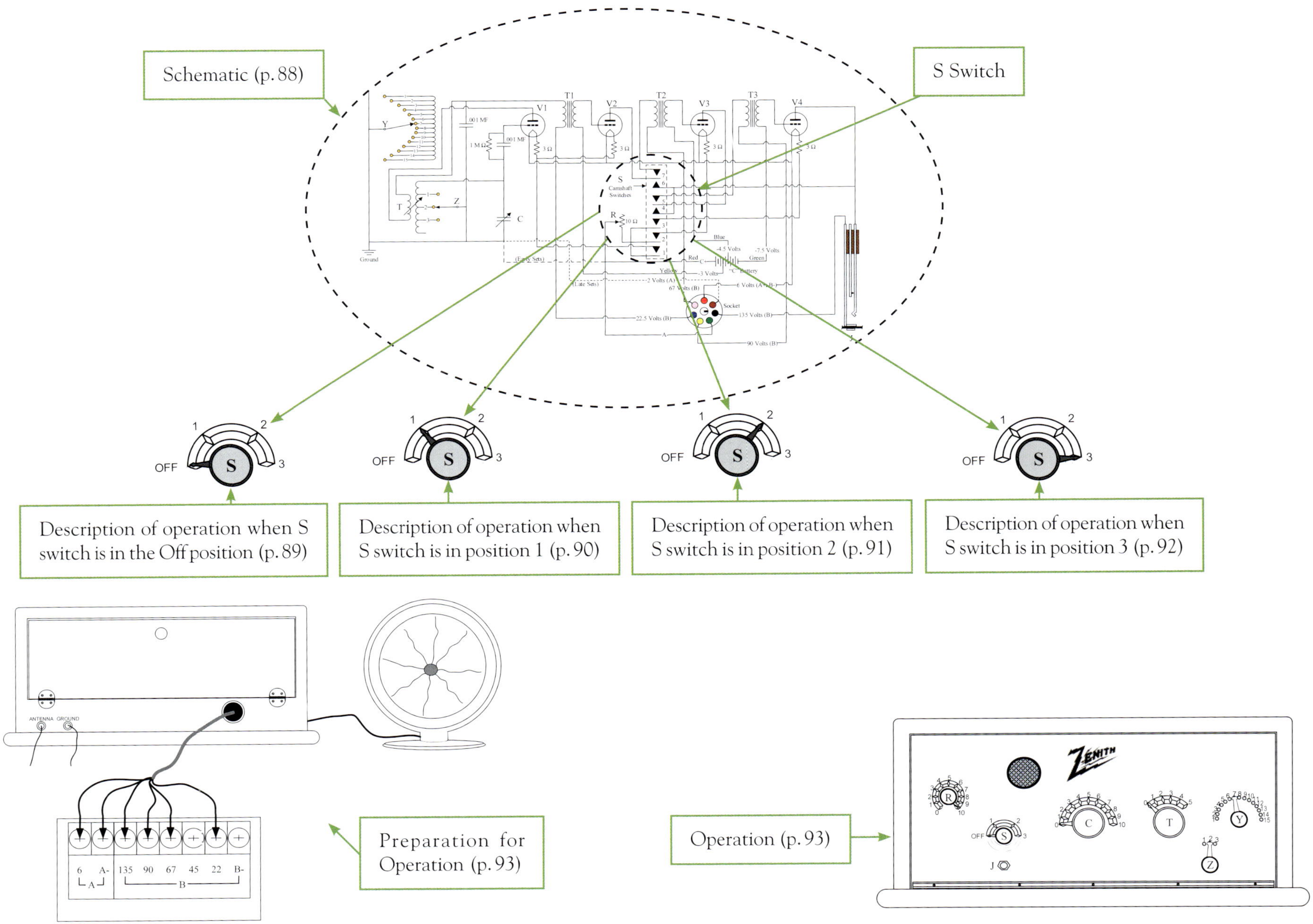

4-R (E) Schematic

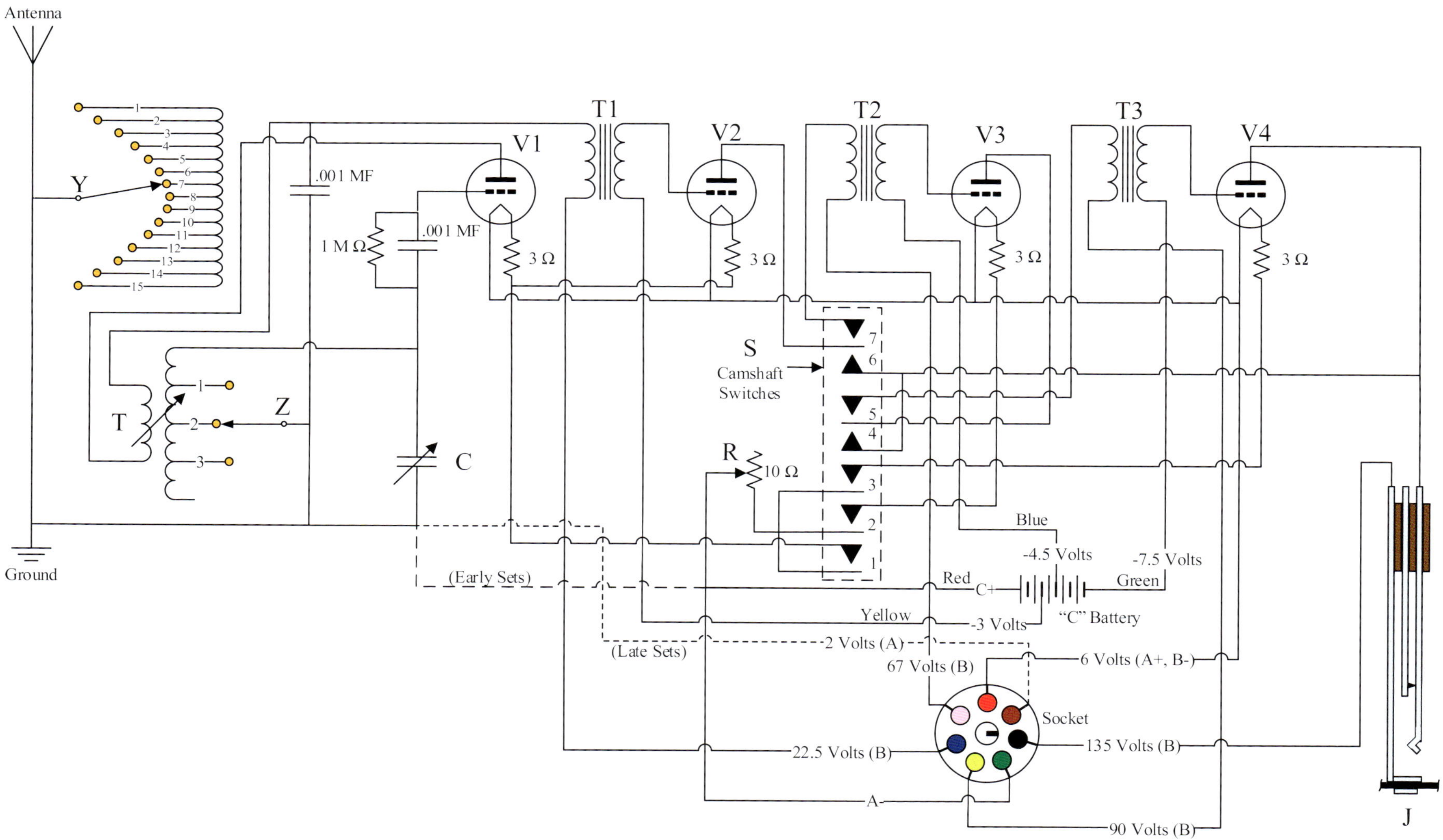

4-R (E) Description of Operation—Switch S in Off Position

When switch S is in the OFF position the filament circuits are open and the radio is inoperable. Plate voltage is applied to the plates of V1, V2, V3 and V4. A negative bias voltage is applied to the grids of V2, V3, and V4. This is considered the initial state of the radio for future discussions of circuit operation.

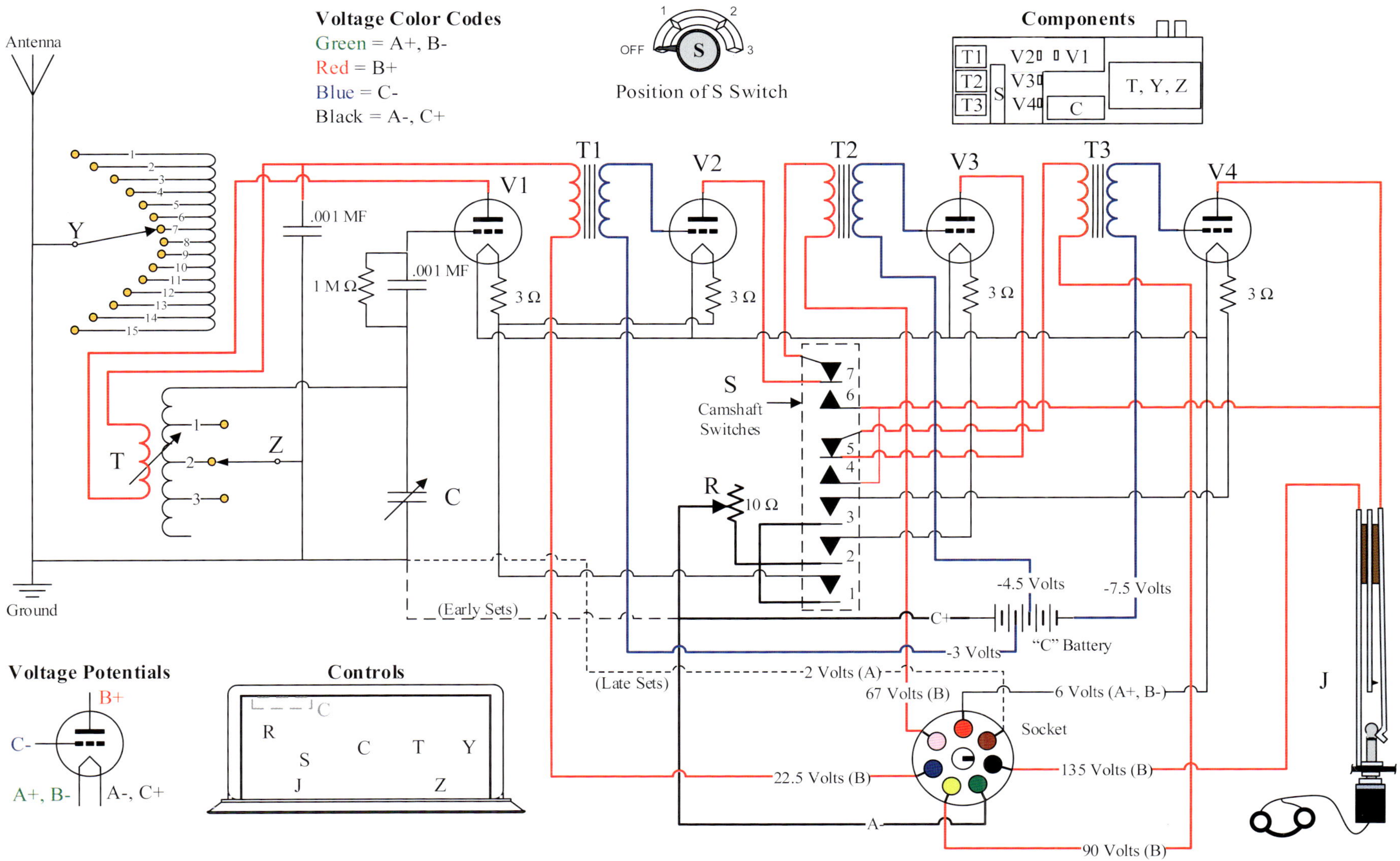

4-R (E) Description of Operation—Switch S in Position 1 (One AF Amplifier Stage)

When switch S is moved to position 1, camshaft switch 1 closes energizing the filament circuits of V1 and V2. At the same time camshaft switch 7 opens disconnecting the plate of V2 from the primary of transformer T2, and camshaft switch 6 closes connecting the plate of V2 to the headphones. The radio is now playing with the detector (V1) and one stage (V2) of AF amplification.

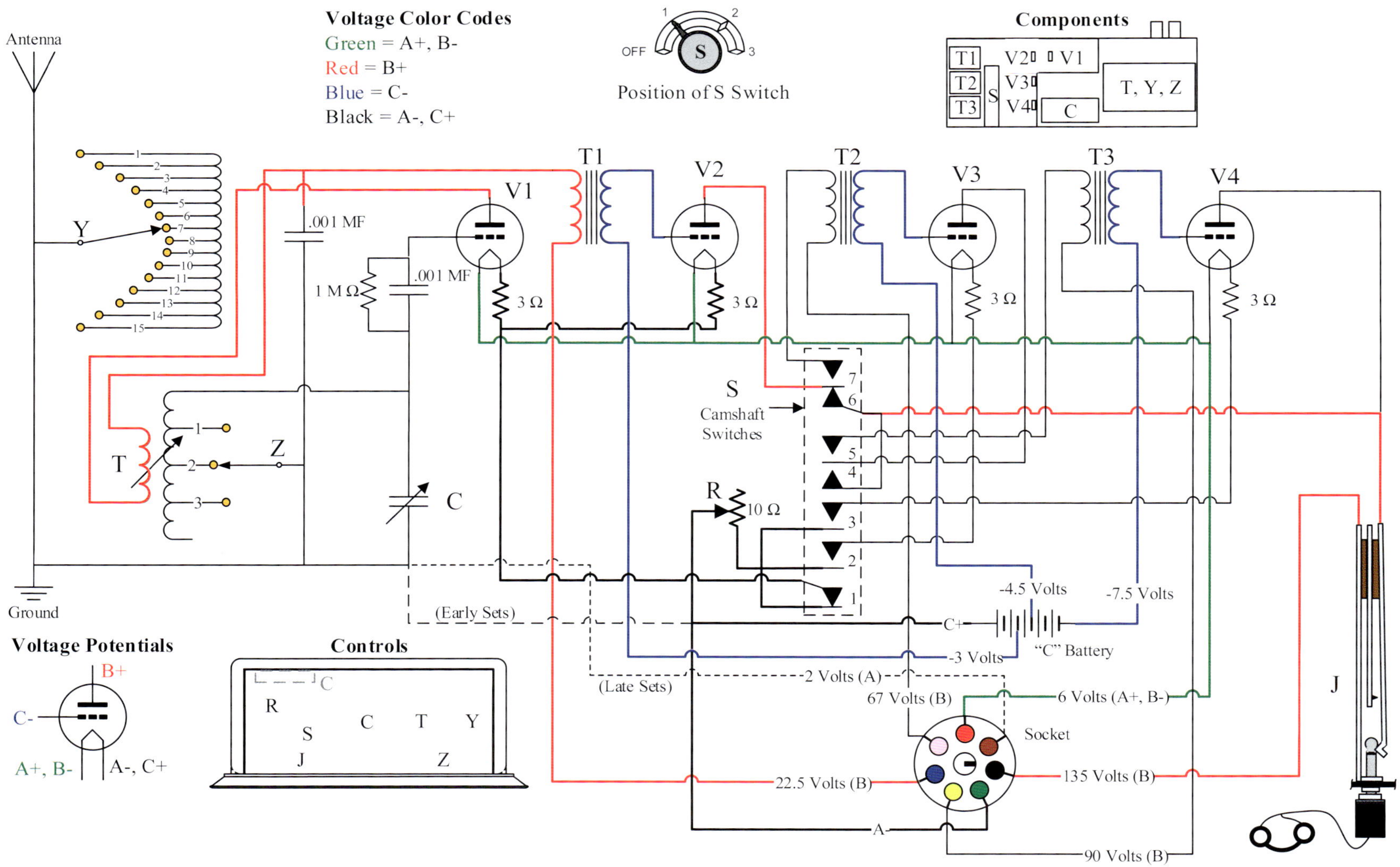

4-R (E) Description of Operation—Switch S in Position 2 (Two AF Amplifier Stages)

When switch S is moved to position 2, camshaft switch 2 closes (camshaft switch 1 remains closed) energizing the filament circuits of V1, V2 and V3. At the same time camshaft switch 6 opens disconnecting the plate of V2 from the headphones circuit, camshaft switch 7 closes connecting the plate of V2 to the primary of transformer T2, and camshaft switch 4 closes connecting the plate of V3 to the headphones circuit. The radio is now playing with the detector (V1) and two stages (V2, V3) of AF amplification.

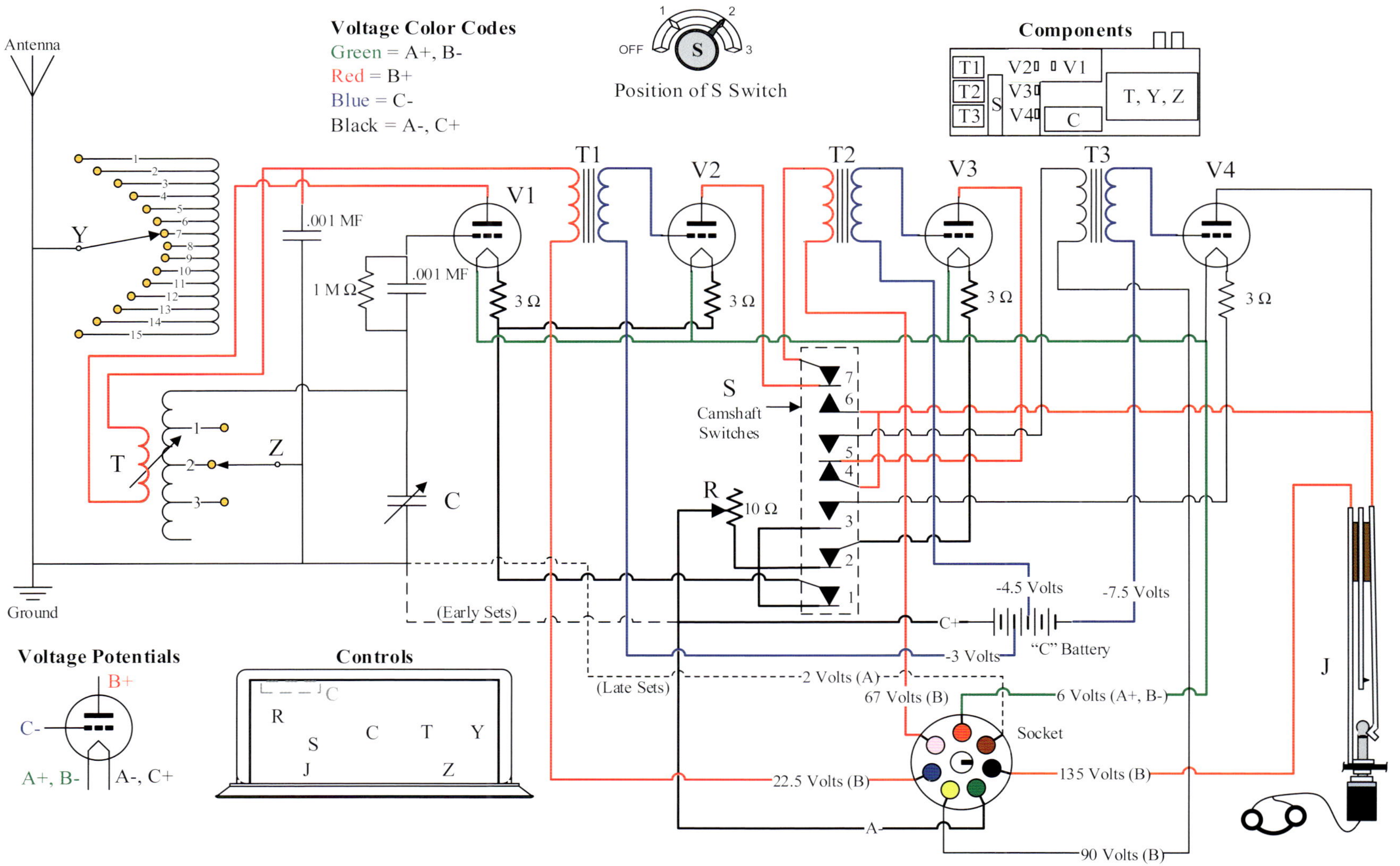

4-R (E) Description of Operation—Switch S in Position 3 (Three AF Amplifier Stages)

When switch S is moved to position 3, camshaft switch 3 closes (camshaft switches 1 and 2 remain closed) energizing the filament circuits of V1, V2, V3 and V4. At the same time camshaft switch 4 opens disconnecting the plate of V3 from the headphones circuit (the plate of V4 is already connected to the headphones circuit), and camshaft switch 5 closes connecting the plate of V3 to the primary of transformer T3. The radio is now playing with the detector (V1) and three stages (V2, V3, V4) of AF amplification.

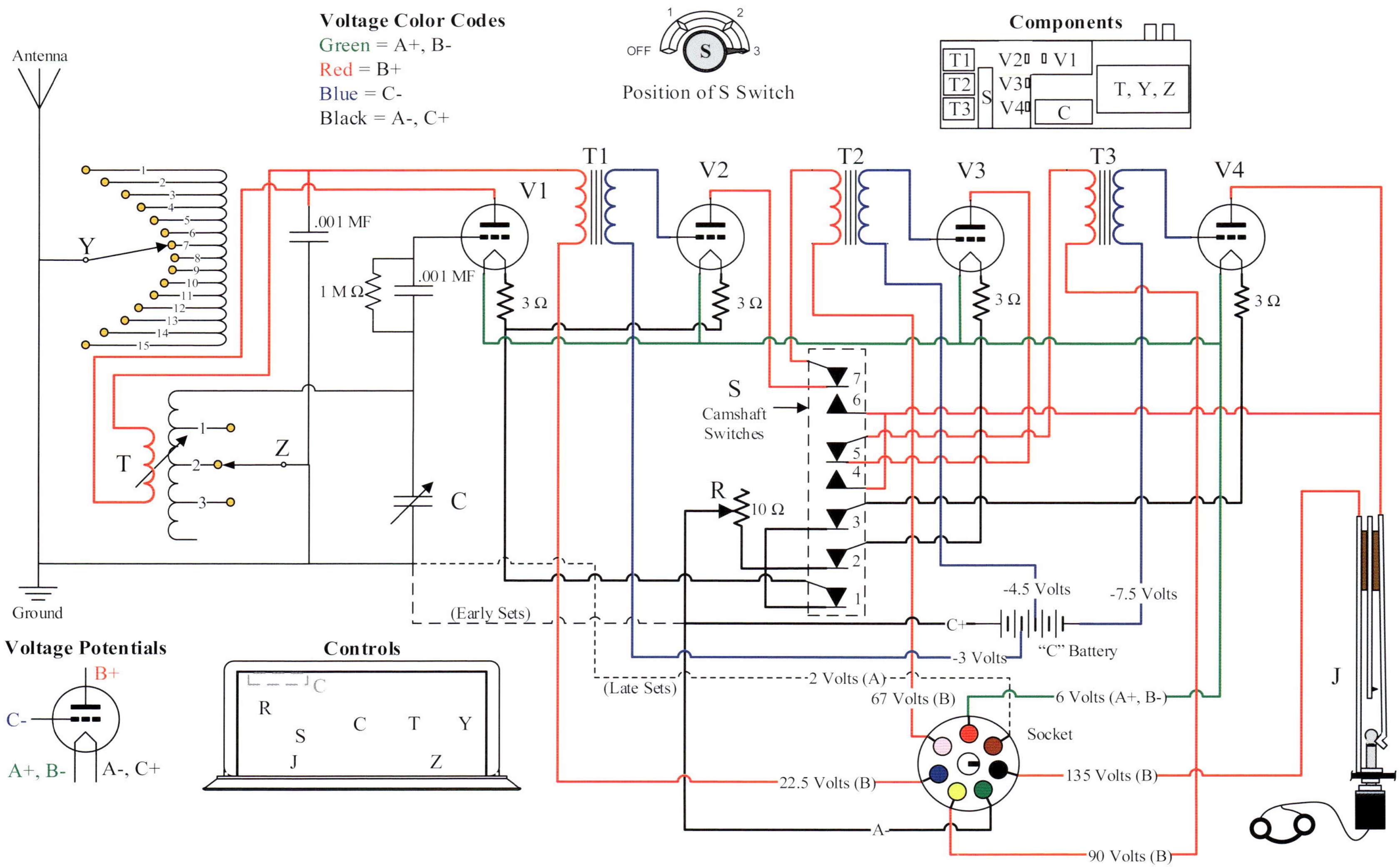

4-R (E) Preparation for Operation

Install the tubes and the "C" battery inside the radio. Connect the antenna and ground to the ANTENNA and GROUND posts. Use the appropriate battery connection diagram, inside the back door of the radio, to make sure the power can be connected properly and agrees with the 4-R (E) Connections diagram below. If the power connections are not correct or otherwise unavailable, utilize the physical and functional descriptions to assist in determining the correct power connections. Hook up the power to the radio. Proceed to 4-R (E) Operation. See Common Radio Attachments (page 8) for additional information related to antennas, loudspeakers and power options.

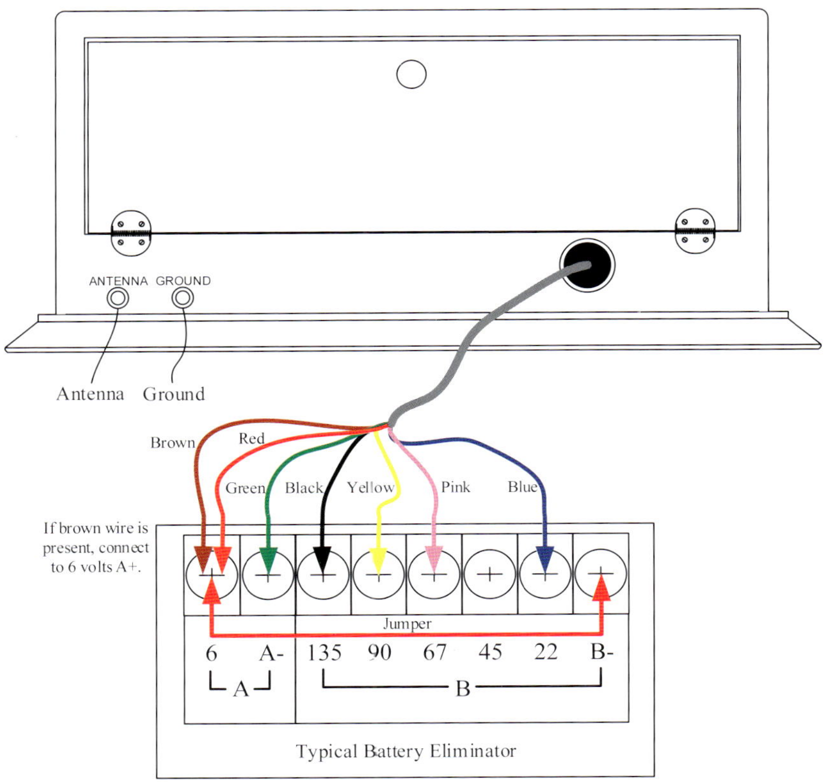

4-R (E) Connections

4-R (E) Operation

Follow the 4-R (E) Preparation for Operation instructions at left. Set the pointers to the positions shown in the Initial Dial Settings diagrams below.

The variable capacitor C selects the stations and the tickler T provides the feedback (regeneration) that increases the volume of the selected signal. Practice using knobs C and T simultaneously.

Insert the headphone plug into jack J. Turn the switch "S" to position 1 or 2.

Gradually move "T" to the right until a hissing sound is heard and hold the pointer to the beginning of that sound. Gradually move "C" to the right until a bird whistle is heard, first high, then descending to low; then a pause will be heard. As you continue to move "C" the whistle is again low, then ascending to high. The station is between these low points or valleys. Move "T" to the beginning of the whistle, while "C" is being moved. If you have C in the valley of the whistle, slightly move "T" to the left to clear the station. For best results operate C and T simultaneously. If an unpleasant howl (oscillations) is heard and the station is lost, turn "T" back to the left until the howl stops.

Adjust the audio level (volume) by setting switch S to position 1, 2 or 3, as appropriate.

Knob Y tunes the antenna circuit. A little manipulation of this switch will enable the operator to obtain improved results.

Knob Z changes the bandwidth of the receiver and is generally left at position 2. Position 1 decreases the bandwidth and position 3 increases the bandwidth.

Record the control settings for each station in the Station Log (next page). Then, to return to a particular station, set the controls to the settings that were previously recorded in the log for that station.

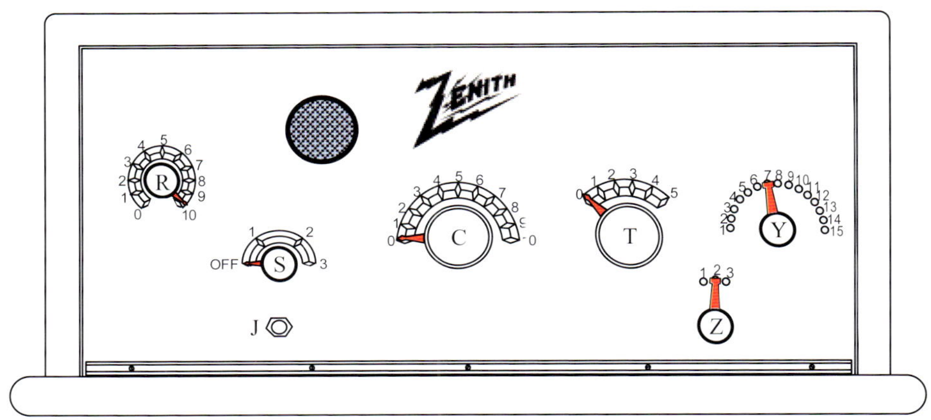

Initial Dial Settings (pointers shown in red for clarity; R=10, S=OFF)

4-R (E) Station Log

Name: ______________________________

Date	Time	Station	Frequency	C	T	Y	Z

From *The Early Zenith Radios: The Battery Powered Table Sets 1922–1927*, © 2014 by Gilbert M. Hedge with Durell M. Roth. ISBN 978-0-7643-4674-3. Schiffer Publishing Ltd. www.schifferbooks.com

Chapter 6: **Zenith Super-Portable**

Super-Portable General Specifications

Company:	Zenith Radio Corporation
Model:	Super-Portable (Companion)
Year Manufactured:	1924
Manufacture Location:	48th and Kedzie, Chicago IL (QRS Factory)
Serial Number:	2085
Price:	$224
Tube:	UV-199 (6), three RF amplifiers, one detector, two AF amplifiers
Circuit Type:	TRF with uncontrolled regeneration

Super-Portable General Information

Zenith claims that the Companion was the first portable radio produced.[1] Collectors disagree as to whether or not it was the very first, but it is unquestionably among the first to hit the broadcast market. The Companion portable radio is a very rare radio and nearly all the radios that were sold were returned to the manufacturer because of a serious technical defect. See Failure of the Zenith Super-Portable, page 192.

The Companion radios had two different styles of decals that were located in the center of the antenna cover. One version was the letters ZEV, and the other version has a decal similar to the Zenith lightning-bolt logo but with a lightning bolt around the outside. This decal is similar to but smaller than the logo used on the top of some of the early versions of the 3-Rs and 4-Rs. See Missing Model Numbers and Super Portable Configurations, page 186.

Restoration details for the set that was used as the example in this chapter. *This particular radio required a complete restoration. The horn speaker was missing, and a new one was constructed of paper mâché. A drawing was made in order to construct the new horn speaker and is included in this chapter. Additionally, the antenna cover was missing and a new one was made of 0.125" black plastic. All "A", "B" and "C" batteries were replaced with reproduced batteries. Too, the horn speaker driver-coil was open and required rewinding. Also, both of the Acme R-2 RF transformers were open and were re-wound. Drawings for both types of Acme RF transformers are included in this chapter. The remainder of the Companion is original, except for the adapter that was built for the "D" battery. The only damaged part that was not replaced was the handle; the repair would have likely resulted in damaging the front of the case.*

Super-Portable Physical Description Quick Reference Guide

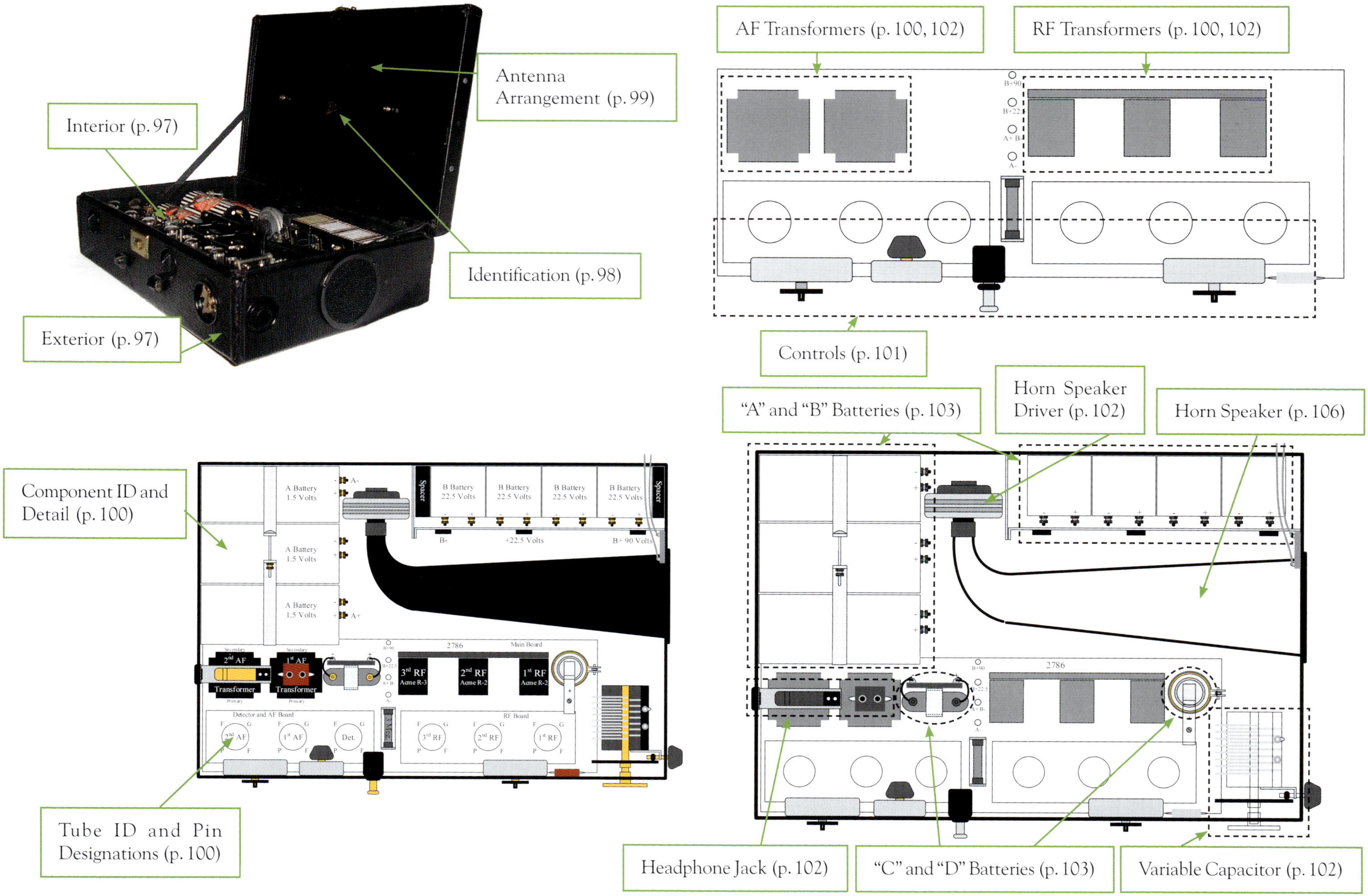

Super-Portable Exterior

The case is black and measures 20.75" by 13.25" by 6". Two rheostat controls are located on the front, along with the ON/OFF switch, tuning indicator, latch and handle. The left side has a jack for a headphone and the right side has a vernier tuning knob and horn speaker cover. There are access holes for connecting an external antenna and ground on the right side of the lid.

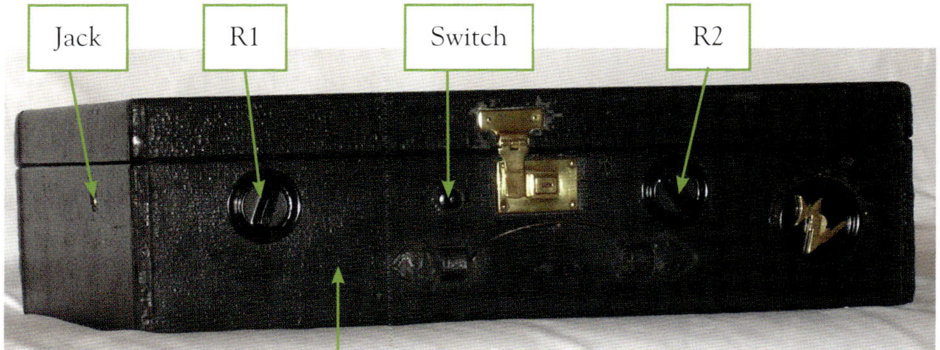

Front View

Control	Function
C1	Variable Capacitor Vernier (right side)
J	Headphone Jack (left side)
Switch	On-Off (front, center)
R1	Rheostat, filament (front, left)
R2	Rheostat, filament (front, right)
R3	Rheostat, filament (internal)

Left Side

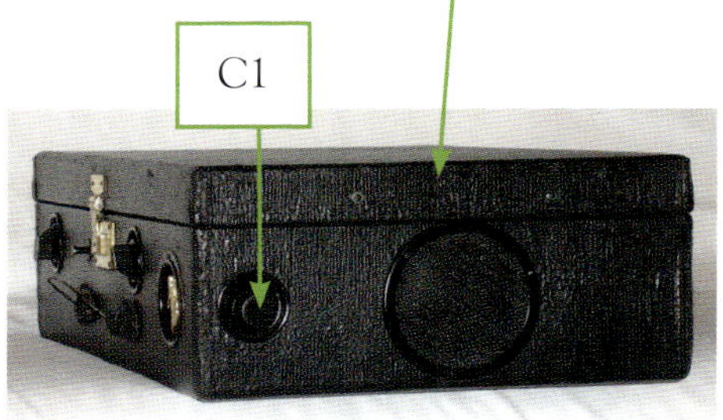

Right Side

Super-Portable Interior

Top View with Lid Open

Super-Portable Identification

This radio does not have the typical identifications that were present in the other early Zenith radios. The only markings are the logo on the antenna cover, a serial number on the circuit board, and the Zenith logo–shaped dial indicator. Examples of each are included below.

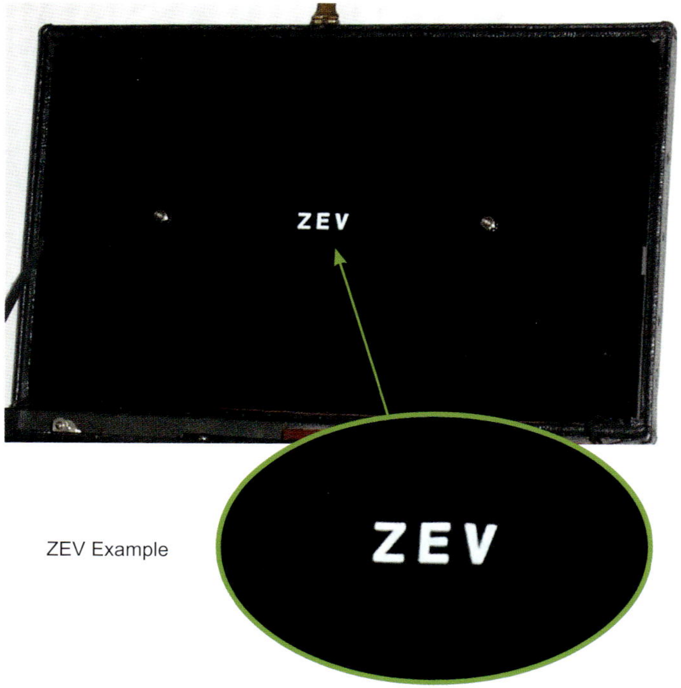

ZEV Example

Zenith Logo Example
(from another Super-Portable radio)

Serial Number Location

Zenith Logo Tuning Indicator

Super-Portable Antenna Arrangement

An external antenna and ground can be connected through holes in the side of the lid. When an external antenna and ground are connected, the RF signal from the external antenna is connected to the antenna coil and then inductively coupled to the loop antenna.

Antenna Cover

Antenna (cover removed)

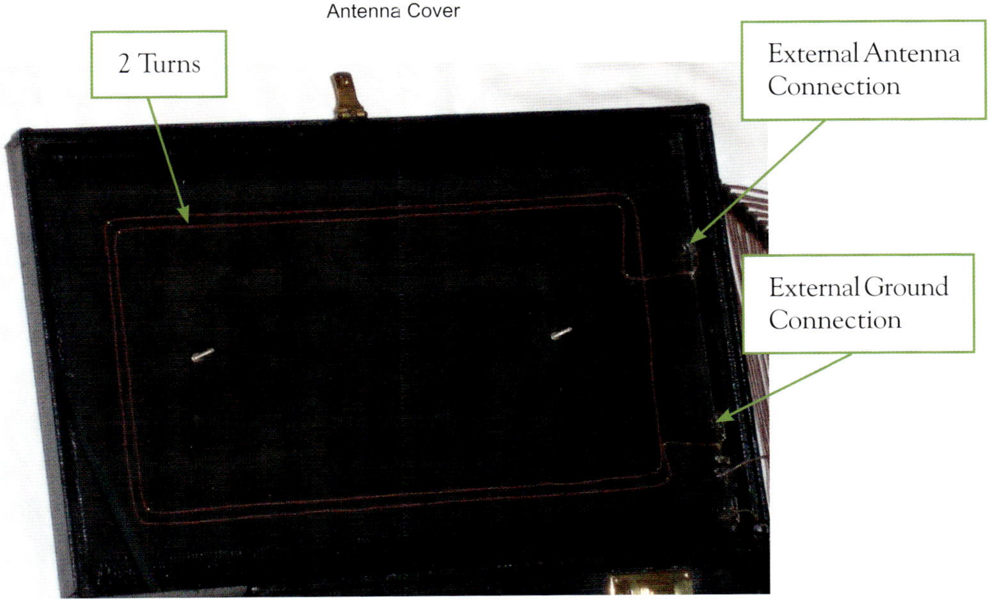

Antenna Coil (loop antenna removed)

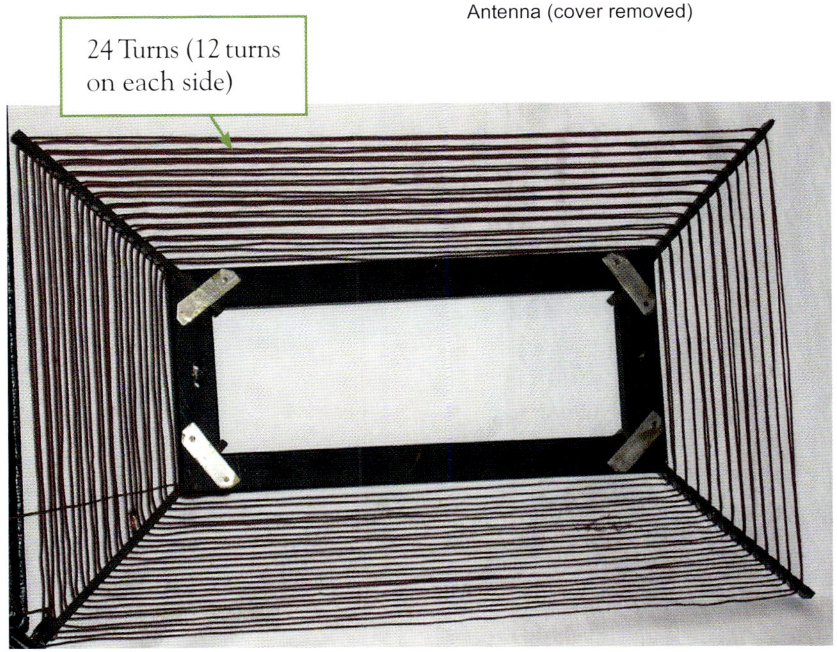

Loop Antenna (removed from lid)

Super-Portable Component Identification and Detail

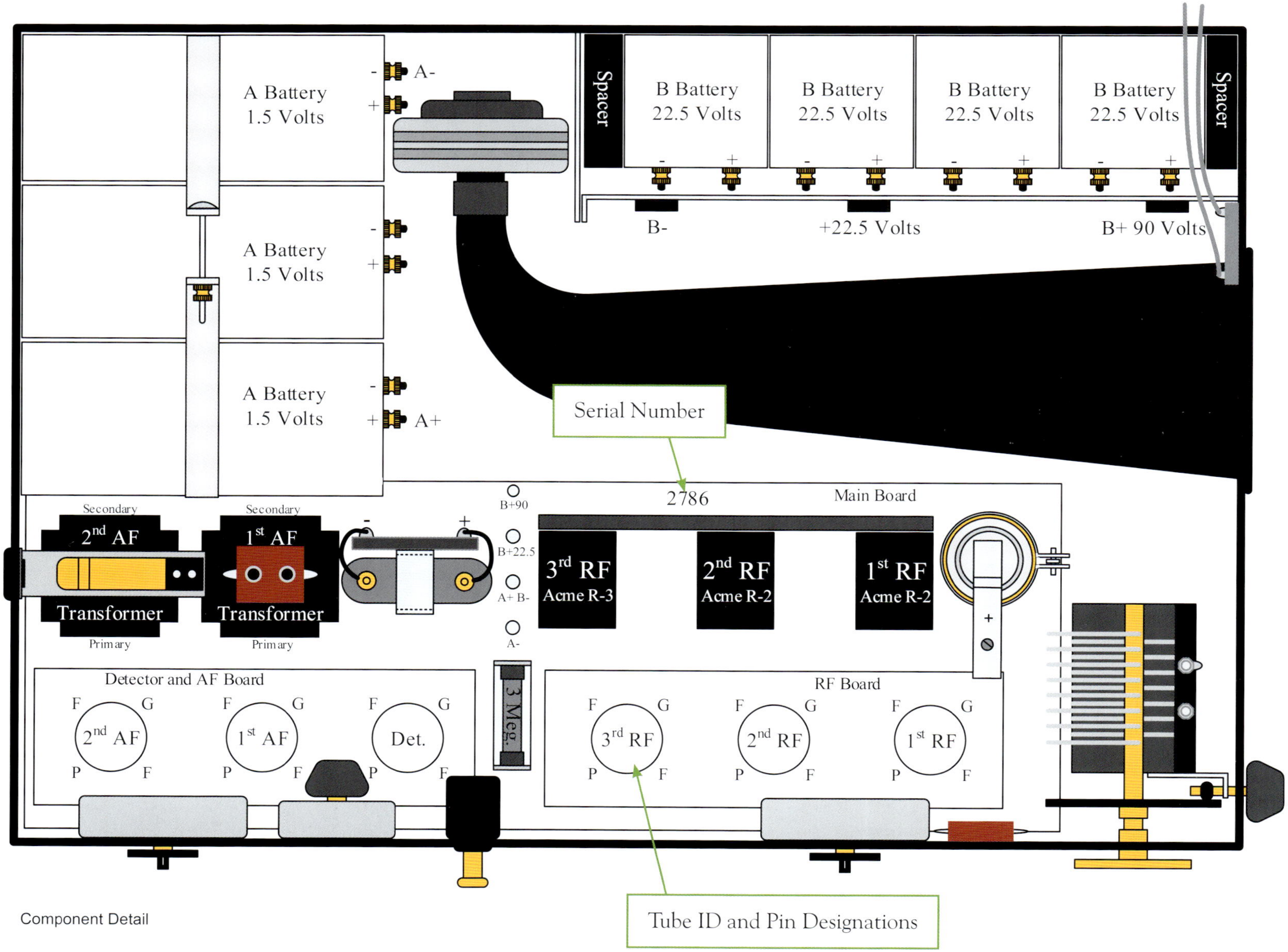

Component Detail

Super-Portable Controls

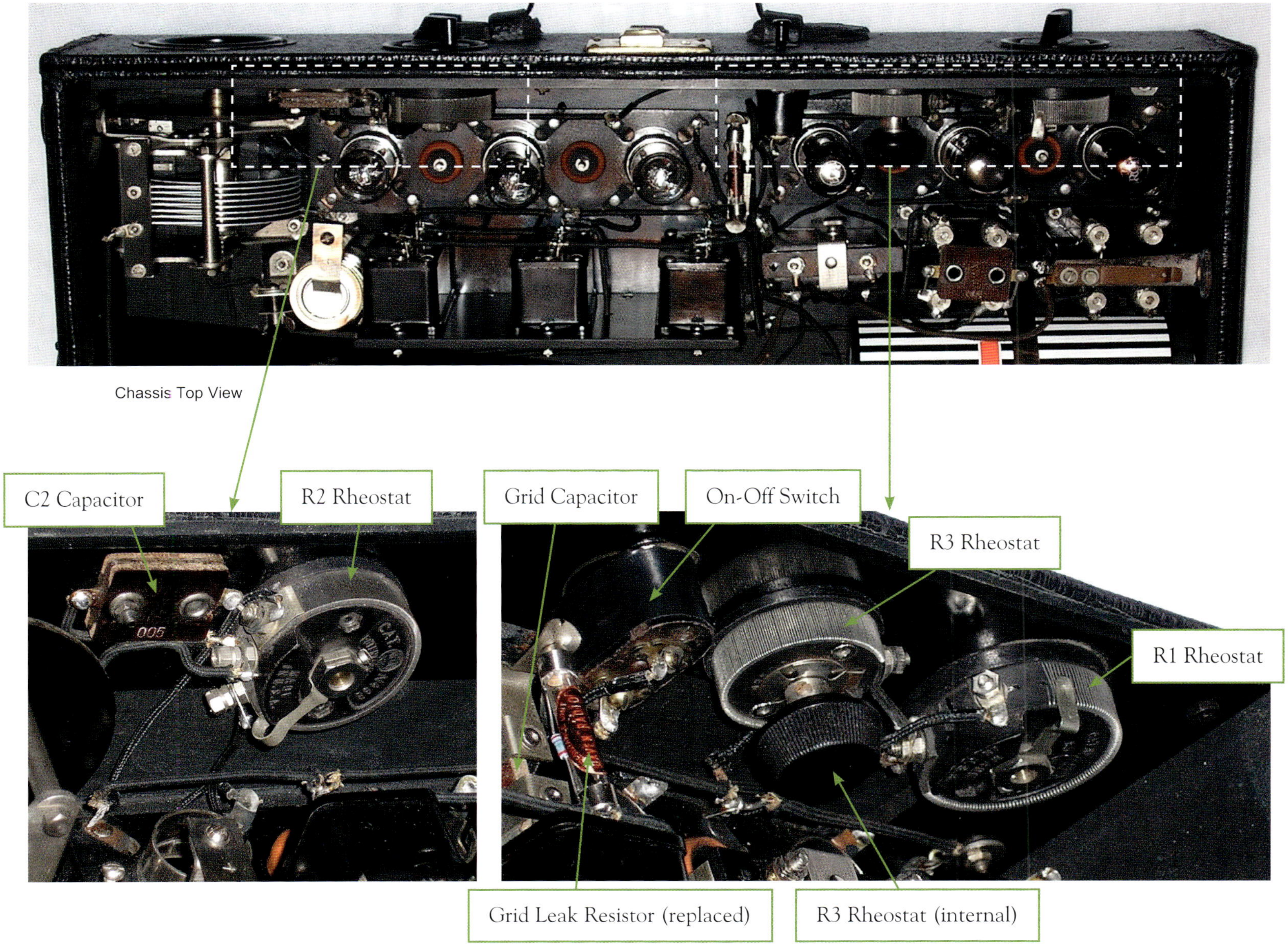

Super-Portable Transformer and Variable Capacitor

Horn Speaker Driver
(horn removed)

Thordarson Audio Frequency Transformers

Transformer DC Resistance

Transformer	Primary	Secondary
Thordarson	≈ 1 KΩ	≈ 5 KΩ
National Baldwin (speaker driver)	≈ 1 KΩ	N/A
Acme R-2 RF	≈ 8.4 Ω	≈ 13 Ω
Acme R-3 RF	≈ 28 Ω	≈ 37 Ω

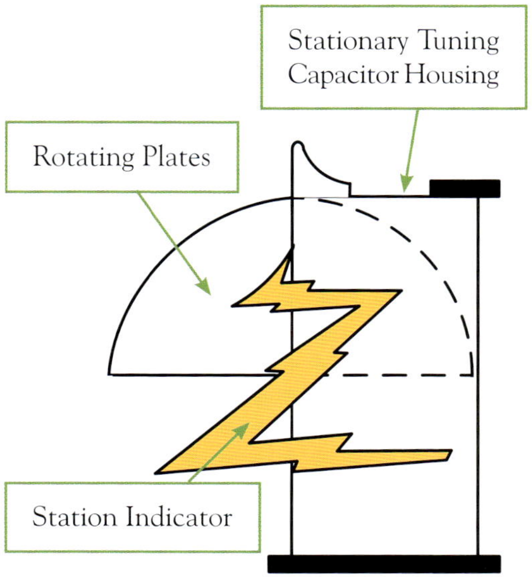

Tuning Capacitor and
Pointer Orientation

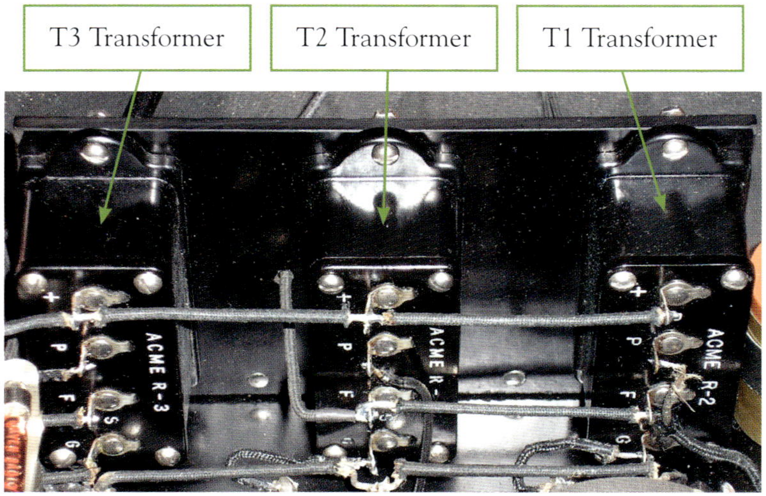

Acme Transformers

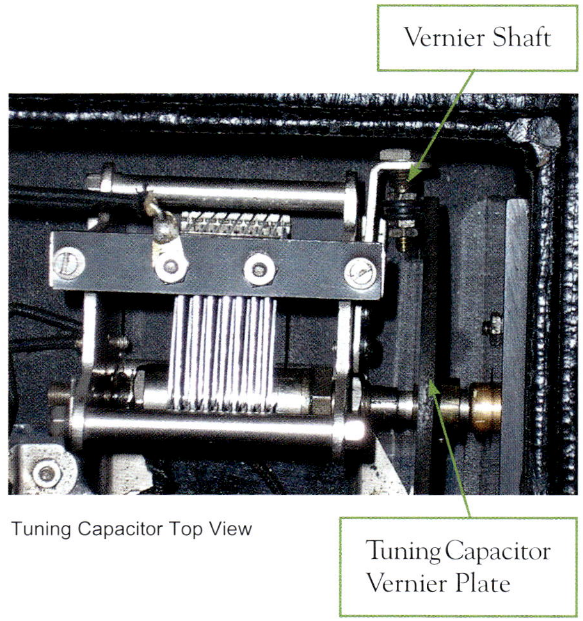

Tuning Capacitor Top View

Super-Portable Batteries

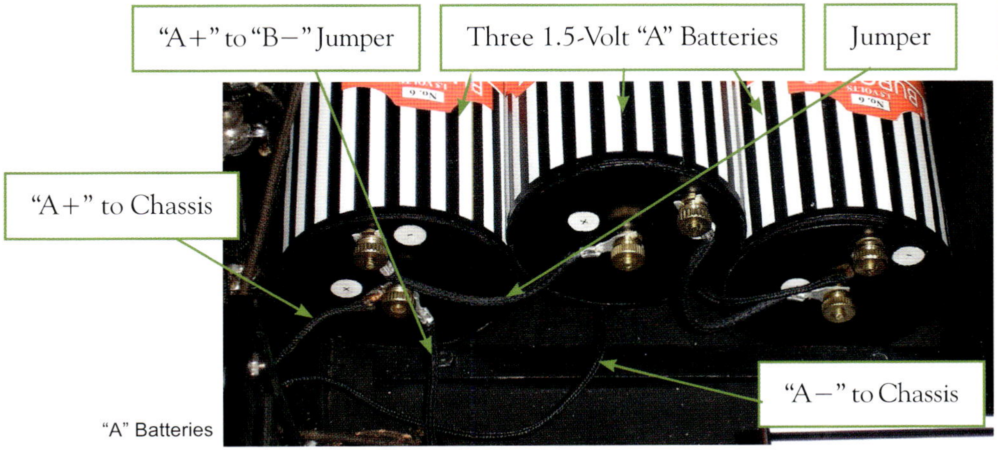

"A" Batteries

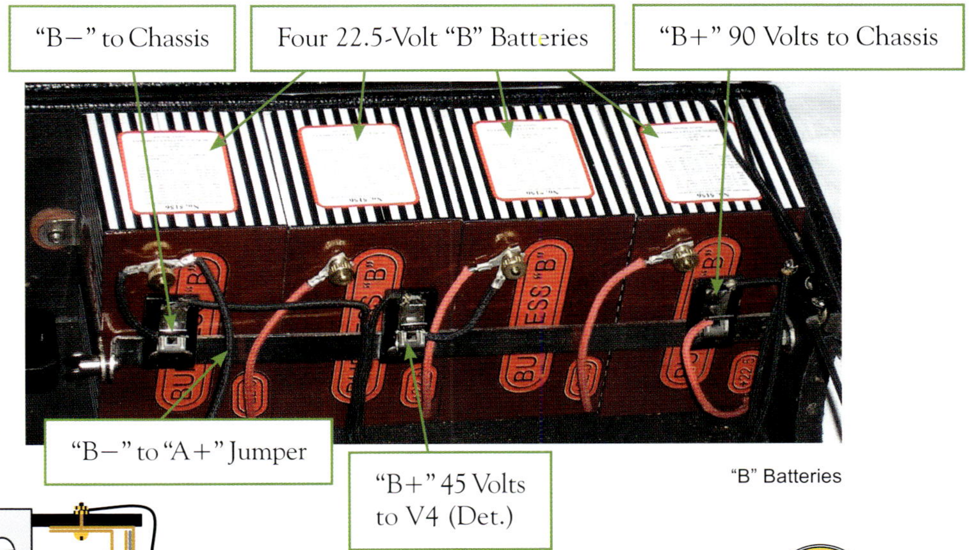

"B" Batteries

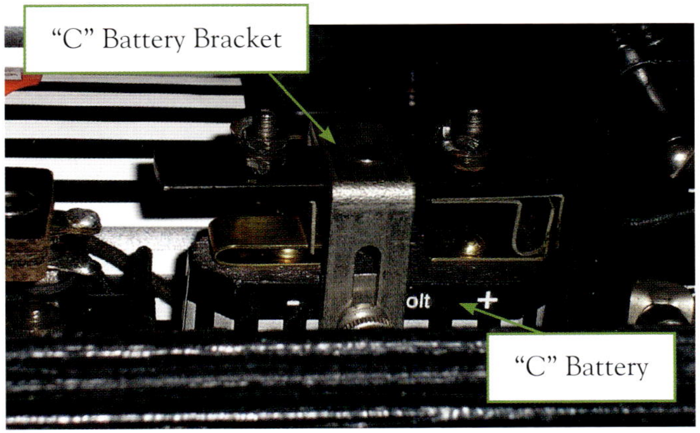

"C" Battery

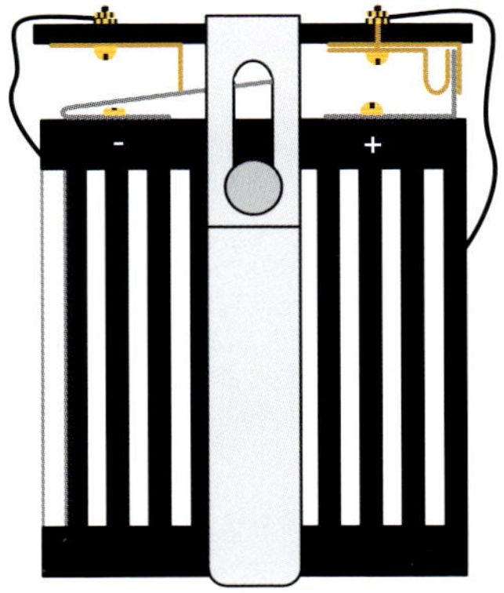

"C" Battery Detail

"D" Battery Adapter Detail

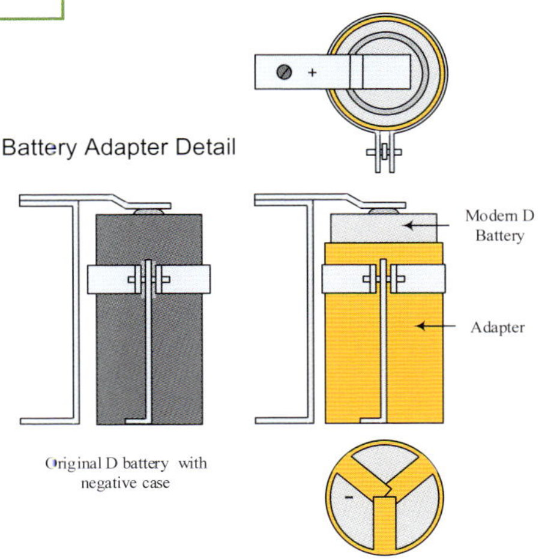

Reproduced Batteries

"D" Battery and Adaptor

Installed "D" Battery and Adaptor

"D" Battery Adapter

An adapter is required because the original D battery had a negative case and the modern D battery has an insulated case. See "D" Battery Adapter Modification, page 244.

Super-Portable Acme R-2 Detail

See ACME RF Transformer Repair, page 243.

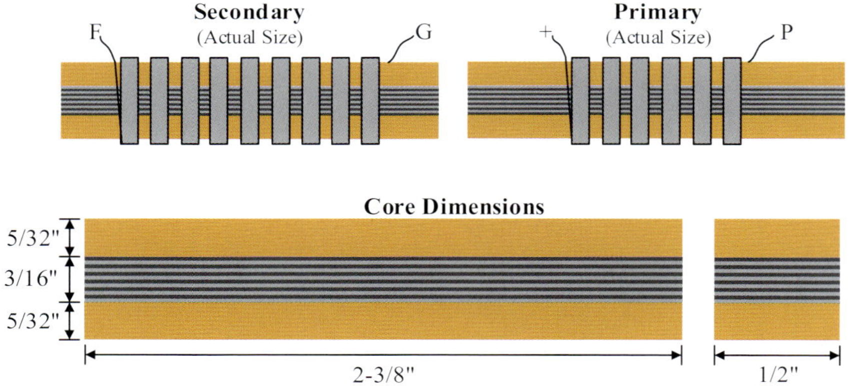

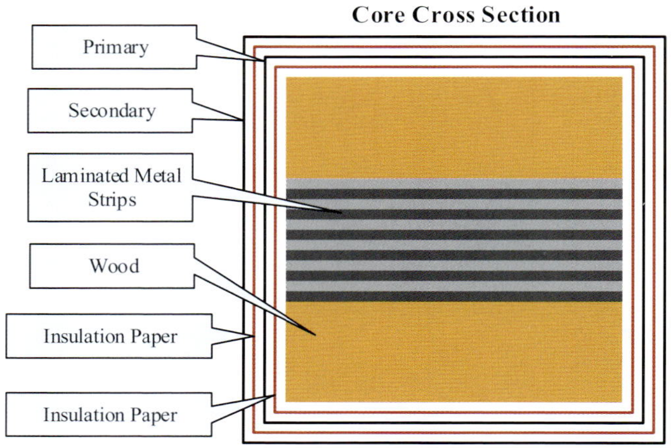

" = Inch

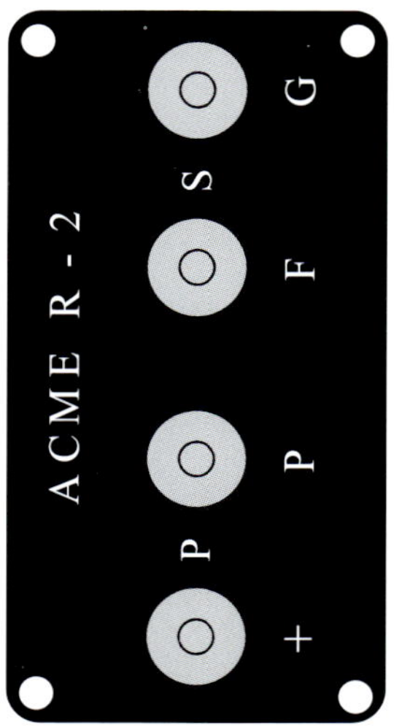

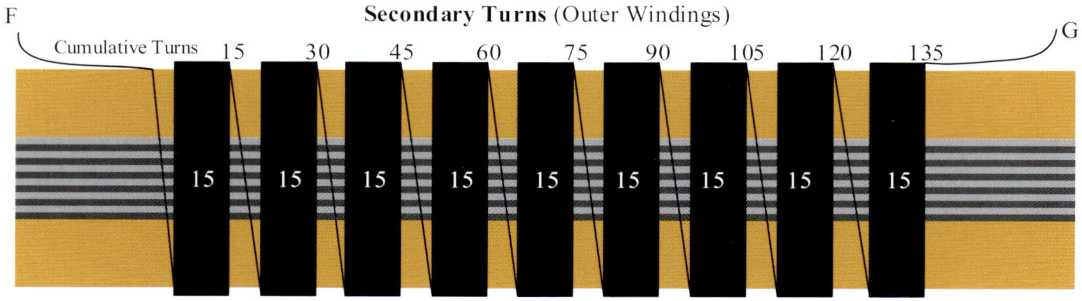

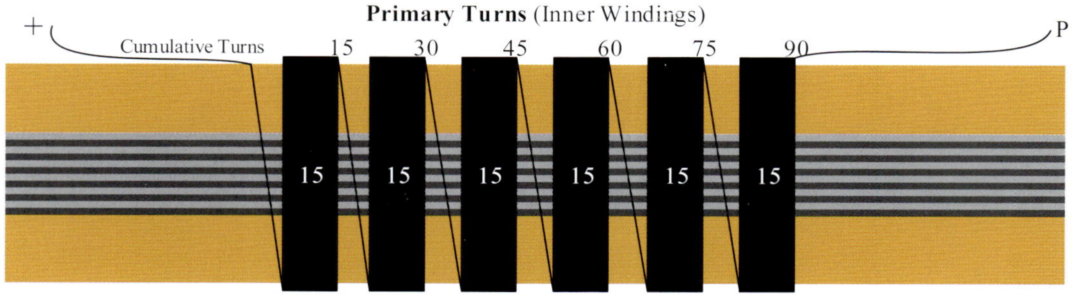

From *The Early Zenith Radios: The Battery Powered Table Sets 1922–1927,* © 2014 by Gilbert M. Hedge with Durell M. Roth. ISBN 978-0-7643-4674-3. Schiffer Publishing Ltd. www.schifferbooks.com

Super-Portable Acme R-3 Detail

See ACME RF Transformer Repair, page 243.

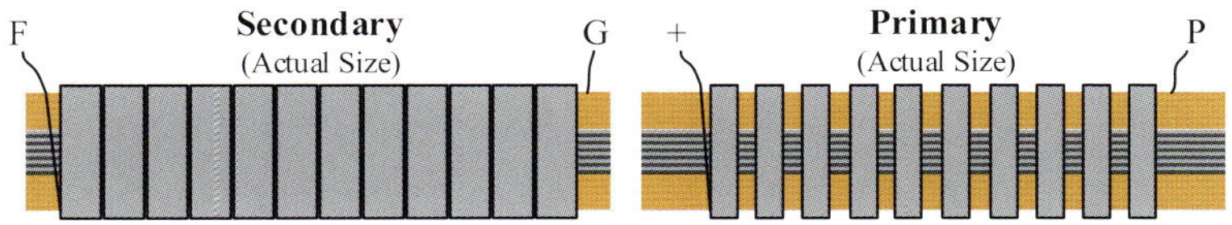

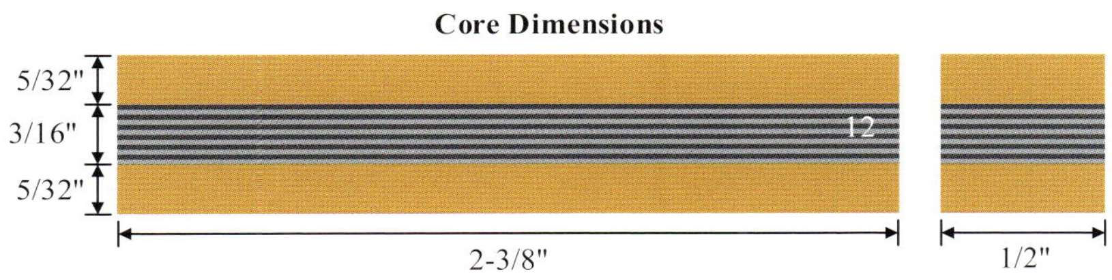

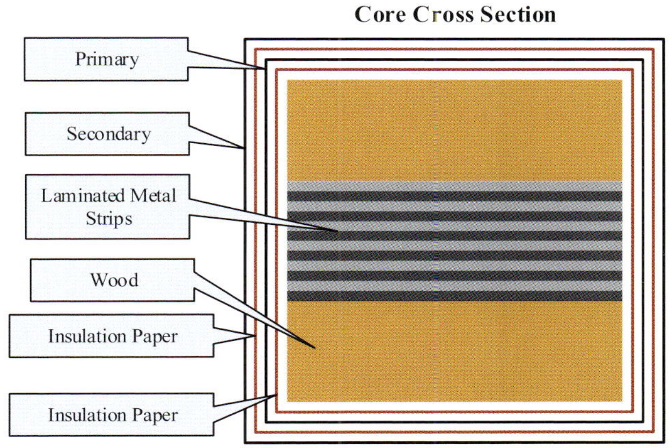

" = Inch

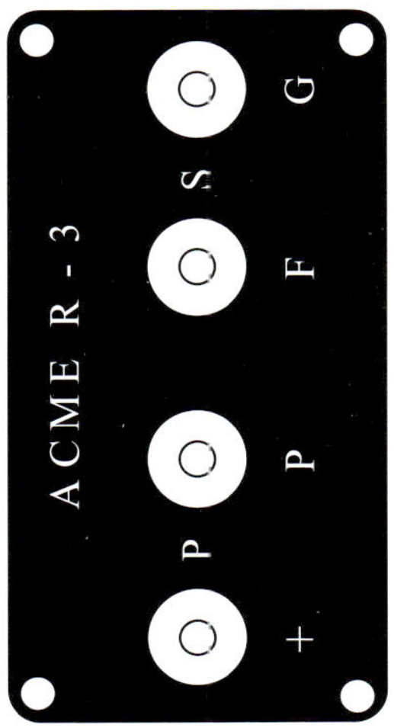

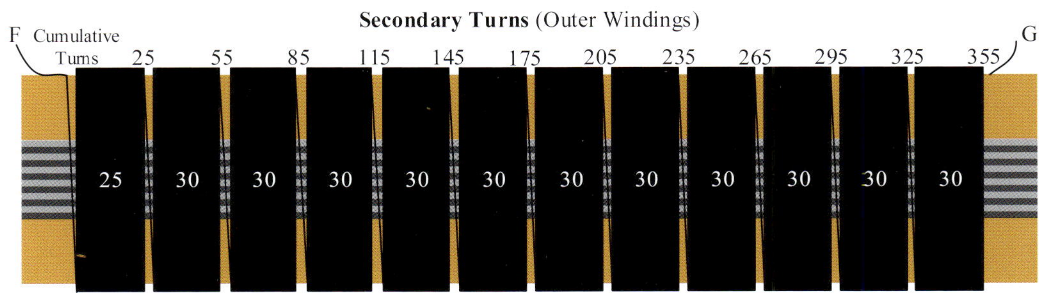

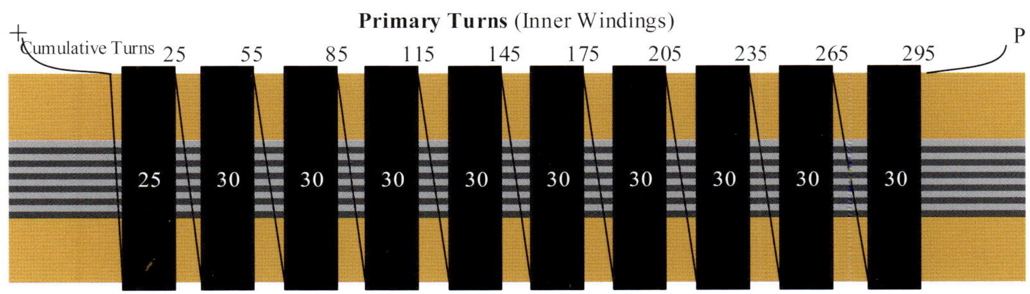

From *The Early Zenith Radios: The Battery Powered Table Sets 1922–1927*, © 2014 by Gilbert M. Hedge with Durell M. Roth. ISBN 978-0-7643-4674-3. Schiffer Publishing Ltd. www.schifferbooks.com

Super-Portable Horn Speaker Detail

The completed horn speaker will be slightly longer than required. Fit check it, and trim it to fit. See Paper Mâché Horn Speaker Repair and Construction, page 234.

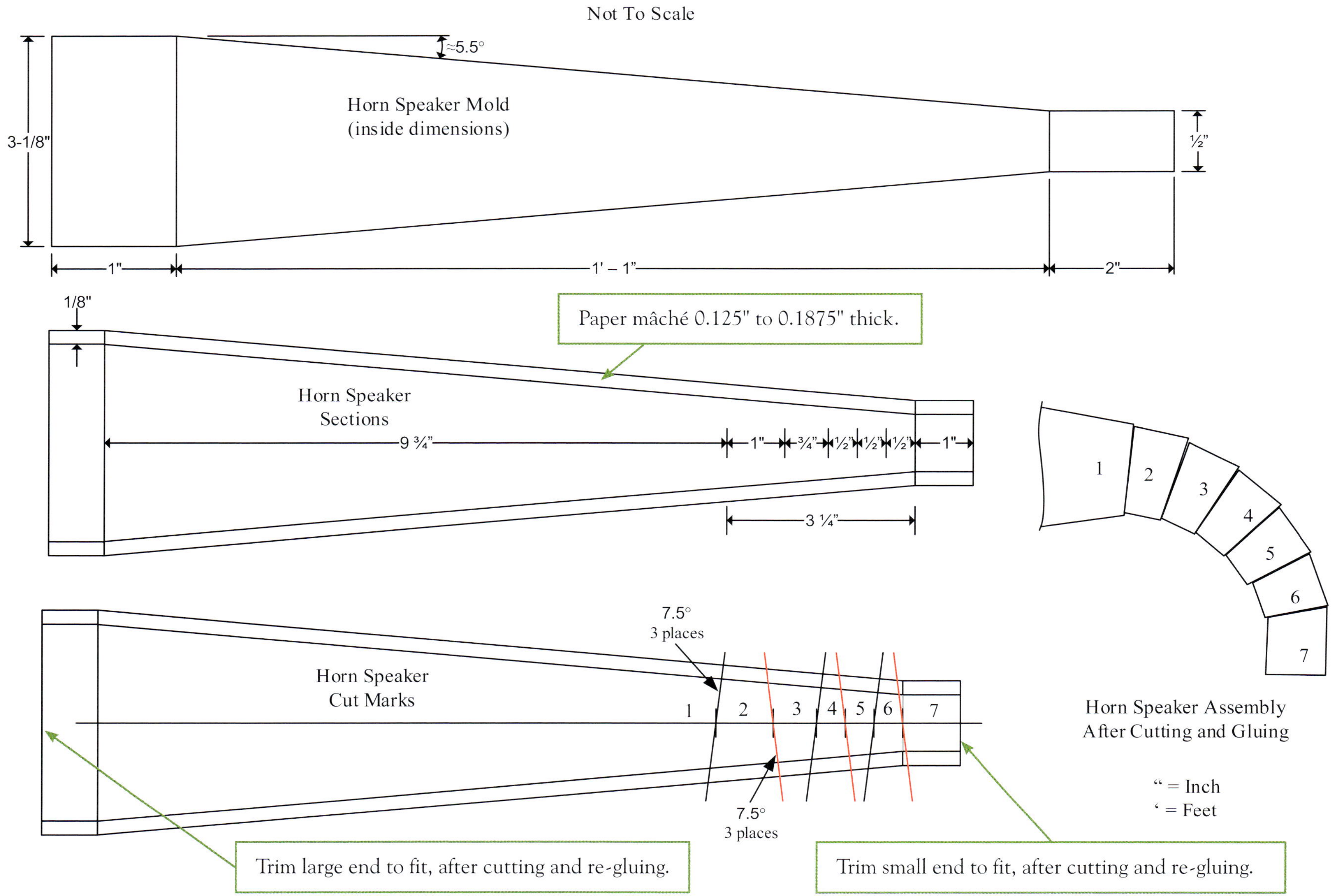

Super-Portable Functional Description Quick Reference Guide

Schematic (p. 108)

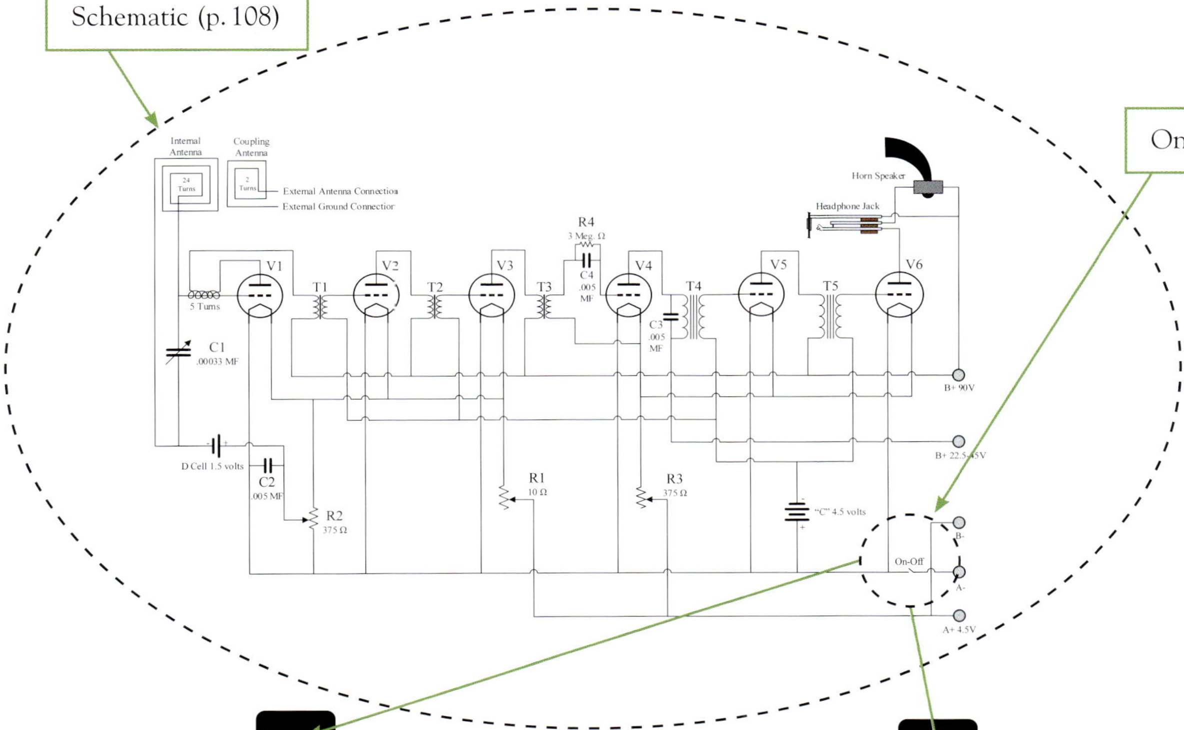

On-Off Switch

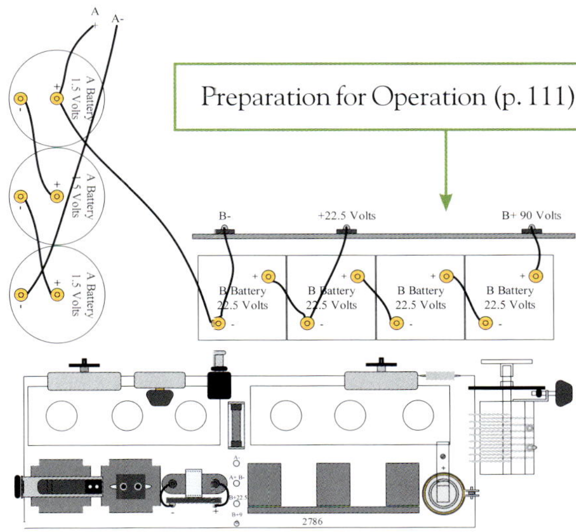

Preparation for Operation (p. 111)

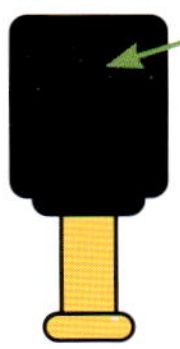

On-Off Switch Off (Push)

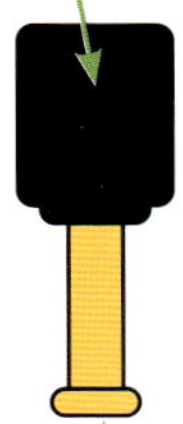

On-Off Switch On (Pull)

Description of operation when switch is in the Off position (p. 109)

Description of operation when switch is in the On position (p. 110)

Operation (p. 112)

Super-Portable Schematic

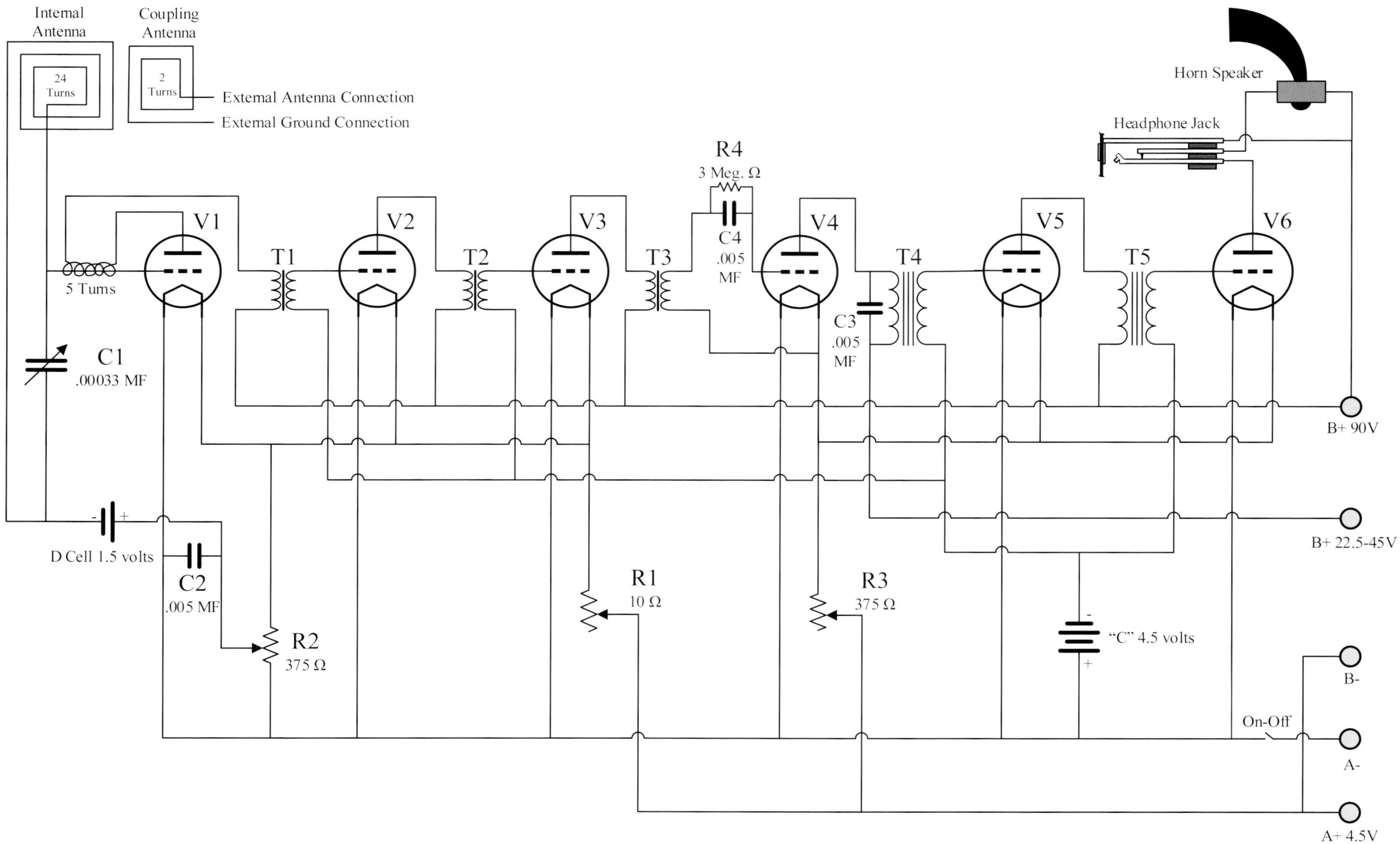

Super-Portable Description of Operation—Switch Off

The On/Off switch of the Super-Portable is a Pull-ON Push-OFF style switch. When pushed to its OFF position the filament circuits are open and the radio is inoperable. Plate voltage is applied to the plate of each tube. A negative voltage is applied to the grid of tubes V1, V2, V3, V5 and V6. This is considered the initial state of the radio for the discussion of circuit operation.

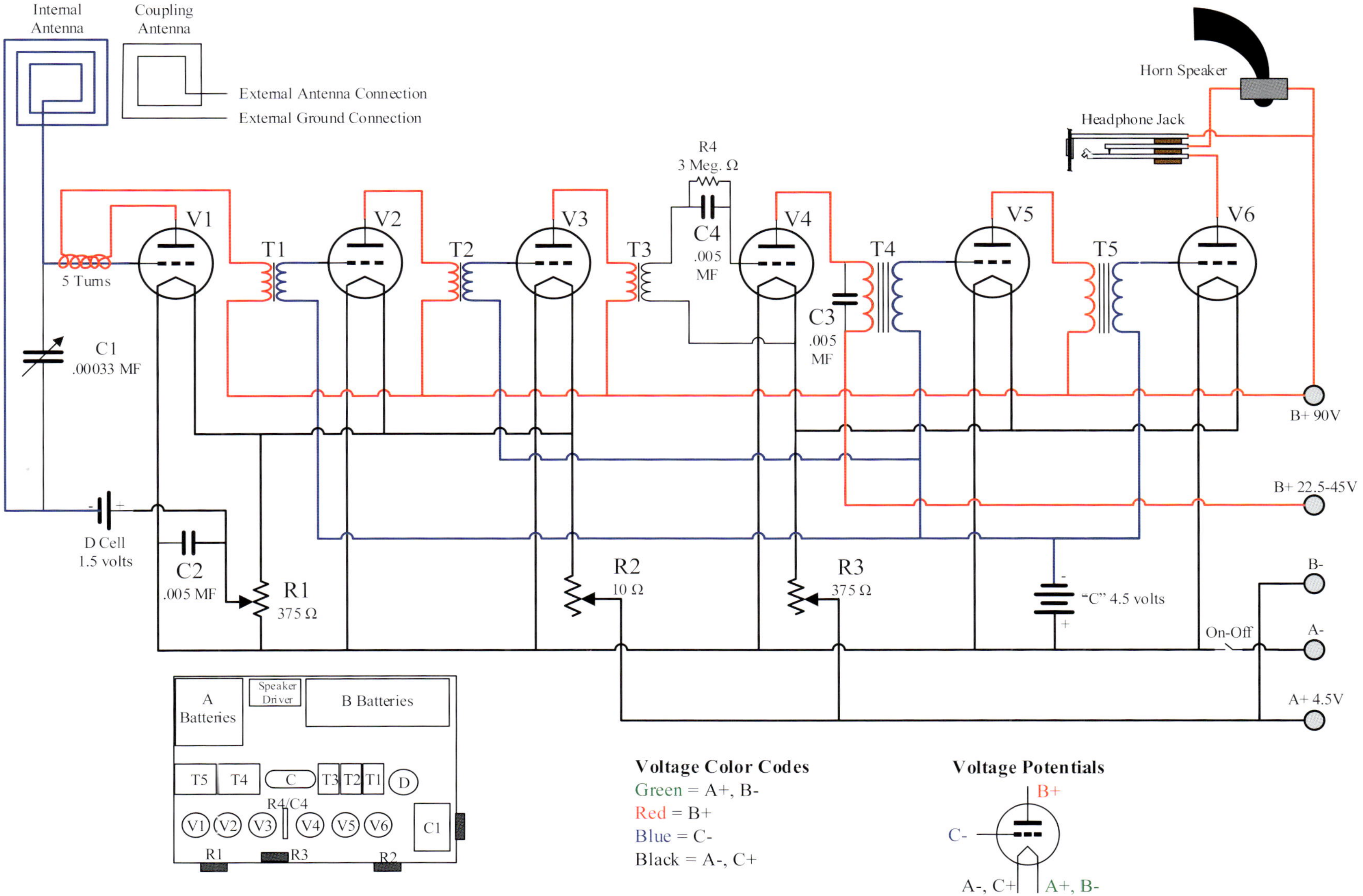

Super-Portable Description of Operation—Switch On

When the On/Off switch is pulled to the ON position, the filament circuit is closed energizing the filament circuits of the tubes. The set is fully operational. The tuning capacitor C1 is rotated (tuned) to select the desired signal which is applied to the grid of the first RF amplifier V1. A small portion of the amplified output signal from the plate of V1 is inductively coupled back to its grid through a 5-turn coil (regeneration), which provides a measure of additional amplification. This amplified output signal from the plate of V1 is transformer coupled, via T1, to the grid of the second RF amplifier V2. In like manner, the output signal of V2 is inductively coupled to third RF amplifier V3 and V3 is inductively coupled to the detector V4. The detector V4 removes the RF portion of the signal, leaving only the AF signal. The AF signal from the plate of V4 is further smoothed by capacitor C3 then coupled through AF transformer T4 to the grid of the first AF amplifier V5. The output of V5 is transformer coupled through T5 to the grid of the second AF amplifier tube V6. The plate of V6 feeds the AF signal to the horn speaker where it is converted to sound.

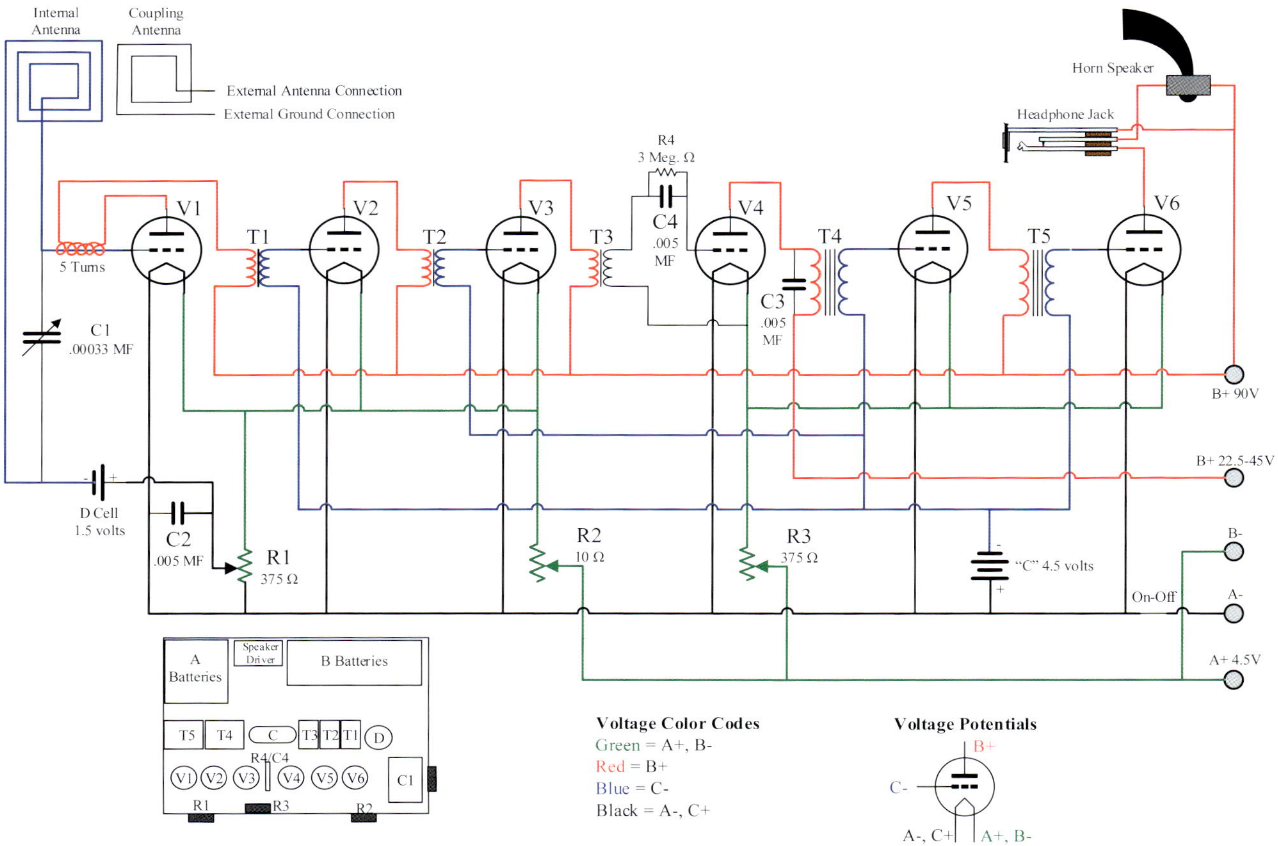

Super-Portable Preparation for Operation

Install and connect the "A", "B" and "C" batteries, as shown. Install the "D" battery and "D" Battery Adapter.

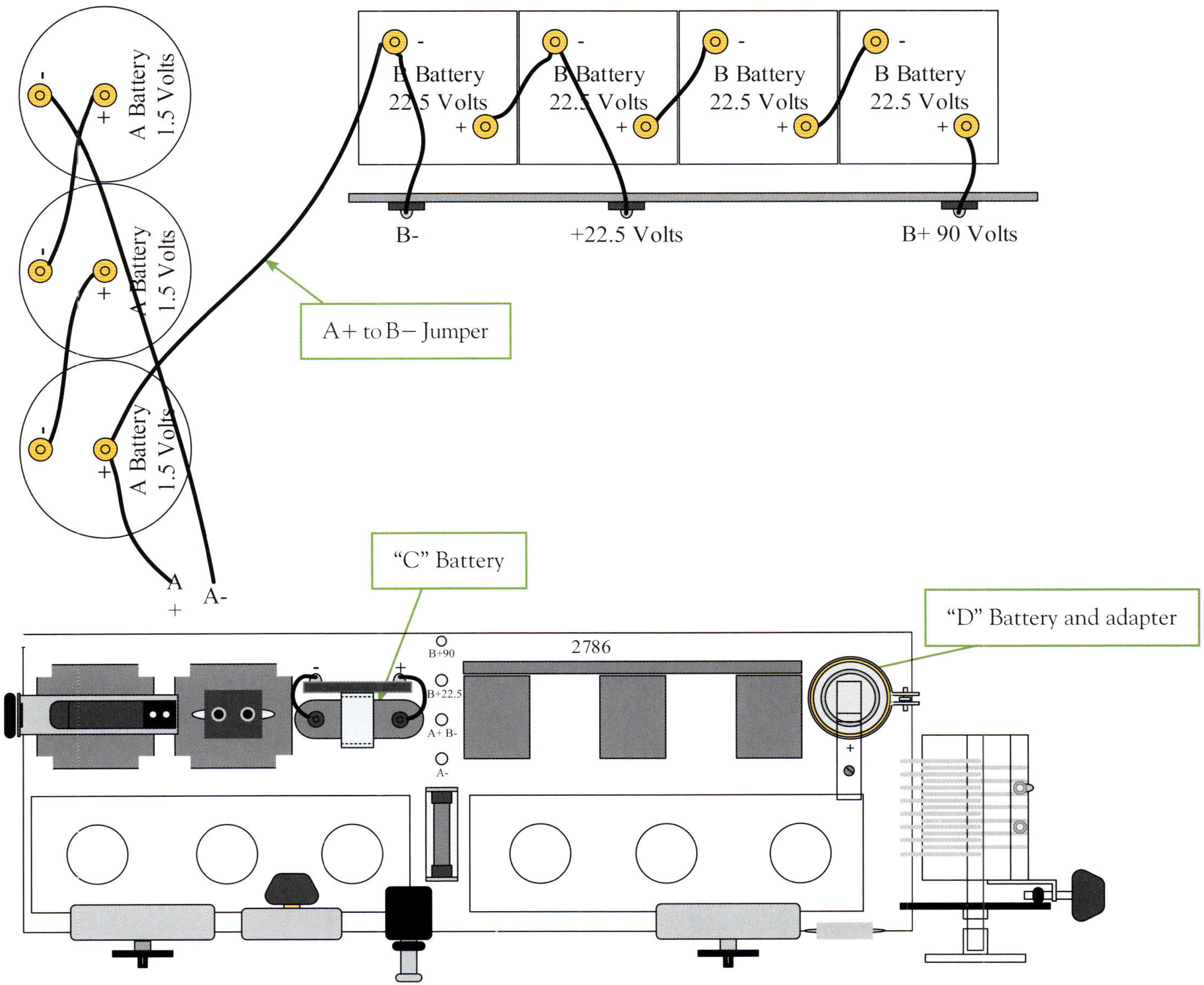

Super-Portable Operation

Follow the Super-Portable Preparation for Operation instructions.

Connect an external antenna, if desired.

Turn the wiper of the R3 (internal) clockwise to approximately 80%, as viewed from the front (see page 101). As the battery voltage decreases over time, slightly increase R3 clockwise, otherwise no adjustments are required. R1 and R2 are very critical to the performance of this radio. Turn "R2" a half-turn to the right. Now turn "R1" to the right three-quarters of a turn. Turn the "C" vernier to obtain a signal. Now readjust R1 and R2 until the station is the loudest and clearest. If the R1 and R2 controls are advanced too far to the right, a growling (oscillations) will be heard, and if they are not turned far enough the weak signals cannot be heard. Because of control interaction, R1 and R2 require adjusting as the stations are tuned. See Common Radio Attachments (page 8) for additional information related to antennas, loudspeakers and power options.

Chapter 7: **Zenith 3-R** (Late Version)

3-R (L) General Specifications

Company:	Zenith Radio Corporation
Model:	3-R (Late Version)
Year Manufactured:	1924
Manufacture Location:	48th and Kedzie, Chicago IL (QRS Factory)
Serial Number:	328984
Price:	$160
Tube:	UV-201-A (4), one detector, three AF amplifiers
Circuit Type:	Regenerative

3-R (L) General Information

The late version of the 3-R had several improvements over the earlier version. No major changes were made to the regenerative circuit, but the voltage to the plates of the AF amplifier tubes (V2, V3, and V4) was changed to make each one operate at 90 volts. Because the multiple plate voltages were eliminated, the power cable was changed from 7 to 5 conductors (wires). The stage selection switching mechanism was changed from a multi-lobed camshaft switch to a multi-contact rotary switch. The order of the AF amplifier stages that could be selected was changed. The peep hole was removed.

Restoration details for the set that was used as the example in this chapter. *This particular radio was extensively affected by the pot metal problem. Five of the seven pointers were replaced with handmade aluminum pointers. The two large pointers were not replaced and showed only a small amount of deterioration. The pulleys were replaced with machined aluminum pulleys. The two small pulleys were replaced with reproduced brass pulleys. The 5 conductor cable was replaced with a reproduced cable. The original "C" battery was replaced with a reproduced "C" battery.*

3-R (L) Physical Description Quick Reference Guide

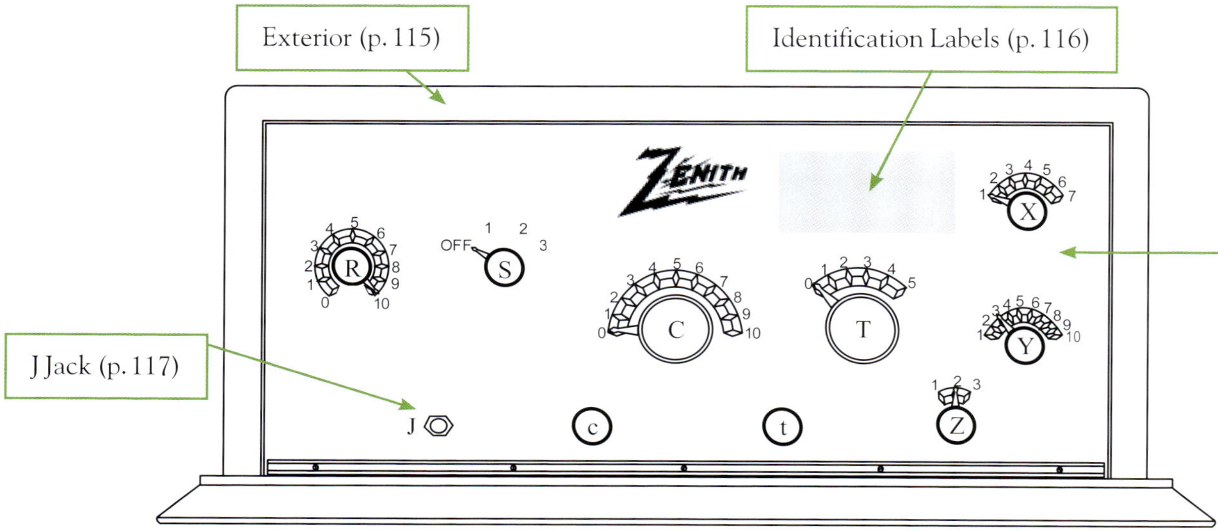

Control	Function
C	Variable Capacitor (tuning)
c	Vernier for Variable Capacitor
J	Headphone Jack
R	Rheostat (filament voltage control)
S	Switch (Off, 1, 2 or 3 AF amp. stages)
T	Tickler (regeneration)
t	Vernier for Tickler
X	Tap Switch (tunes the antenna circuit)
Y	Tap Switch (tunes the antenna circuit)
Z	Tap Switch (selects the bandwidth)

Note: Shaded item is located inside set.

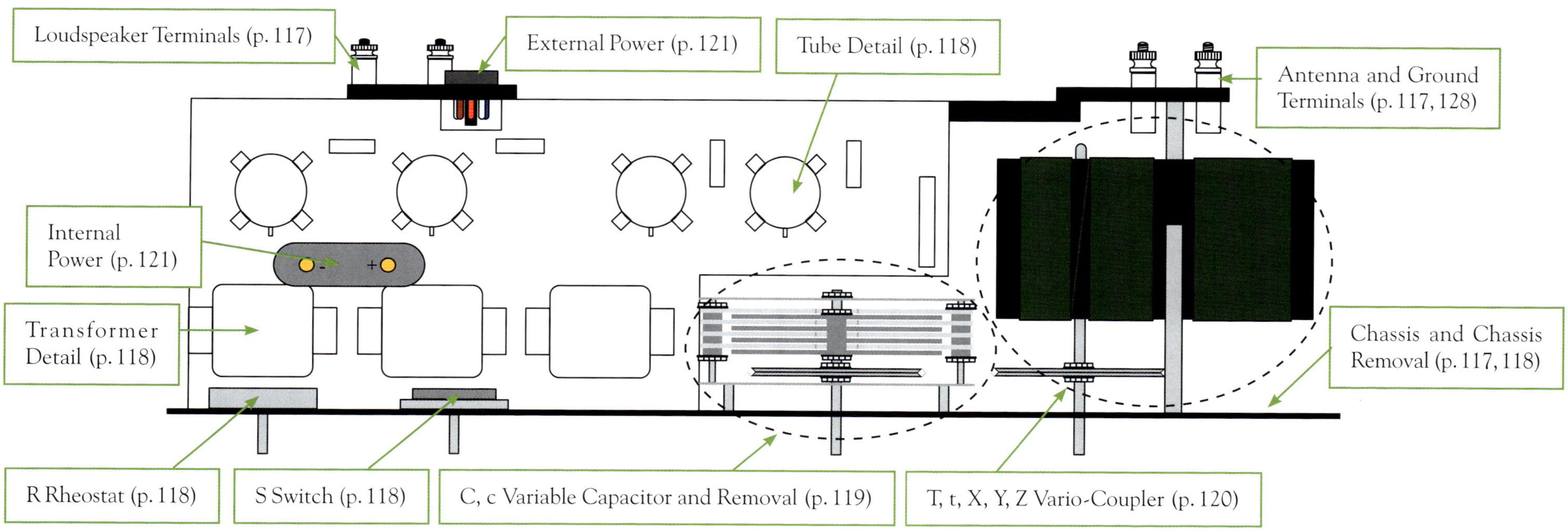

3-R (L) Exterior

The cabinet is made of mahogany and measures 25" by 10.25" by 9.5". The front panel slopes at an angle of approximately 10° and is made of brown Bakelite with faux wood grain.

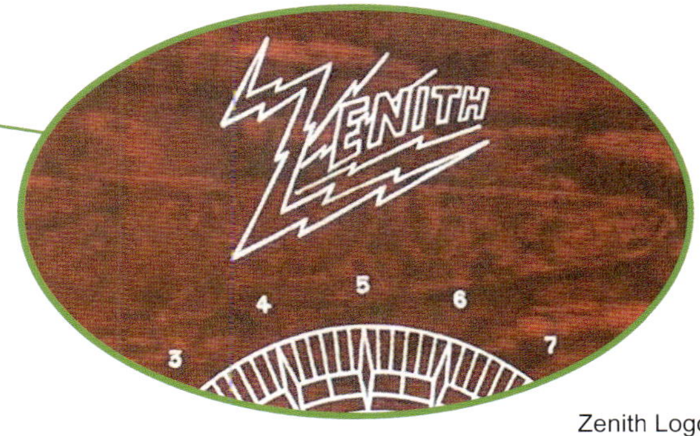

Zenith Logo

Front View

Chicago Radio Laboratory
logo and Armstrong patent information.

Rear View

3-R (L) Identification Labels

Rear View (Armstrong License and Patent label is missing)

Inspection and test sticker (partially missing).

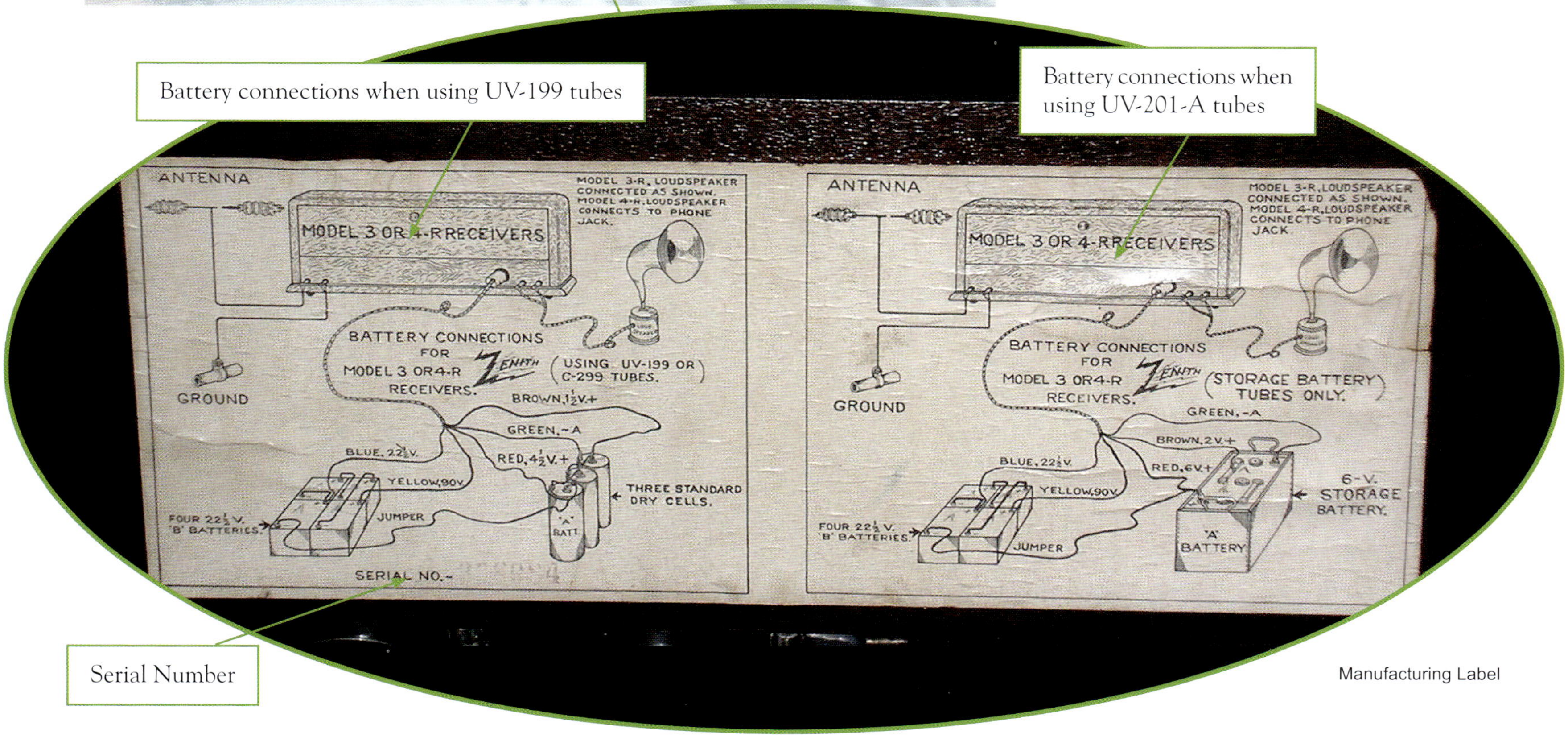

Battery connections when using UV-199 tubes

Battery connections when using UV-201-A tubes

Serial Number

Manufacturing Label

3-R (L) Chassis

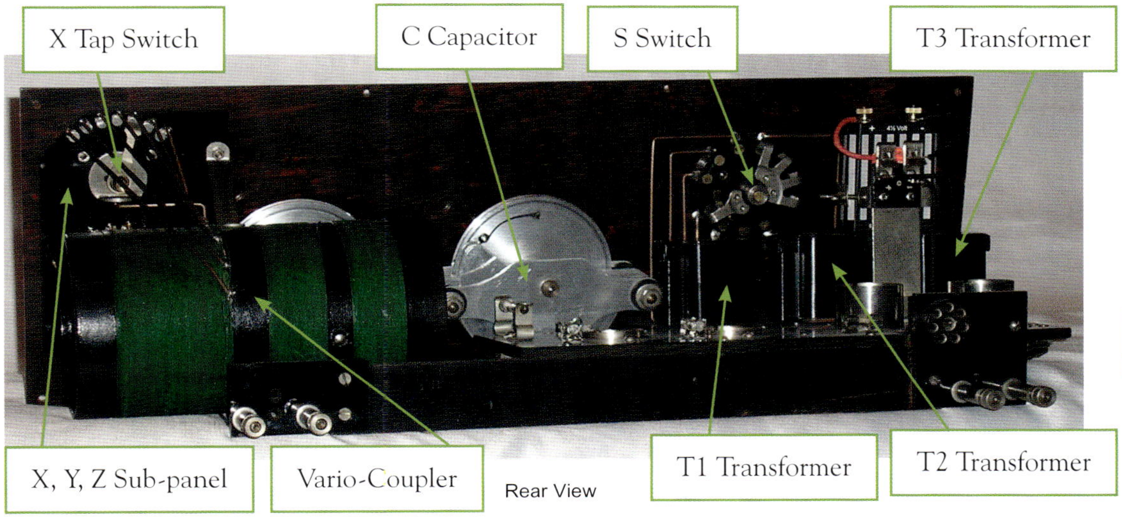

Rear View

Front View

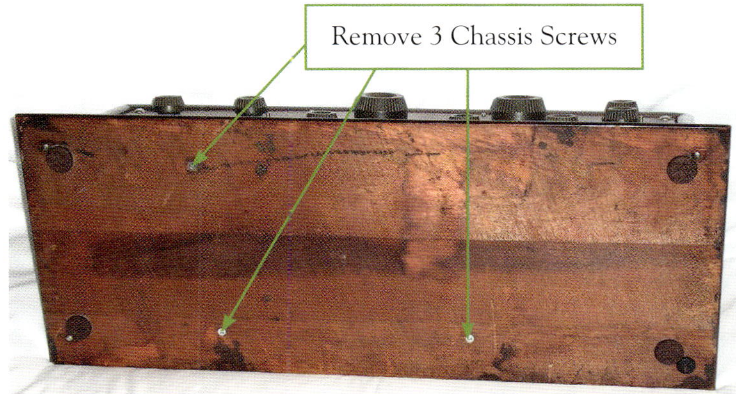

Bottom View

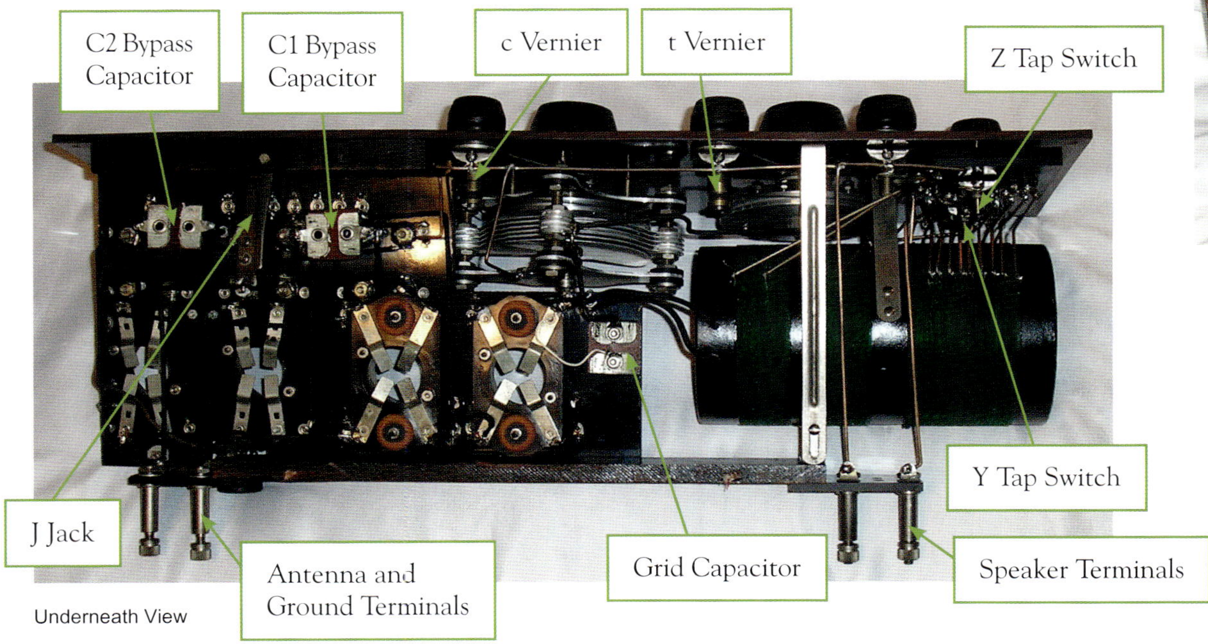

Underneath View

Removal of Chassis from Cabinet

1. Remove three screws from the underside of cabinet. See Bottom View.
2. On the front panel, remove trim molding, and the six screws. See Front View.
4. The whole chassis slides out the front of the cabinet. Gently push the chassis forward from the rear.

To reinstall the chassis into the cabinet, reverse the removal process.

3-R (L) Chassis (continued)

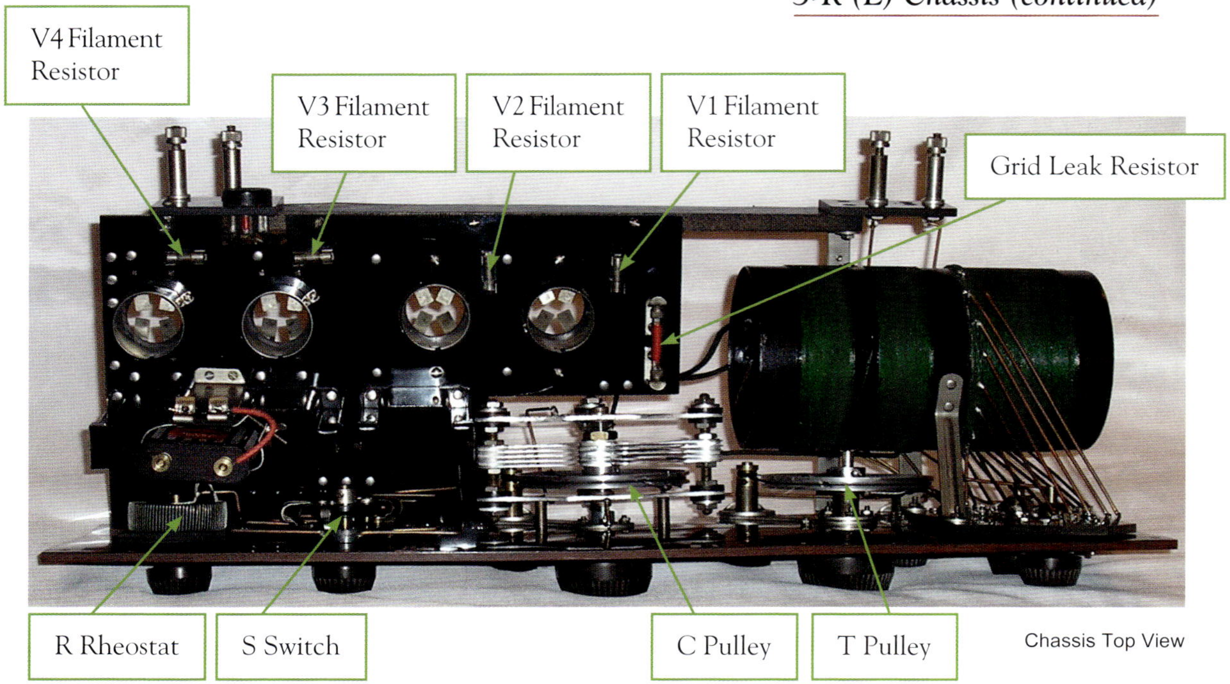

Chassis Top View

Chassis Right Side View

Chassis Left Side View

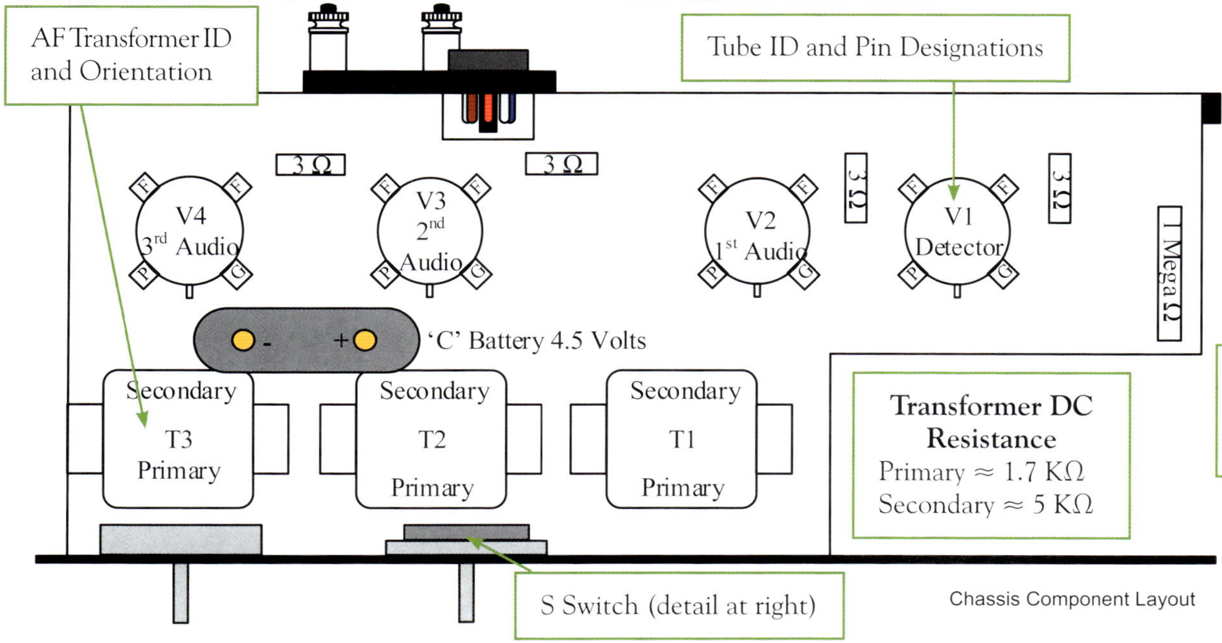

Chassis Component Layout

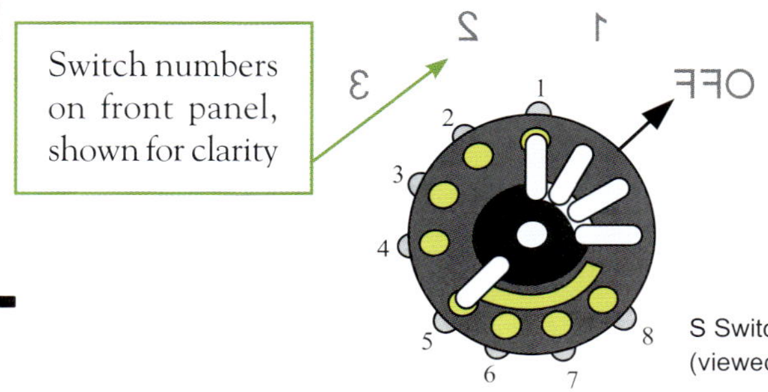

S Switch Detail (viewed from rear)

3-R (L) Variable Capacitor

The variable capacitor housing holds a shaft (rotor) that has 5 plates and a large C pulley that is driven by the smaller vernier c pulley. The shaft rotates the 5 plates, merging them with the 5 fixed plates (stator). The large pulley, small pulley and shaft were made of pot metal. See the replacement pulley in the Top View and Underside View, below.

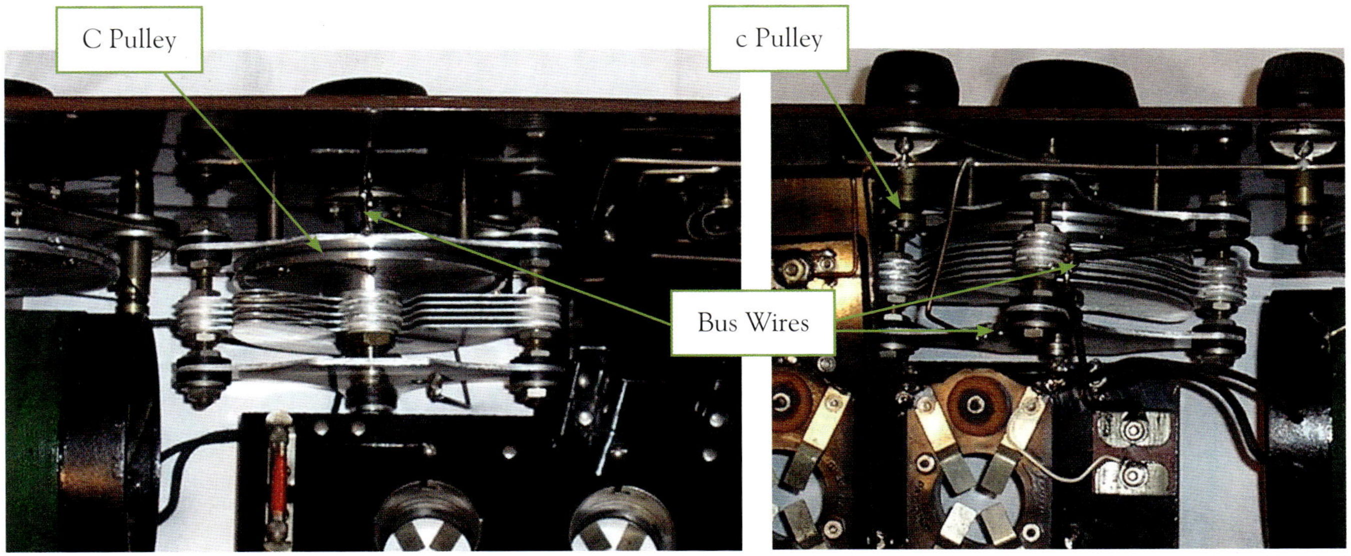

Top View

Underside View

Front View

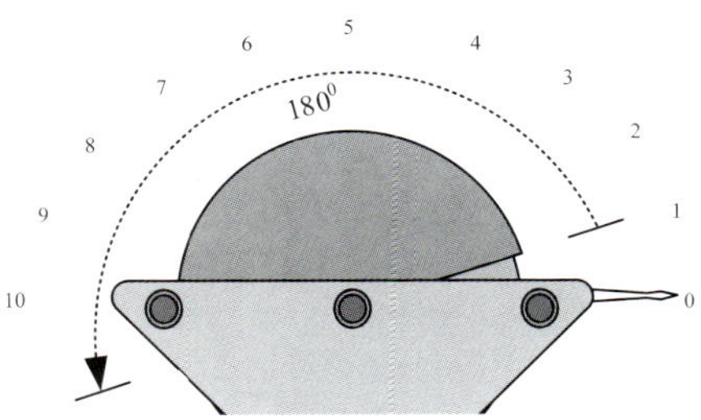

Variable Capacitor Clocking (Viewed from Rear)

Initial starting position of the rotating plates and C pointer in relation to the numbers on the front panel. As the Variable Capacitor rotates 180° right-to-left from the position shown the pointer rotates from 0 to 10, respectively. The pointer and logging numbers are located on front panel, but are shown for clarity.

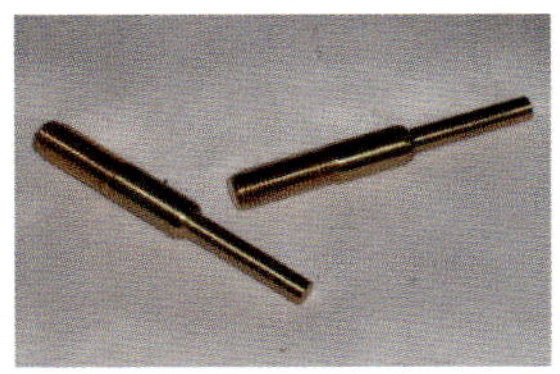

Reproduced Variable Capacitor Shafts

Removal of Variable Capacitor from Chassis

1. Disconnect the bus wires attached to the variable capacitor housing. See Top View and Underside View.
2. On the front panel, remove the two C pointer stops and screw directly below the C knob. See Front View.
4. Gently remove the variable capacitor assembly from the chassis.

To reinstall the variable capacitor, reverse the removal process.

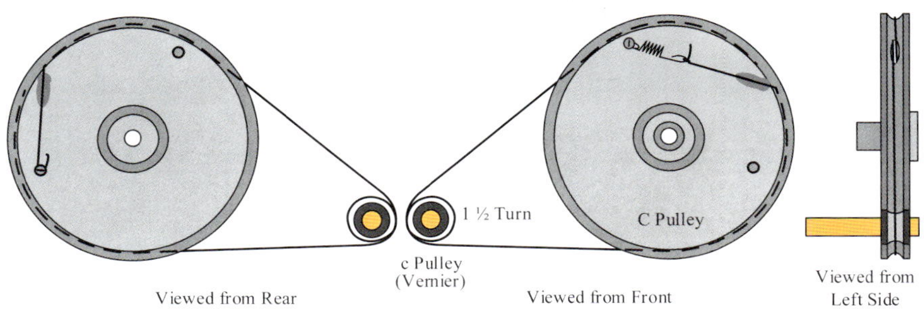

C Pulley Detail

3-R (L) Vario-Coupler

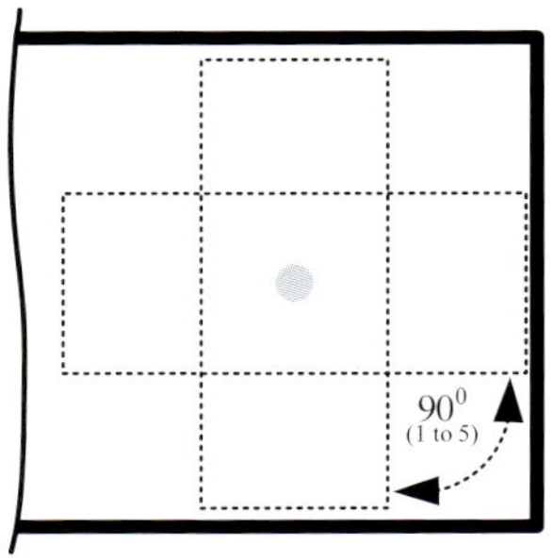

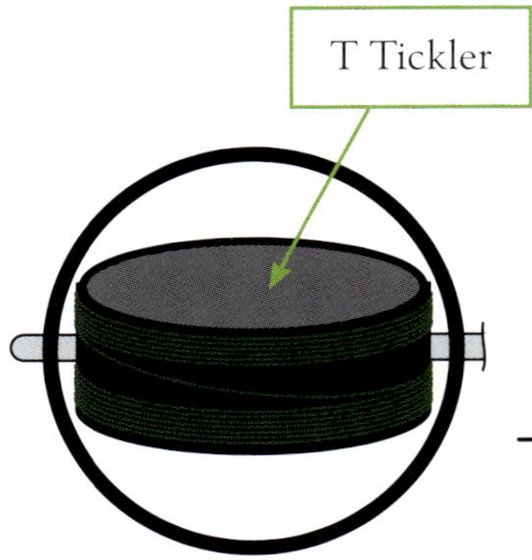

Tickler Clocking

Tickler rotates down, from horizontal to vertical, with T pointer going from 5 to 0.

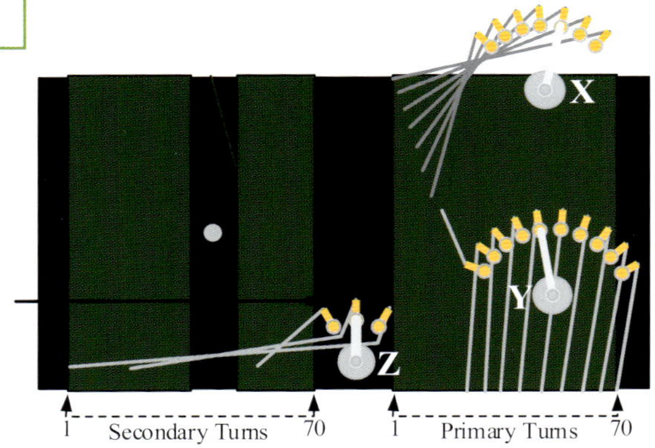

Vario-Coupler Front View

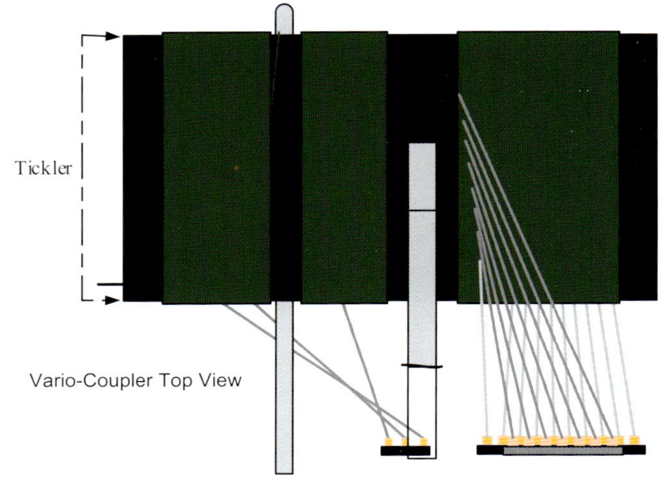

Vario-Coupler Top View

Top View

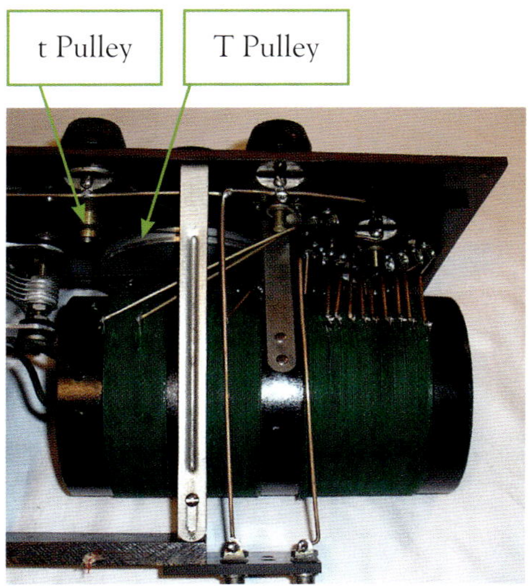

Underside View

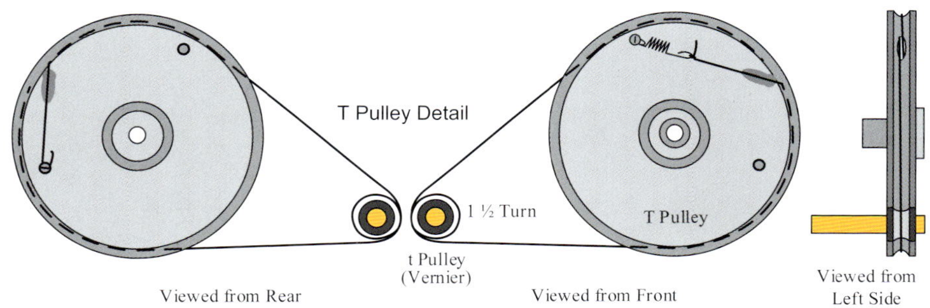

Vario-Coupler Windings and Tap Switch Connections

The main coil of the 3-R (L) consists of a primary and secondary coil each having 70 turns. The secondary coil is physically wound as two coils separated by ½". The coil is constructed in this manner to facilitate the mounting of the small tickler coil inside the main coil. The X-switch tap points 1 through 7 are connected to turns 7 through 1 of the primary coil. In a like manner, Y-switch tap points 1 through 10 are connected to turns 8, 15, 22, 29, 36, 43, 50, 56, 63, and 70 of the primary coil. Additionally, the Y-switch tap point 1 is connected to X-switch tap point 1. The Z-switch taps are connected to the turns of the secondary coil as follows: 1 (tap 3), 20 (tap 2) and 50 (tap 1). The T tickler coil has two connected sets of windings, separated by 5/8", with each coil section having 10 turns (no taps).

3-R (L) Internal Power

The "C" battery provides a negative bias to the grids of tubes (V2, V3, V4).

Internal "C" Battery

Reproduced "C" Battery

3-R (L) External Power

The 3-R (L) uses a five conductor power cable to connect to the external batteries. The power cable has a keyed multi-plug on one end, so that it only goes into the socket one way. This keying method ensures that the proper voltages are routed to the proper circuits. The loose wires, at the other end, are color coded and should be connected per the manufacturer's instructions, which are located on the inside of the back door. Instructions are provided for the UV-201-A tubes and the UV-199 tubes.

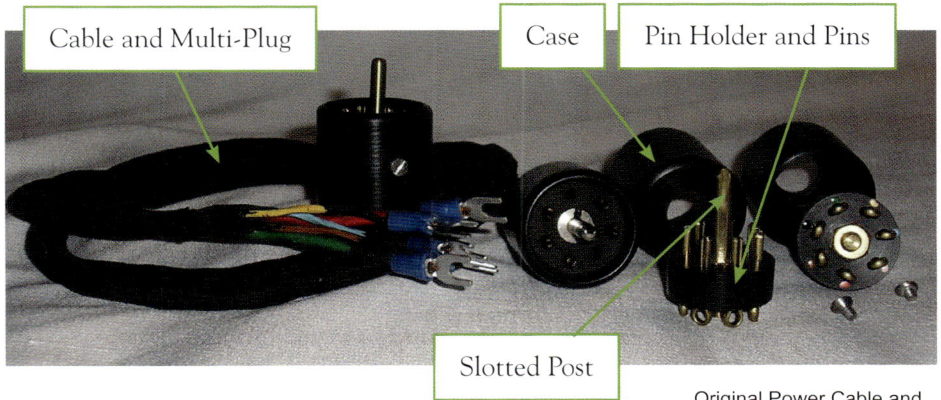

Original Power Cable and Reproduced Multi-Plugs

Power Cable Socket

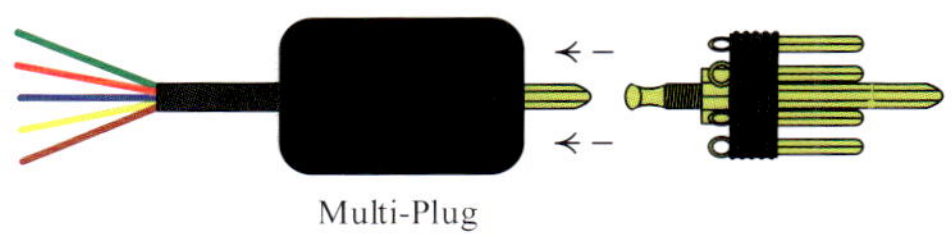

Power Cable and Multi-Plug

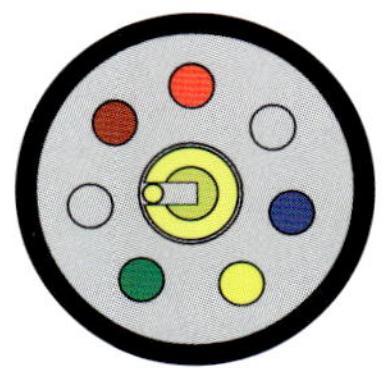

Multi-Plug Color Coding and Clocking

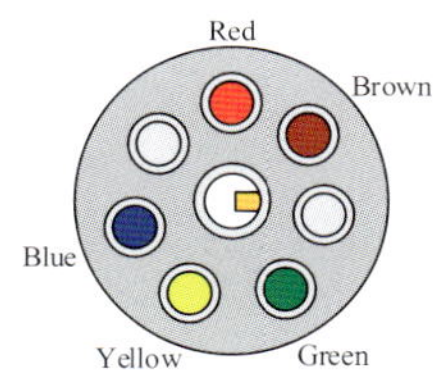

Socket Color Coding and Clocking

3-R (L) Functional Description Quick Reference Guide

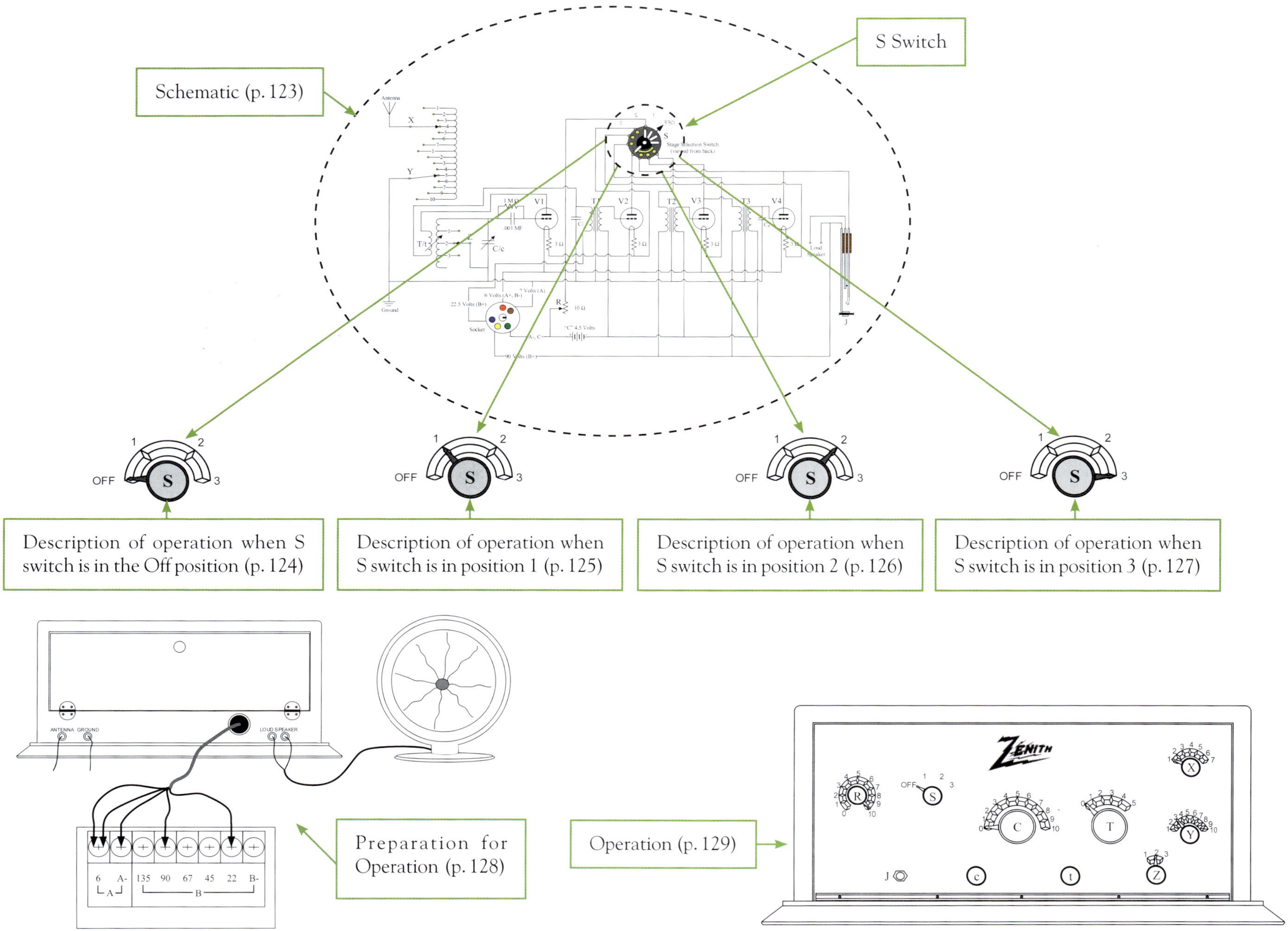

3-R (L) Schematic

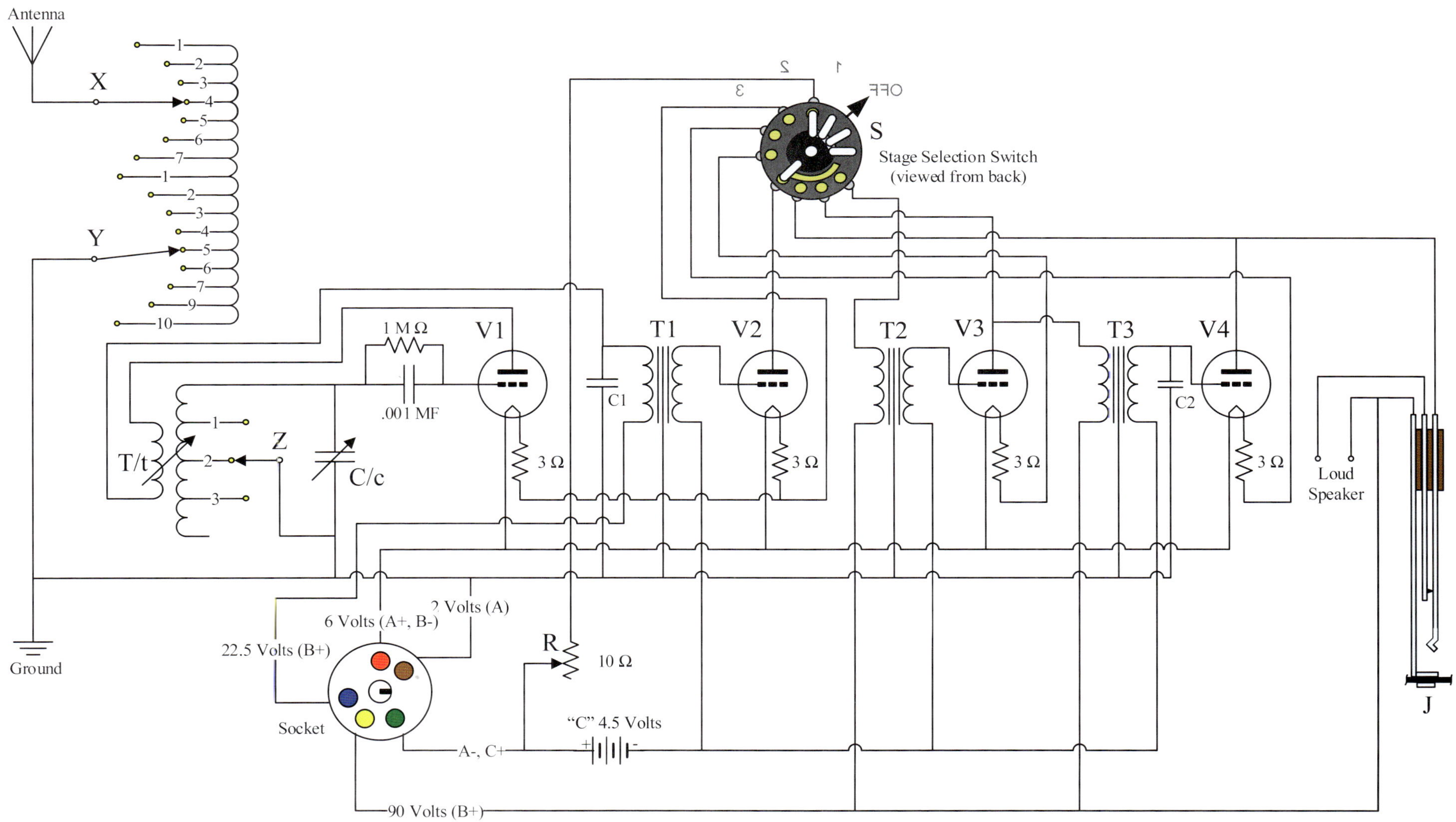

3-R (L) Description of Operation—Switch S in Off Position

When the switch S is in the OFF position the filament circuits are open and the radio is inoperable. Plate voltage is applied to the plate of V1. A negative bias voltage is applied to the grids of V2, V3, and V4. This is considered the initial state of the radio for future discussions of circuit operation.

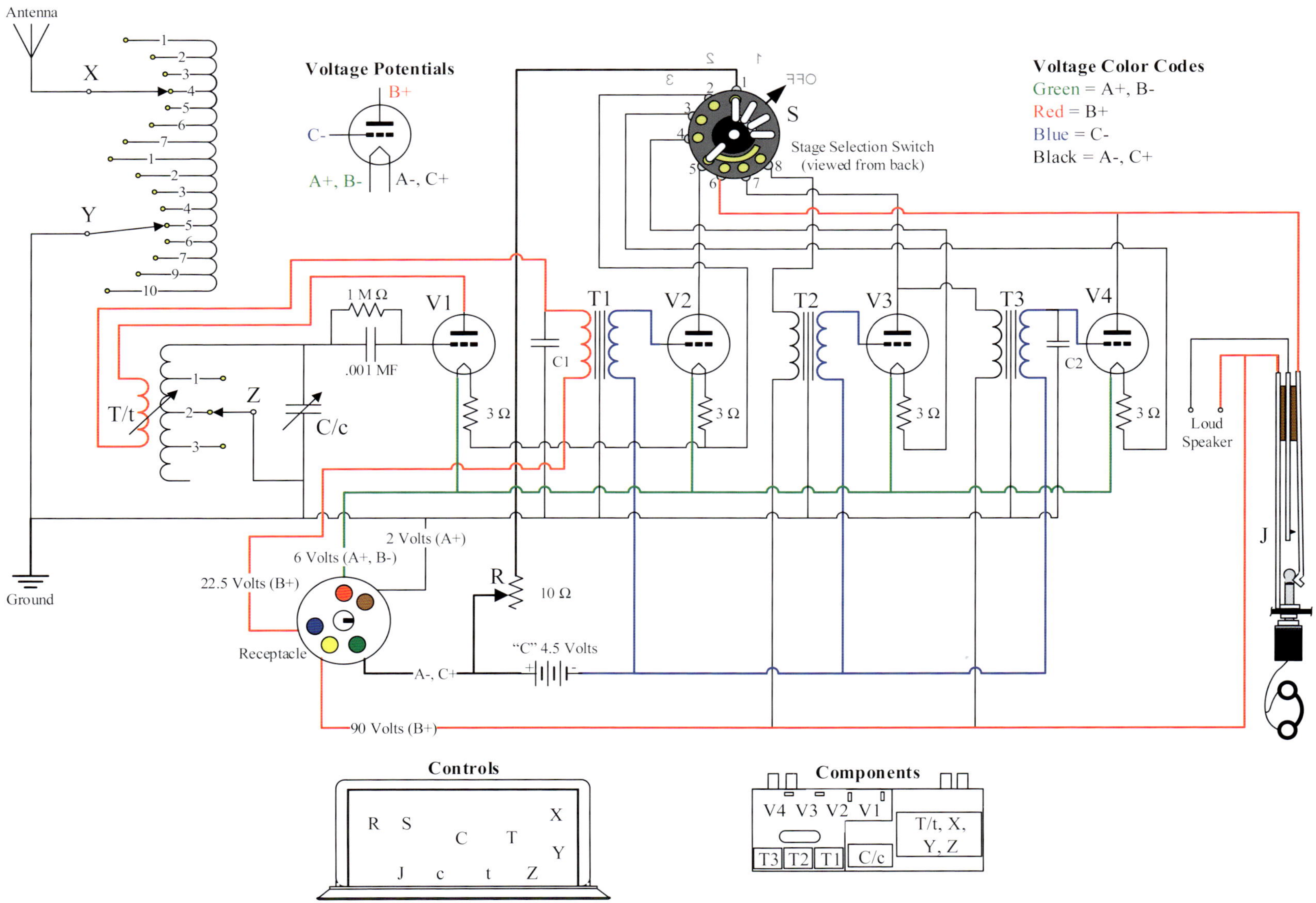

3-R (L) Description of Operation—Switch S in Position 1 (One AF Amplifier Stage)

When switch S is moved to position 1, the double wiper closes contact points 1 and 2 energizing the filament circuits to V1 and V2. At the same time the lower wiper connects contact points 5 and 6, which routes the plate of V2 to the headphones circuit. The radio is now playing with the detector (V1) and one stage (V2) of AF amplification.

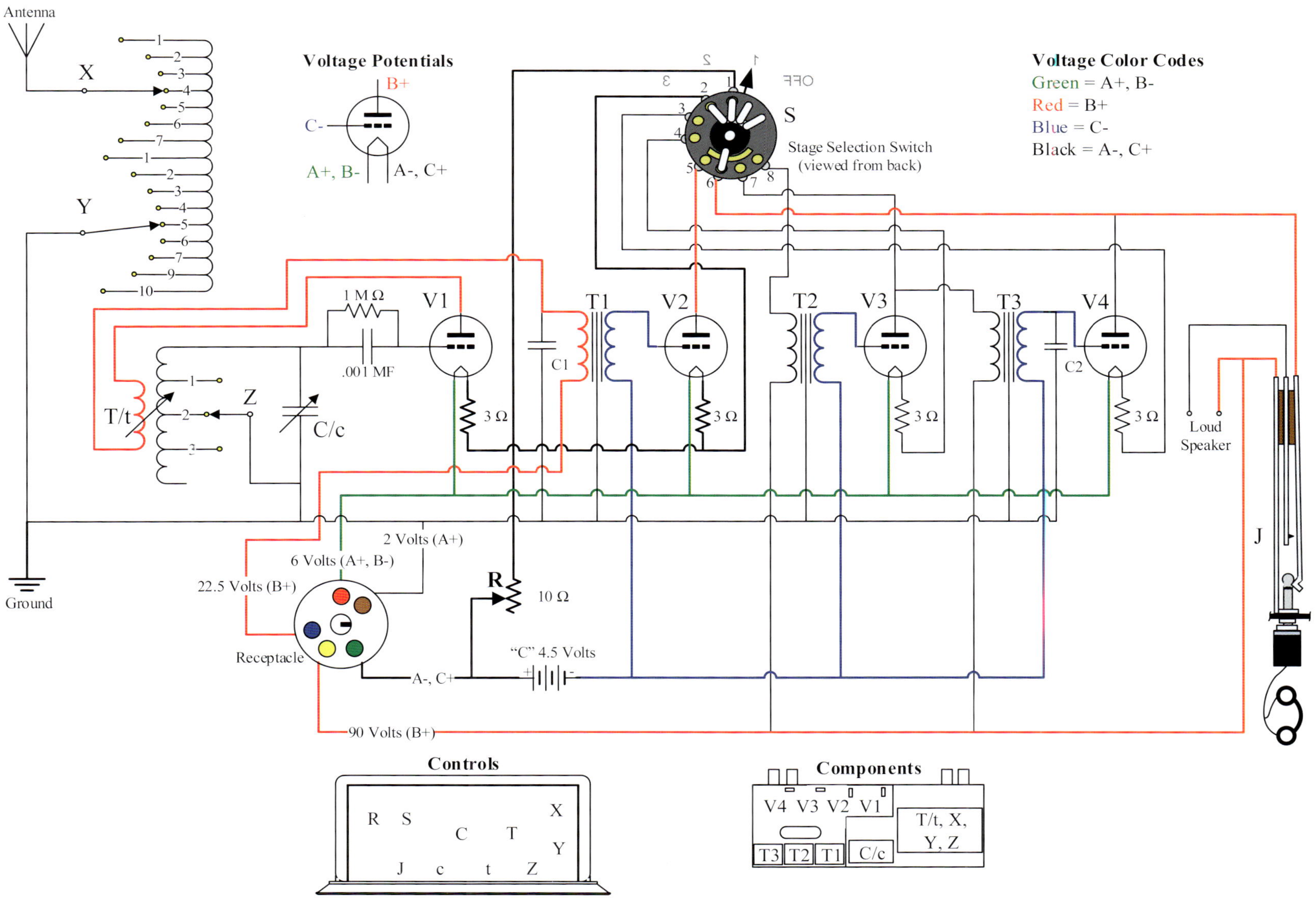

3-R (L) Description of Operation—Switch S in Position 2 (Two AF Amplifier Stages)

When switch S is moved to position 2, the triple wiper closes contact points 1, 2 and 3 energizing the filament circuits to V1, V2 and V4. At the same time, the single wiper disconnects contact points 5 and 6, and connects contact points 5 and 7, which routes the plate of V2 to the primary of transformer T3. The radio is now playing with the detector (V1) and two stages (V2, V4) of AF amplification.

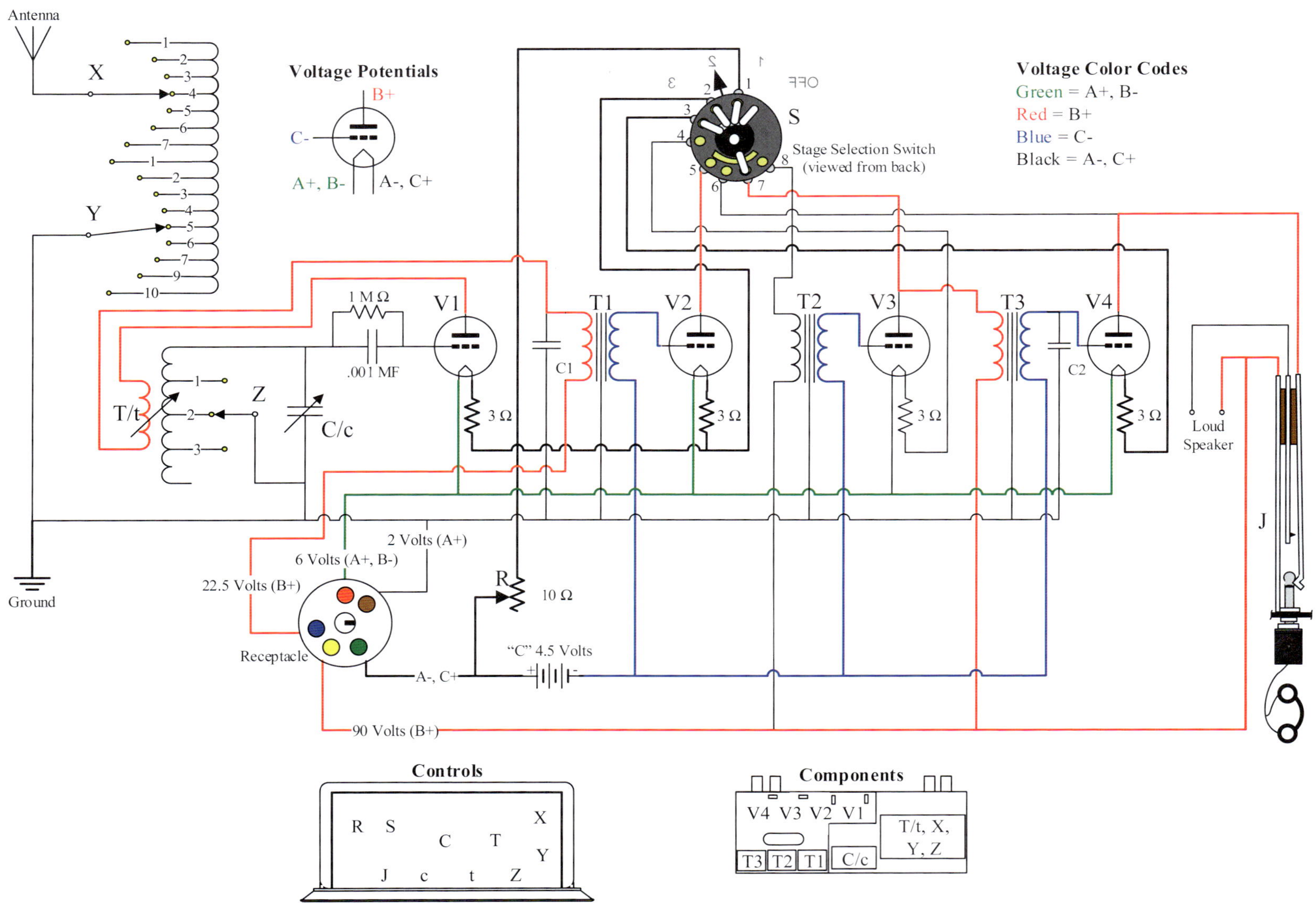

3-R (L) Description of Operation—Switch S in Position 3 (Three AF Amplifier Stages)

When switch S is moved to position 3, the quadruple wiper closes contact points 1, 2, 3 and 4 energizing the filament circuits to V1, V2, V3 and V4. At the same time, the single wiper disconnects contact points 5 and 7, and connects contact points 5 and 8, which routes the plate of V2 to the primary of transformer T2. The radio is now playing with the detector (V1) and three stages (V2, V3, V4) of AF amplification.

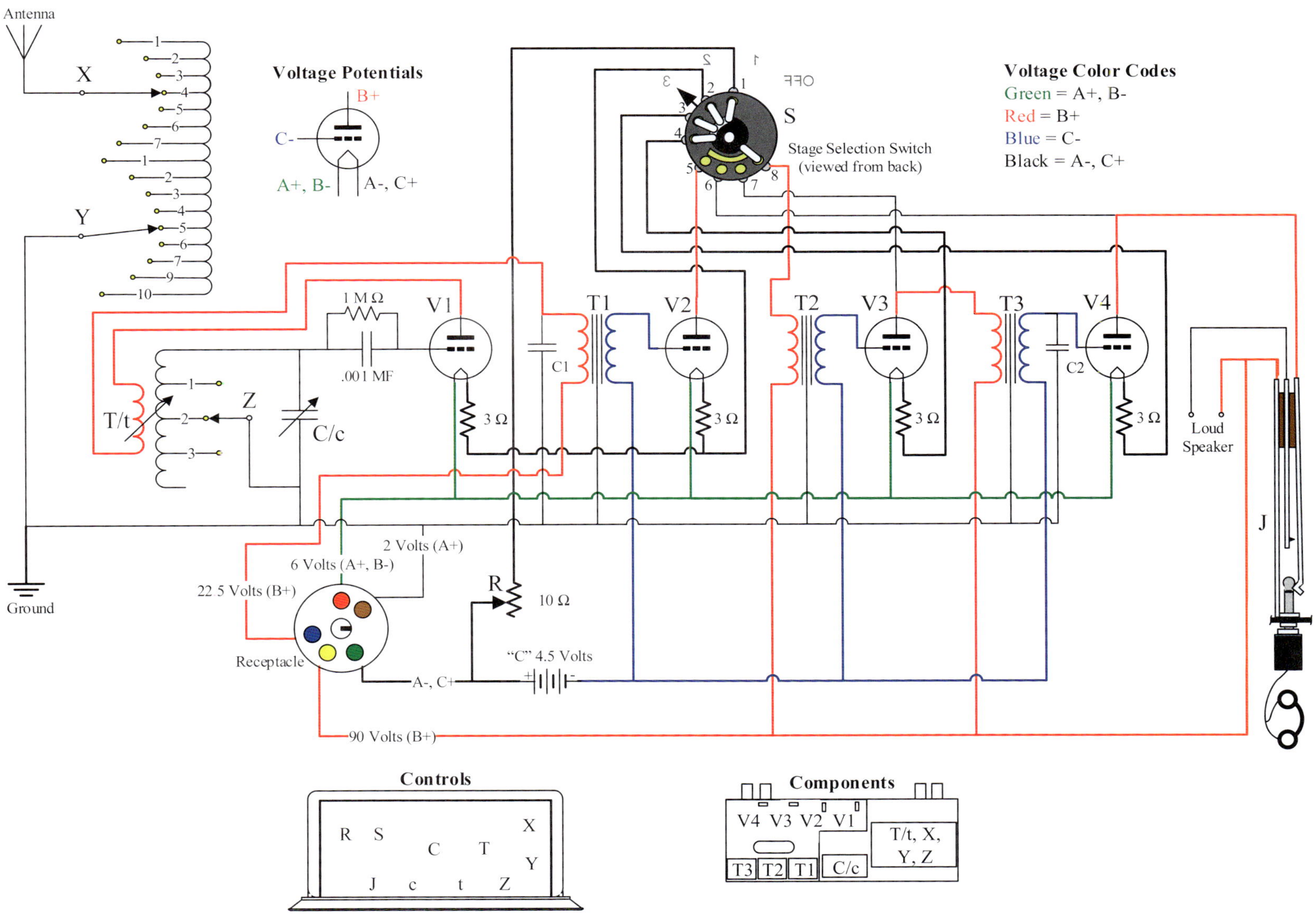

3-R (L) Preparation for Operation

Install the tubes and the "C" battery inside the radio. Connect the antenna and ground to the ANTENNA and GROUND posts. Use the appropriate battery connection diagram, inside the back door of the radio, to make sure the power can be connected properly and agrees with the 3-R (L) Connections diagram below. If the power connections are not correct or otherwise unavailable, utilize the physical and functional descriptions to assist in determining the correct power connections. Hook up the power to the radio. Proceed to 3-R (L) Operation. See Common Radio Attachments (page 8) for additional information related to antennas, loudspeakers and power options.

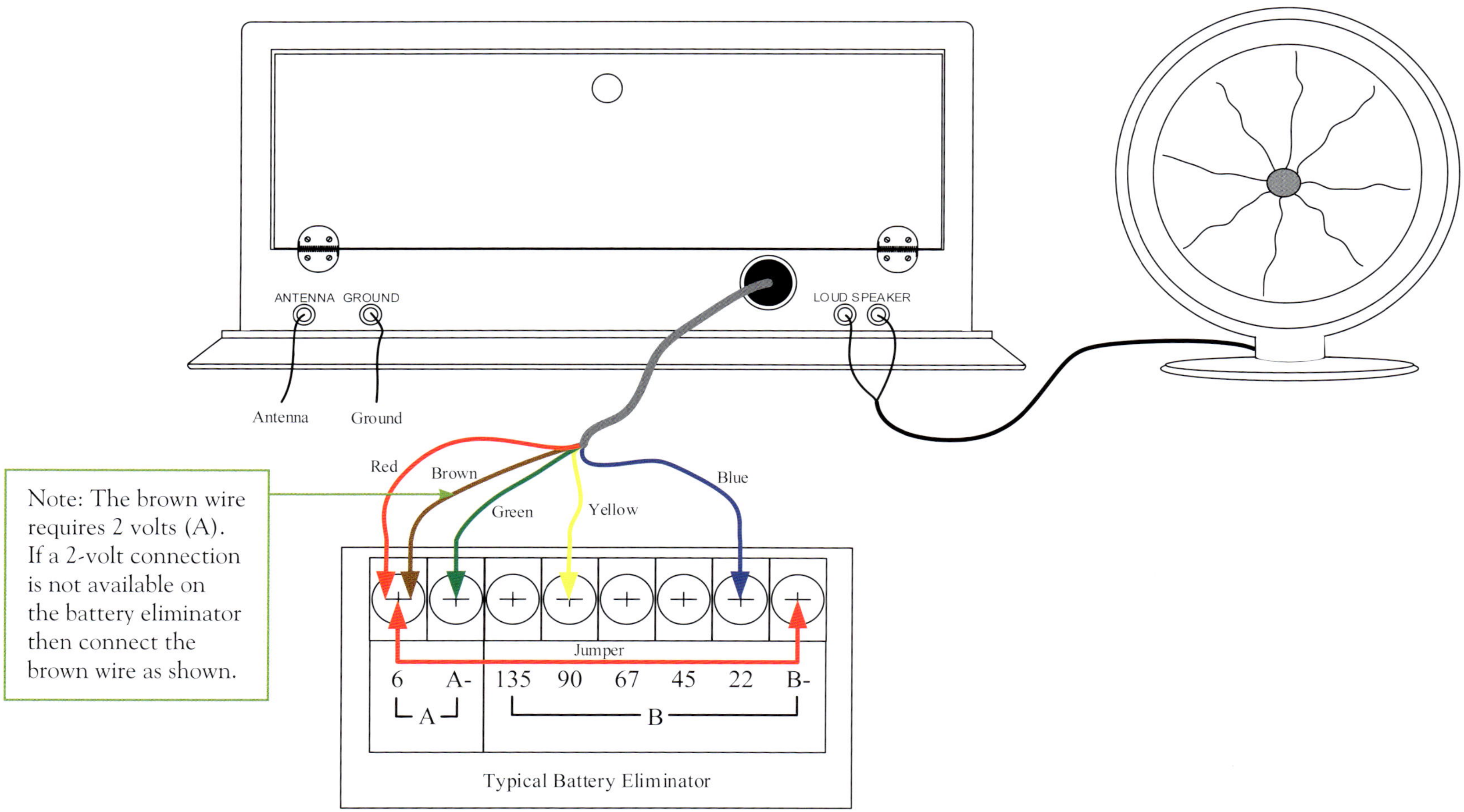

3-R (L) Connections

3-R (L) Operation

Follow the 3-R (L) Preparation for Operation instructions. Set the pointers to the positions shown in the diagrams below.

The tuning capacitor C selects the stations and the tickler T provides the feedback (regeneration) that increases the volume of the selected signal. Practice using knobs C and T simultaneously. Utilize the small knobs c and t to make fine adjustments to the mechanisms C and T.

Insert the headphone plug into jack J. Turn the "S" switch to position 1 or 2.

Gradually move "T" to the right until a hissing sound is heard and hold the pointer to the beginning of that sound. Gradually move "C" to the right until a bird whistle is heard, first high, then descending to low; then a pause will be heard. As you continue to move "C" the whistle is again low, then ascending to high. The station is between these low points or valleys. Move "T" to the beginning of the whistle, while "C" is being moved. If you have C in the valley of the whistle, slightly move "T" to the left to clear the station. For best results operate C and T simultaneously. If an unpleasant howl (oscillations) is heard and the station is lost, turn "T" back to the left until the howl stops.

Adjust the audio level (volume) by setting switch S to position 1, 2 or 3, as appropriate.

If a loudspeaker is connected to the rear terminals and a headphone plug is inserted into the J jack, the loudspeaker will be automatically turned OFF, and when the plug is pulled out of the J jack, the loudspeaker will be automatically turned ON.

Knobs X and Y tune the antenna circuit. A little manipulation of these switches will enable the operator to obtain improved results.

Knob Z changes the bandwidth of the receiver and is generally left in the position 2. Position 1 decreases the bandwidth and position 3 increases the bandwidth.

Record the control settings for each station in the Station Log (next page). Then, to return to a particular station, set the controls to the settings that were previously recorded in the log for that station.

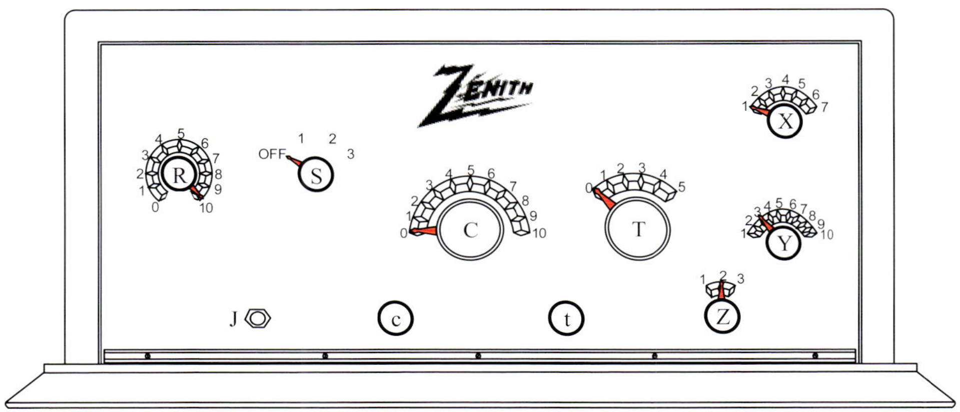

Initial Dial Settings (pointers shown in red for clarity)

3-R (L) Station Log

Name: ____

Date	Time	Station	Frequency	C	T	X	Y	Z

From *The Early Zenith Radios: The Battery Powered Table Sets 1922–1927,* © 2014 by Gilbert M. Hedge with Durell M. Roth. ISBN 978-0-7643-4674-3. Schiffer Publishing Ltd. www.schifferbooks.com

Chapter 8: **Zenith 4-R** (Late Version)

4-R (L) General Specifications

Company:	Zenith Radio Corporation
Model:	4-R (Late Version)
Year Manufactured:	1924
Manufacture Location:	48th and Kedzie, Chicago IL (QRS Factory)
Serial Number:	417406
Price:	$95
Tube:	UV-201-A (4), one detector, three AF amplifiers
Circuit Type:	Regenerative

4-R (L) General Information

The late version of the 4-R had several improvements over the earlier version. No major changes were made to the regenerative circuit, but the voltage to the plates of the AF amplifier tubes (V2, V3, and V4) was changed to make each one operate at 90 volts. Because the multiple plate voltages were eliminated, the power cable was changed from 7 conductors (wires) to 5. Stage switching was changed from the lobed camshaft to a multi-contact rotary switch. The order of the AF amplification tubes that could be selected was changed. The peep hole was removed.

Restoration details for the set that was used as the example in this chapter. *This particular radio was affected by the pot metal problem. The pointers were replaced with handmade aluminum pointers, with the exception of the two large ones, which showed very little effects of the pot metal; they had just a small amount of warping on the variable capacitor pointer. A replacement cable was also reproduced with a multi-plug and a 5 conductor cable. The original "C" battery was replaced with a reproduced "C" battery.*

4-R (L) Physical Description Quick Reference Guide

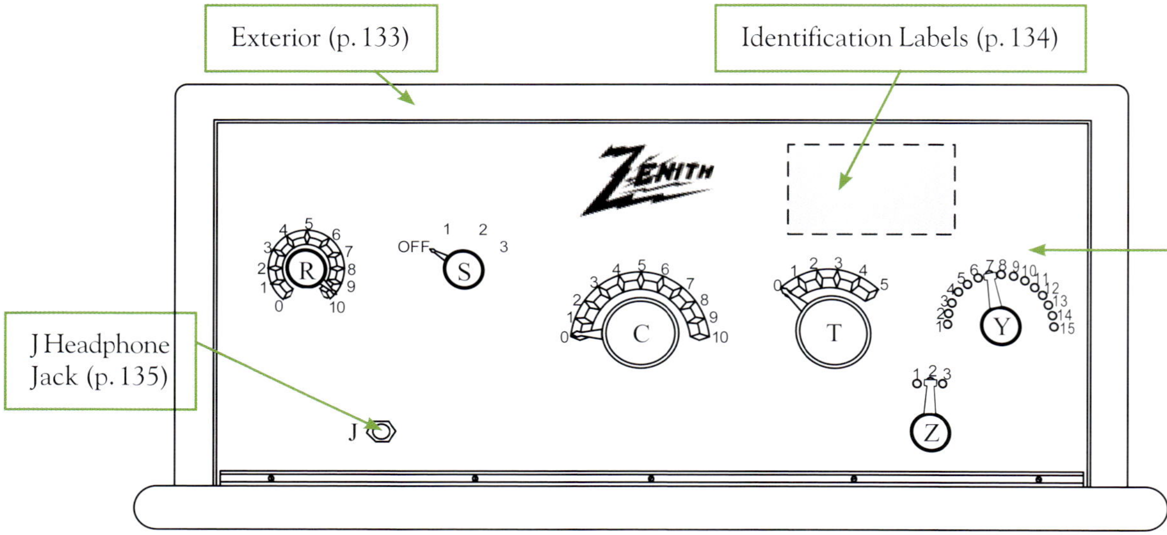

Note: Shaded item is located inside set.

Control	Function
C	Variable Capacitor (tuning)
J	Headphone Jack
R	Rheostat (filament voltage control)
S	Switch (Off, 1, 2 or 3 AF amp. stages)
T	Tickler (regeneration)
Y	Tap Switch (tunes the antenna circuit)
Z	Tap Switch (selects the bandwidth)

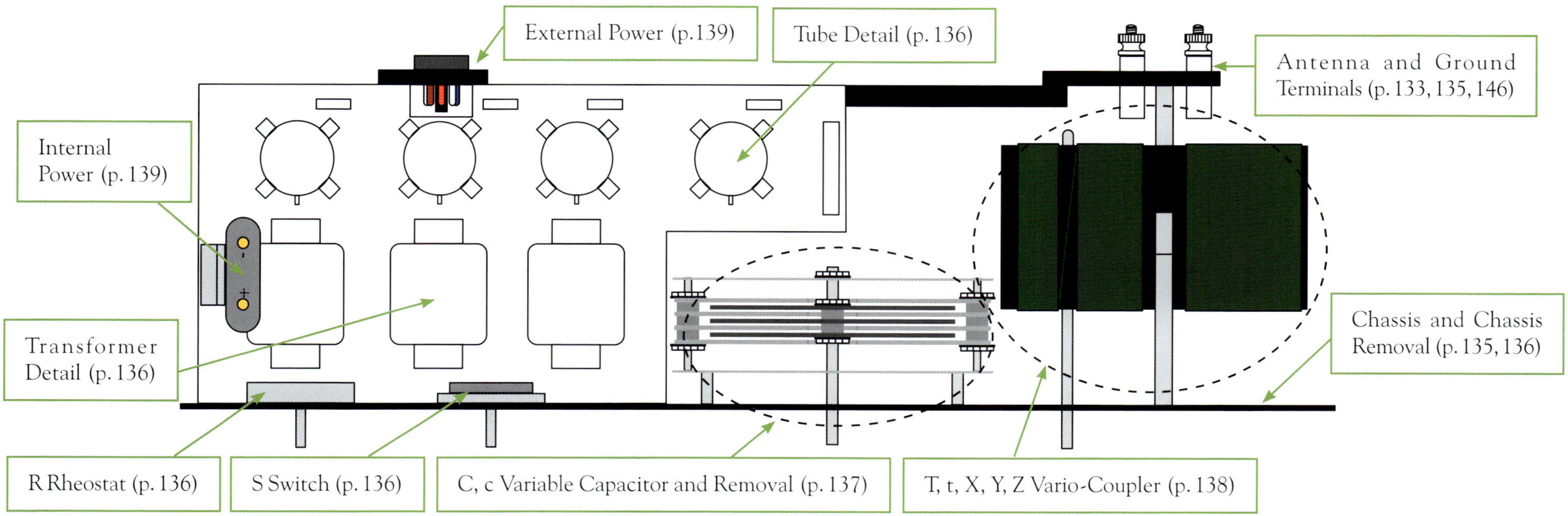

4-R (L) Exterior

The cabinet is made of dark red mahogany and measures 21.75" by 10" by 9.75". The front panel is made of brown Bakelite.

Front View

Zenith Logo

Chicago Radio Laboratory logo and Armstrong patent information.

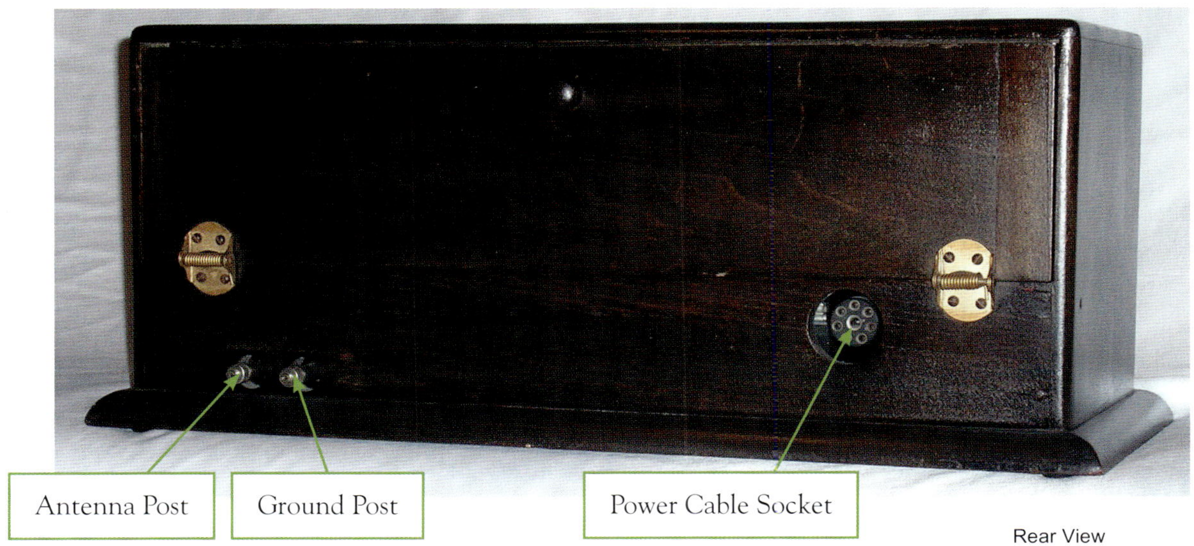

Rear View

4-R (L) Identification Labels

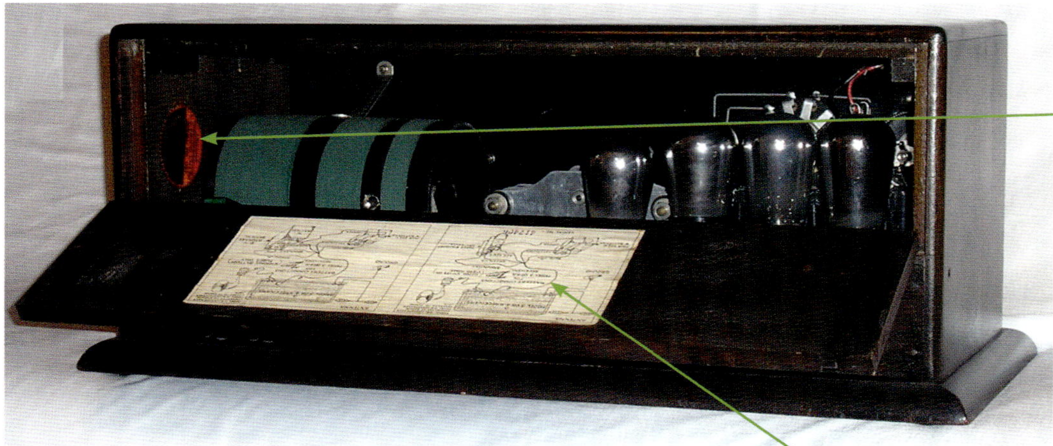

Rear View

Armstrong License and Patent Label

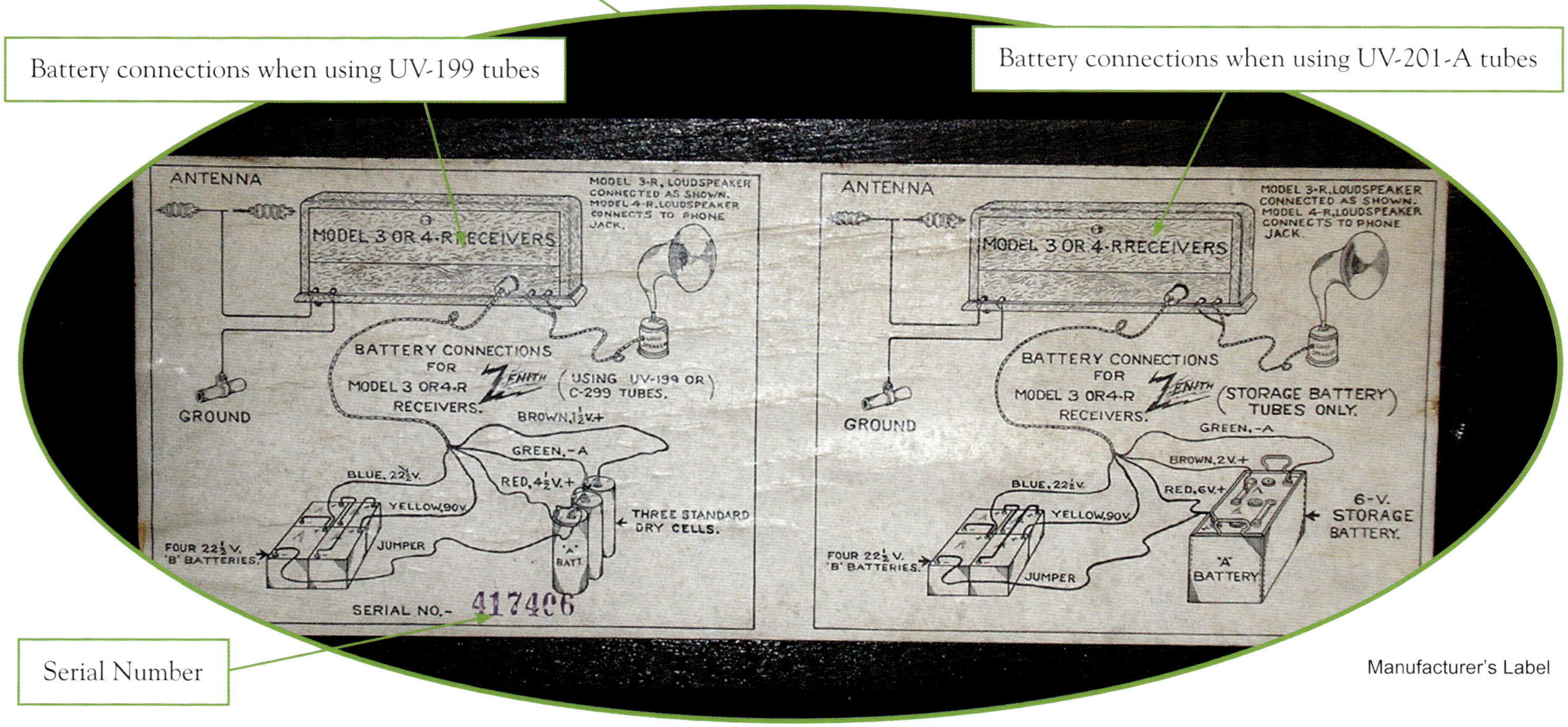

Manufacturer's Label

4-R (L) Chassis

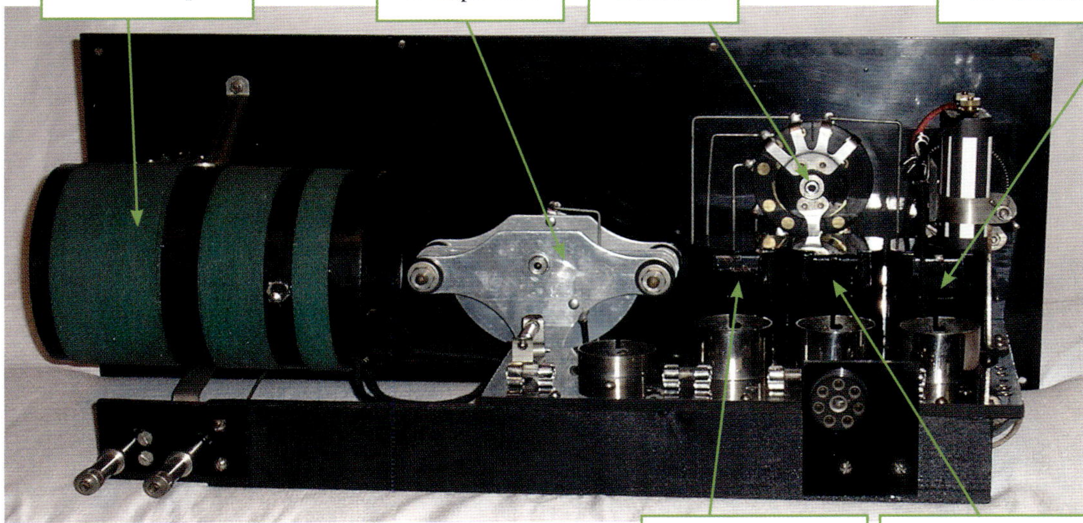

Rear View

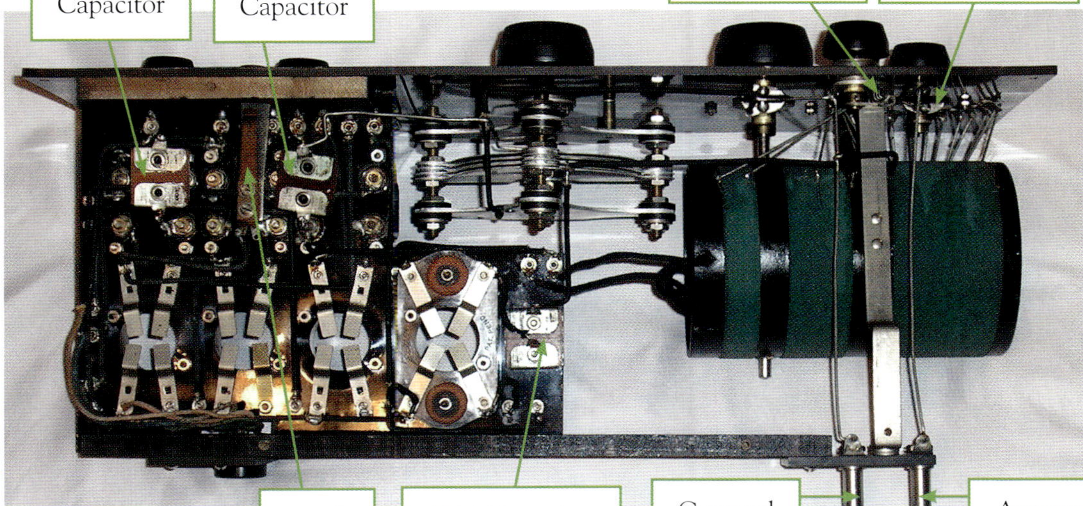

Underneath View

Front View

Bottom View

Removal of Chassis from Cabinet

1. Remove the three screws from the underside of cabinet. See Bottom View.
2. On the front panel, remove trim molding and the six screws. See Front View.
4. The whole chassis slides out the front of the cabinet. Gently push the chassis forward from the rear.

To reinstall the chassis into the cabinet, reverse the removal process.

4-R (L) Chassis (continued)

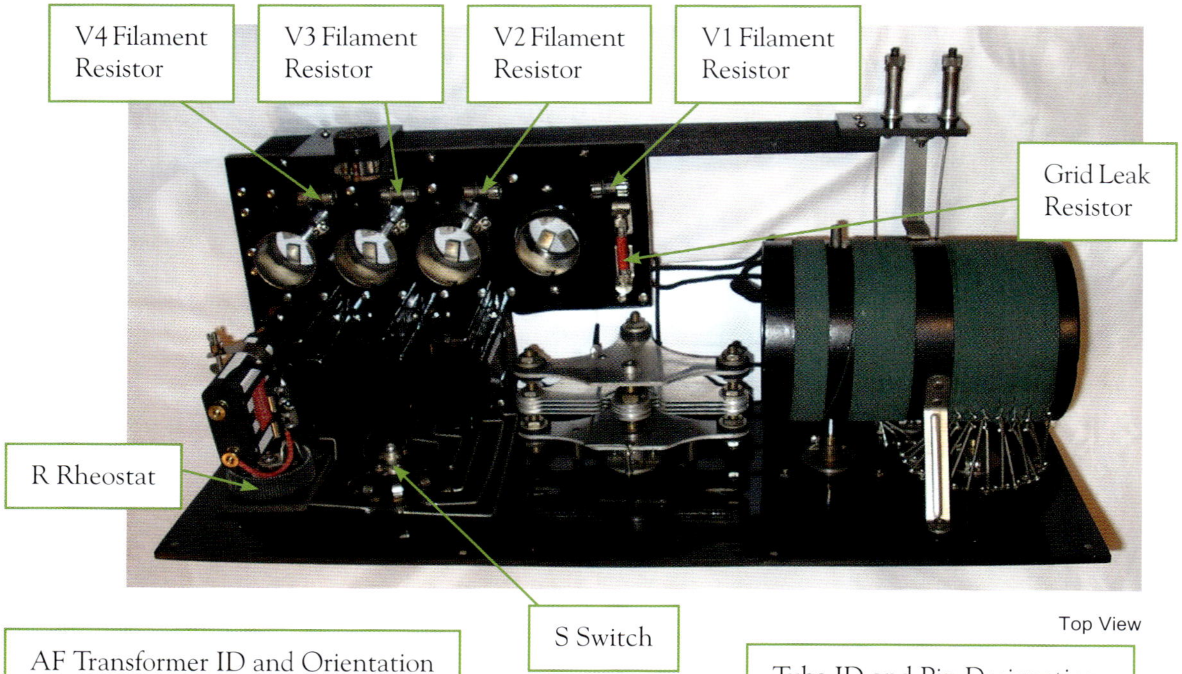

Top View

Right Side View

Left Side View

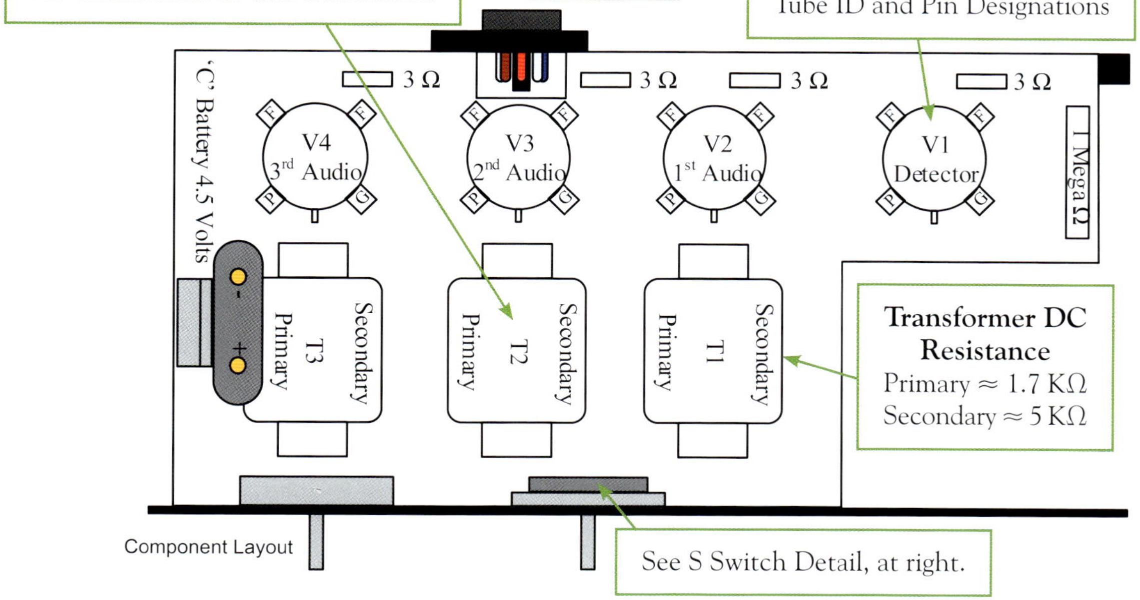

Component Layout

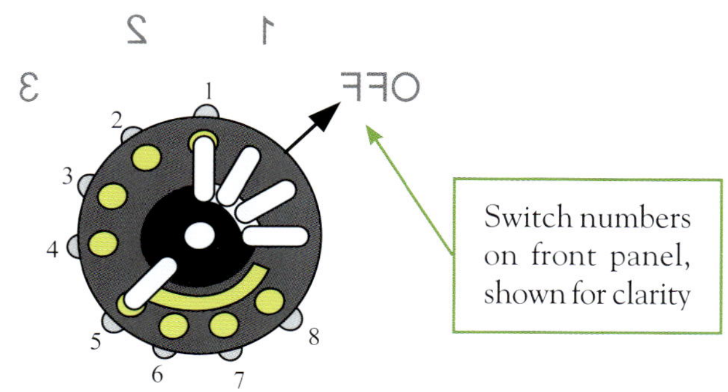

S Switch Detail (viewed from rear)

4-R (L) Variable Capacitor

The variable capacitor housing holds a shaft (rotor) that has 3 plates. The shaft rotates the 3 plates, merging them with the 4 fixed plates (stator).

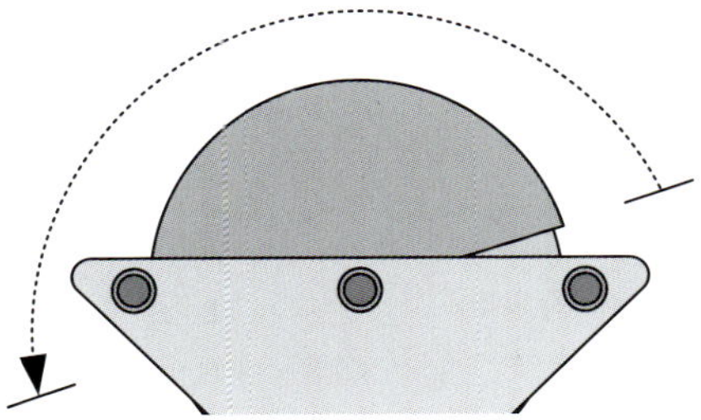

Variable Capacitor Clocking (Viewed from Rear)

Clocking of the movable plates in relation to the C pointer and fixed plates. As the Variable Capacitor rotates 180° right-to-left from the position shown, the pointer rotates from 0 to 10, respectively. Pointer and numbers are shown for clarity and are located on the front panel.

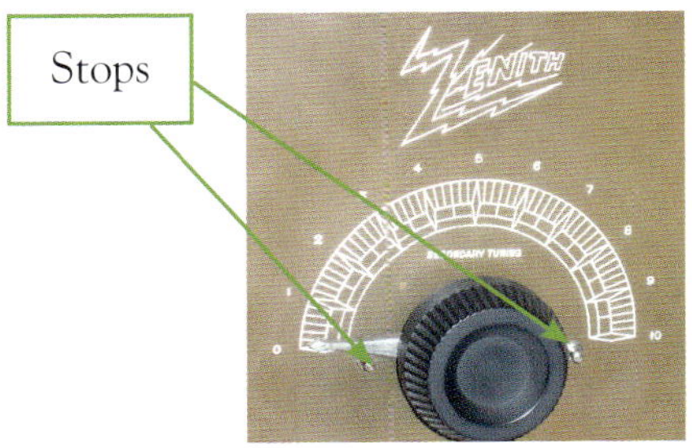

Front Panel

Removal of Variable Capacitor from Chassis

1. Disconnect the bus wires attached to the variable capacitor housing. See Top View and Underside View.
2. On the front panel, remove the two C pointer stops for the C pointer and screw directly below the C knob. See Front Panel.
4. Gently remove the variable capacitor assembly from the chassis.

To reinstall the variable capacitor, reverse the removal process.

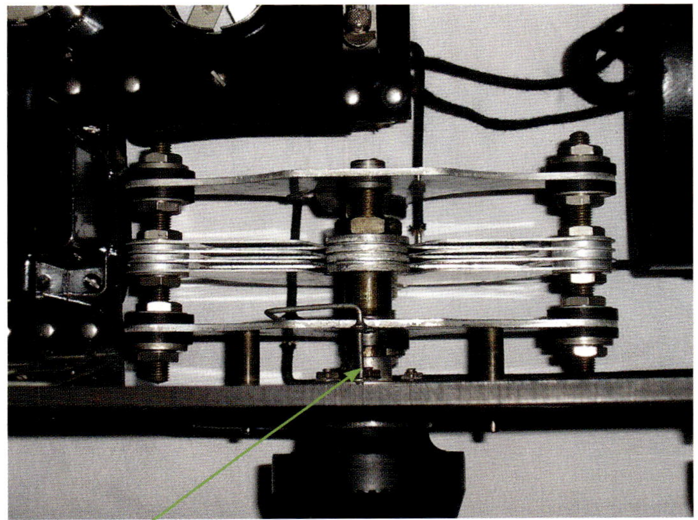

Top View

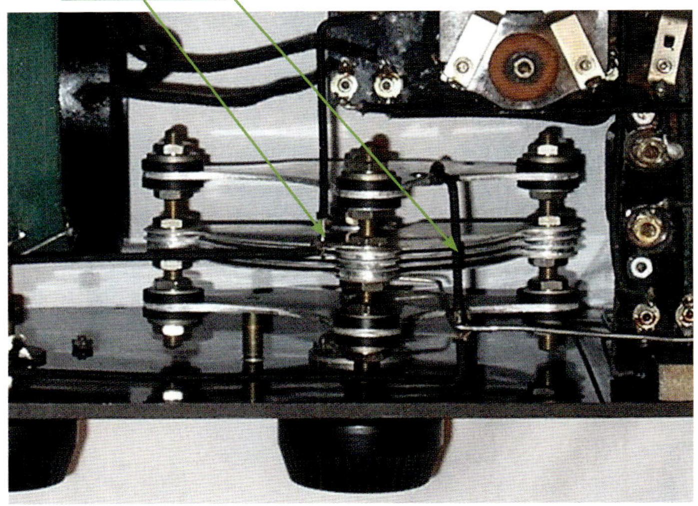

Underside View

4-R (L) Vario-Coupler

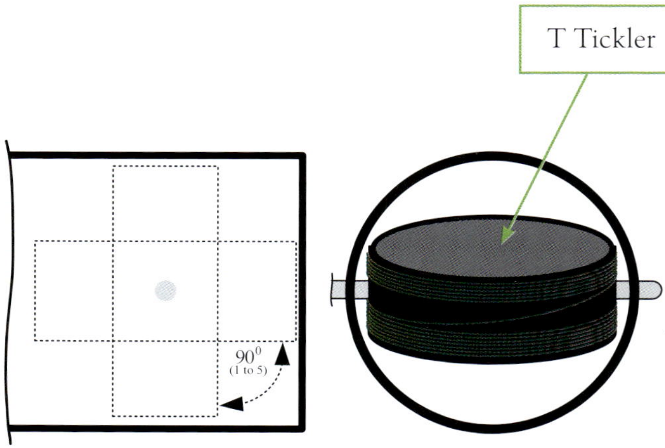

Tickler Clocking

When viewed from the front the tickler rotates from horizontal, down to vertical and T pointer moves from 5 to 0, respectively.

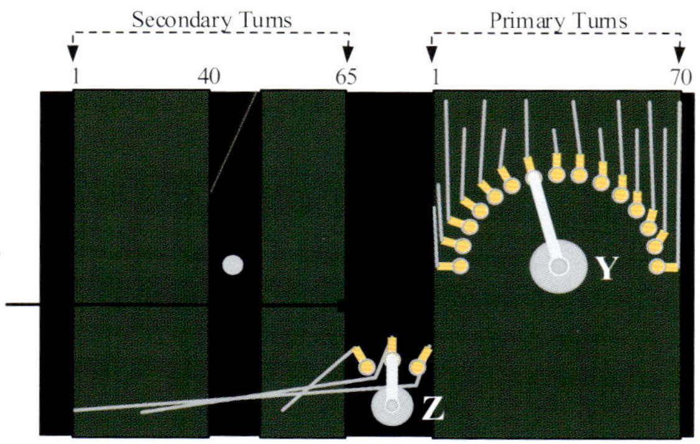

Front View

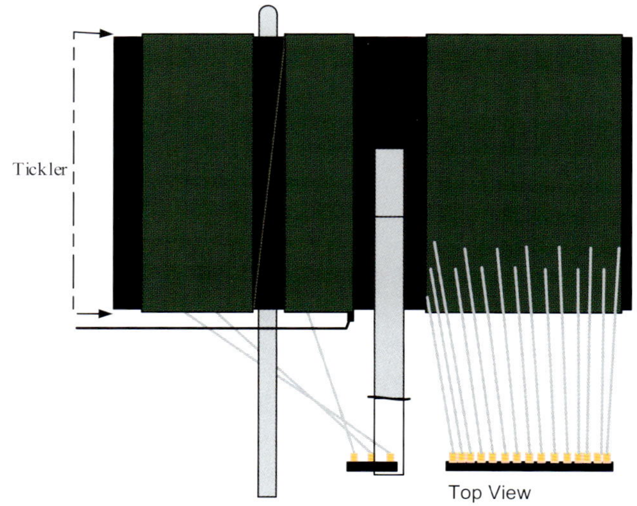

Top View

Vario-Coupler Windings and Tap Switch Connections

The main coil of the 4-R (L) consists of a primary coil with 70 turns and secondary coil with 65 turns. The secondary coil is physically wound as two coils separated by ¾". The coil is constructed in this manner to facilitate the mounting of the small tickler coil inside the main coil. The Y-switch tap points 1 through 15 are connected to turns 1, 2, 8, 10, 15, 20, 25, 30, 35, 40, 45, 50, 55, 60, and 70 of the primary coil. In a like manner the Z-switch taps are connected to the turns of the secondary coil as follows: 1 (tap 3), 20 (tap 2) and 44 (tap 1). The tickler T coil has two connected sets of windings, separated by 5/8", with each coil section having 8 turns (no taps).

4-R (L) Internal Power

The "C" battery provides a negative bias to the grids of tubes V2, V3 and V4.

Internal "C" Battery

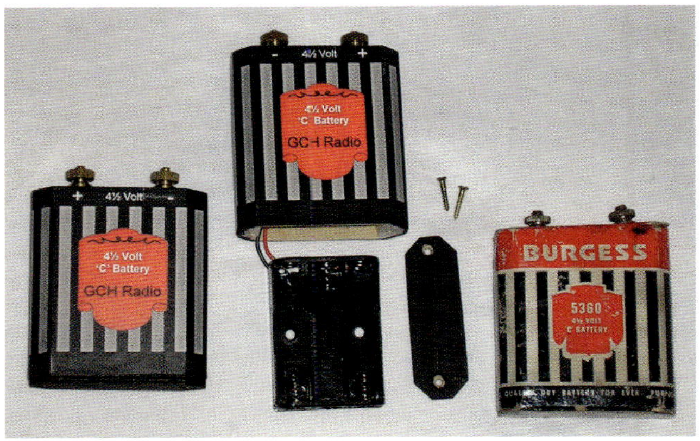

Reproduced and Original "C" Batteries

4-R (L) External Power

The 4-R (L) uses a five conductor power cable to connect to the external batteries. The power cable has a multi-plug on one end, which is keyed, so that it only goes into the socket one way. This keying method ensures that the proper voltages are routed to the proper circuits. The loose wires, at the other end, are color coded and should be connected per the manufacturer's instructions, which are located on the inside of the back door. Instructions are provided for the UV-201-A tubes and the UV-199 tubes.

Original Power Cable and Reproduced Multi-Plugs

Power Cable Socket

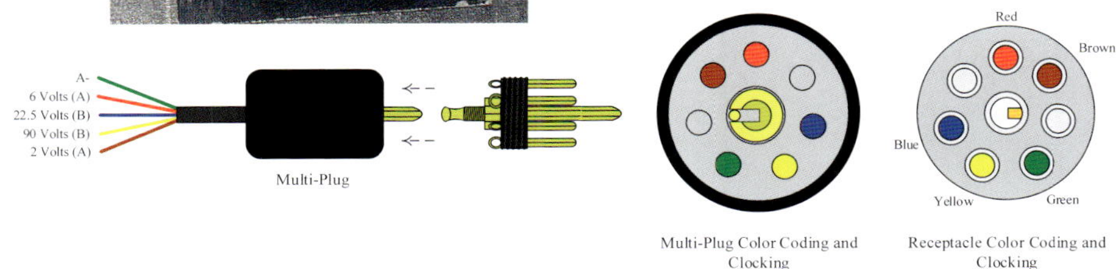

Power Cable and Multi-Plug

4-R (L) Functional Description Quick Reference Guide

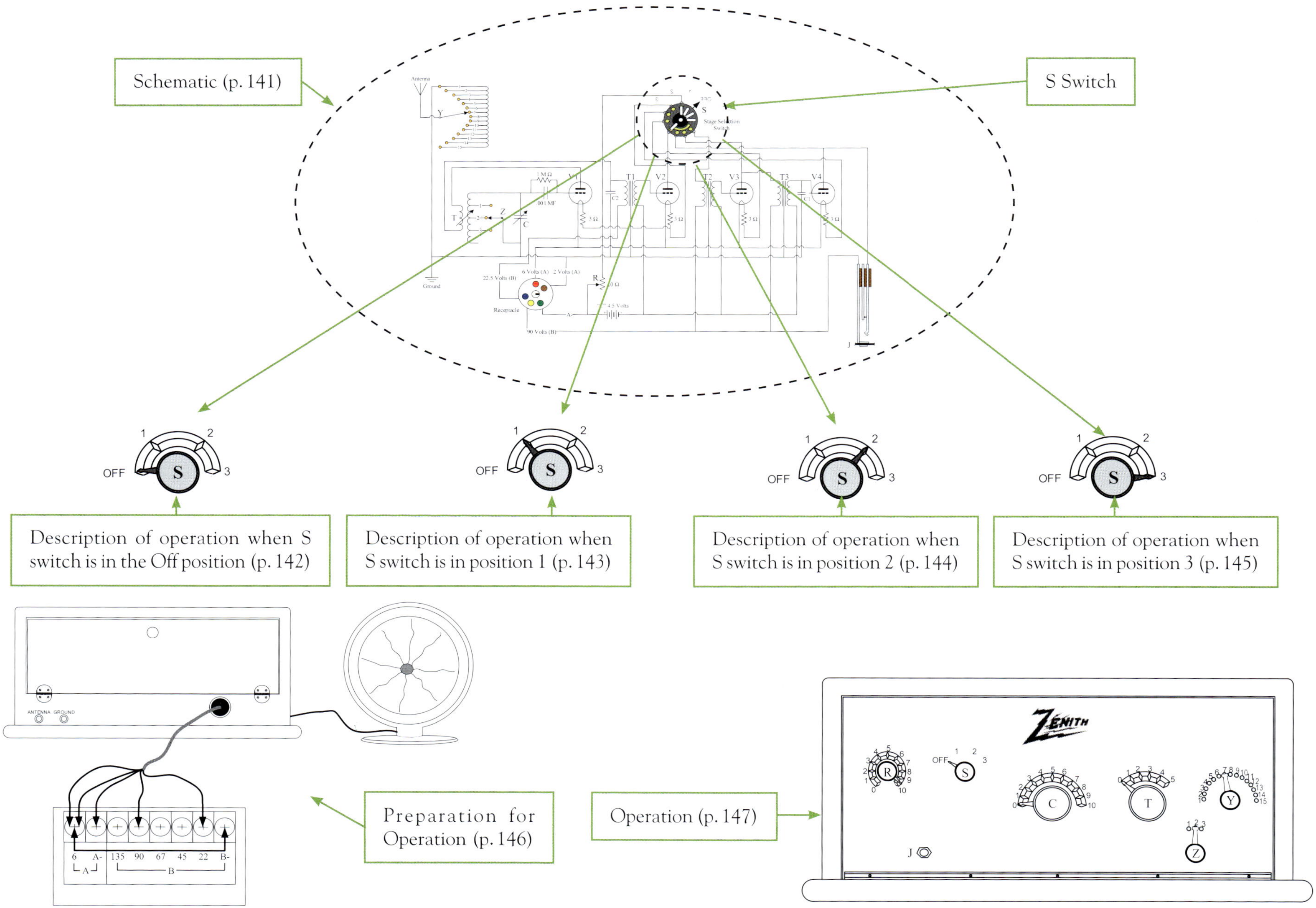

4-R (L) Schematic

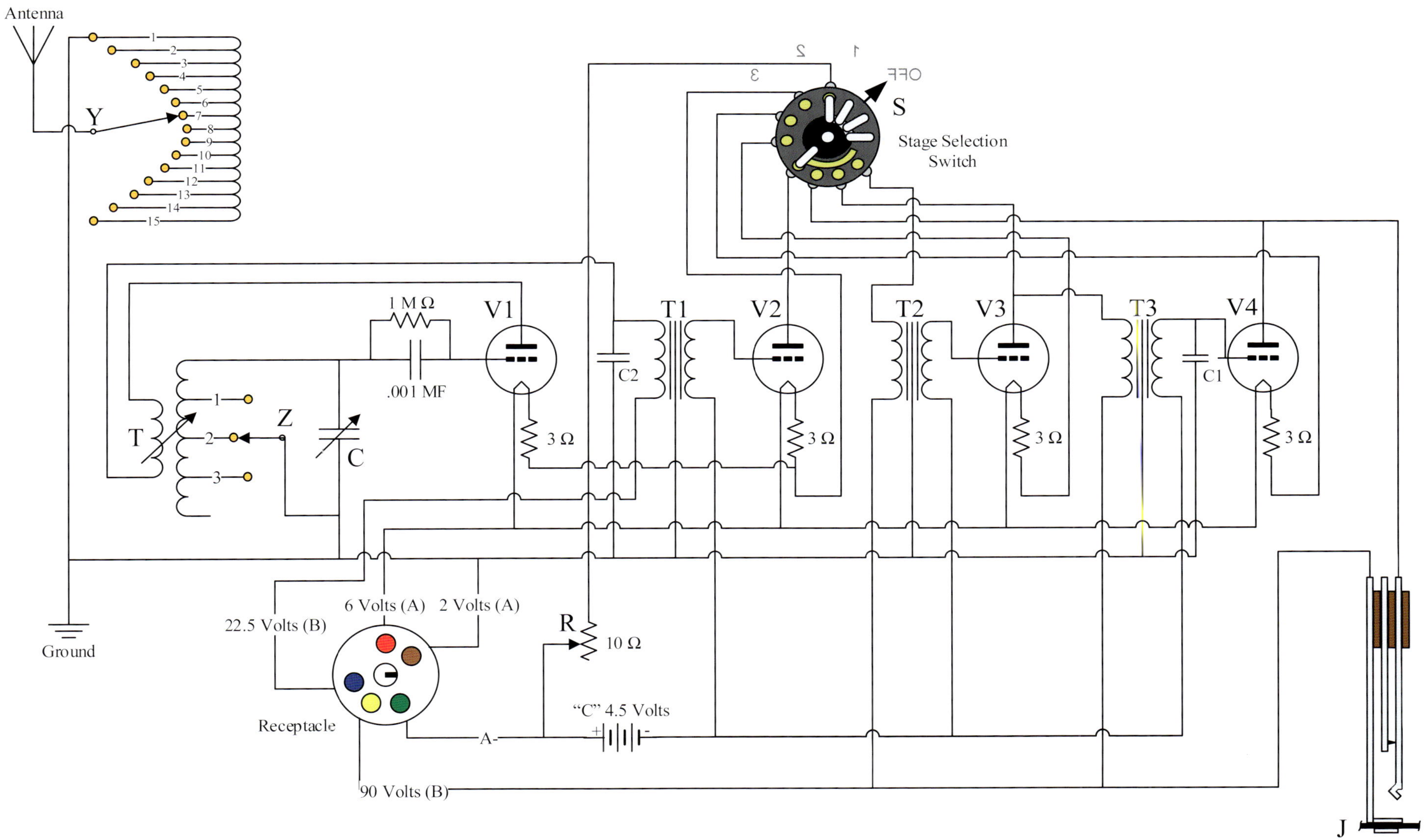

4-R (L) Description of Operation—Switch S in Off Position

When the switch S is in the OFF position the filament circuits are open and the radio is inoperable. Plate voltage is applied to the plate of V1. A negative bias voltage is applied to the grids of V2, V3, and V4. This is considered the initial state of the radio for future discussions of circuit operation.

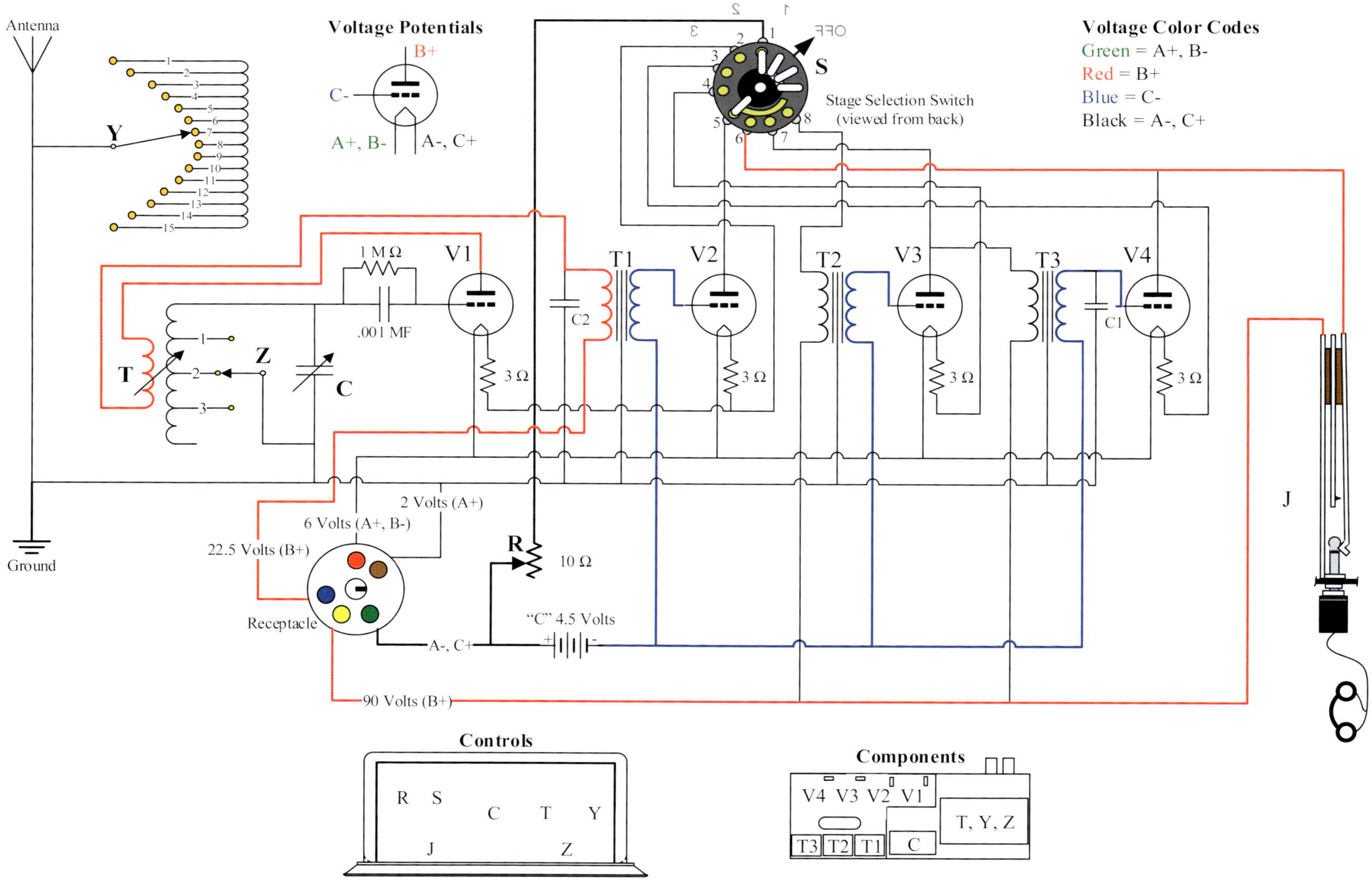

4-R (L) Description of Operation—Switch S in Position 1 (One AF Amplifier Stage)

When switch S is moved to position 1, the double wiper closes contact points 1 and 2 energizing the filament circuits to V1 and V2. At the same time the lower wiper connects contact points 5 and 6, which connects the plate of V2 to the headphones circuit. The radio is now playing with the detector (V1) and one stage (V2) of AF amplification.

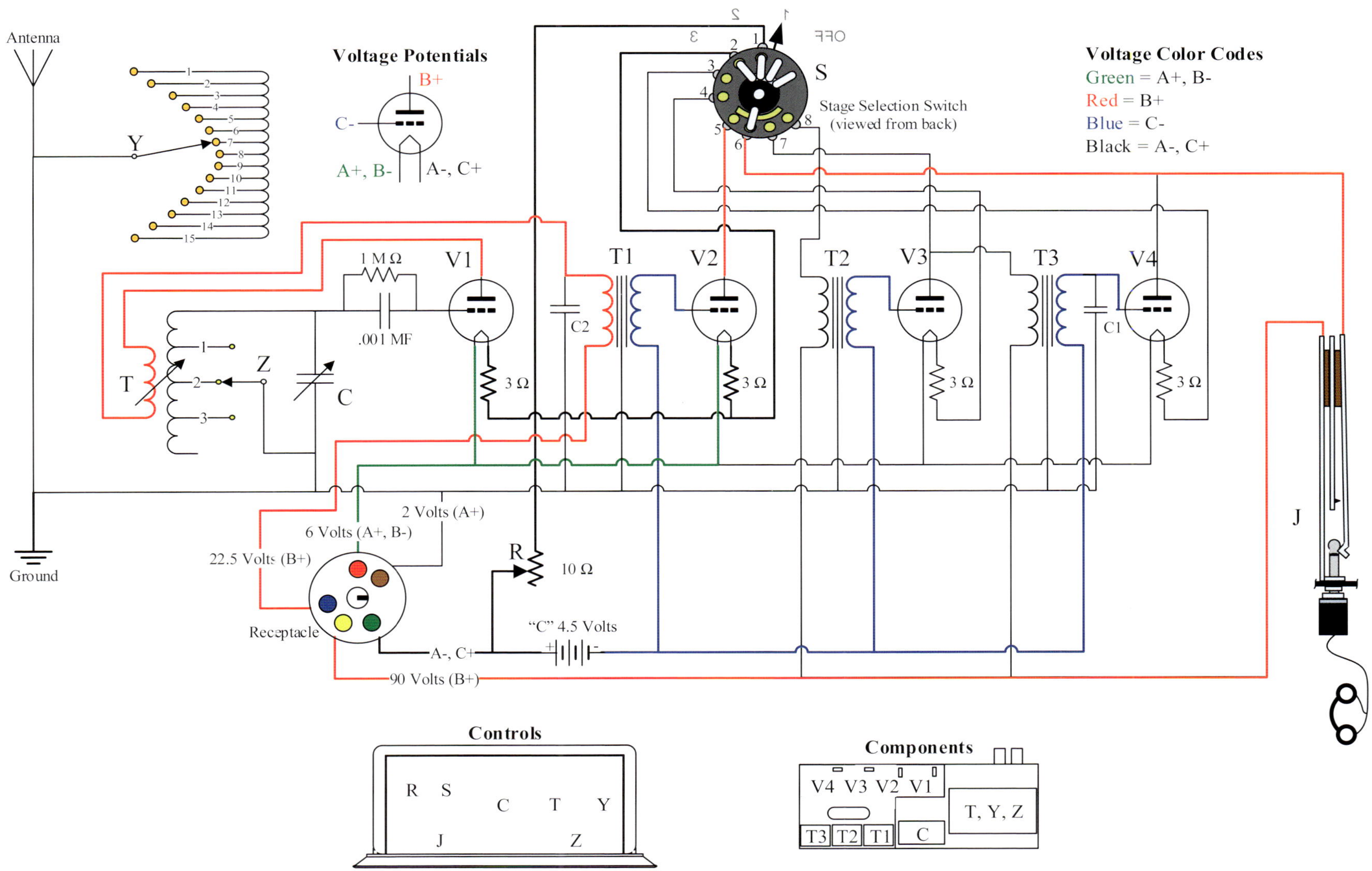

4-R (L) Description of Operation—Switch S in Position 2 (Two AF Amplifier Stages)

When switch S is moved to position 2, the triple wiper closes contact points 1, 2 and 3 energizing the filament circuits to V1, V2 and V4. At the same time, the single wiper disconnects contact points 5 and 6, and connects the contact points 5 and 7, which connects the plate of V2 to the primary of transformer T3. The radio is now playing with the detector (V1) and two stages (V2, V4) of AF amplification.

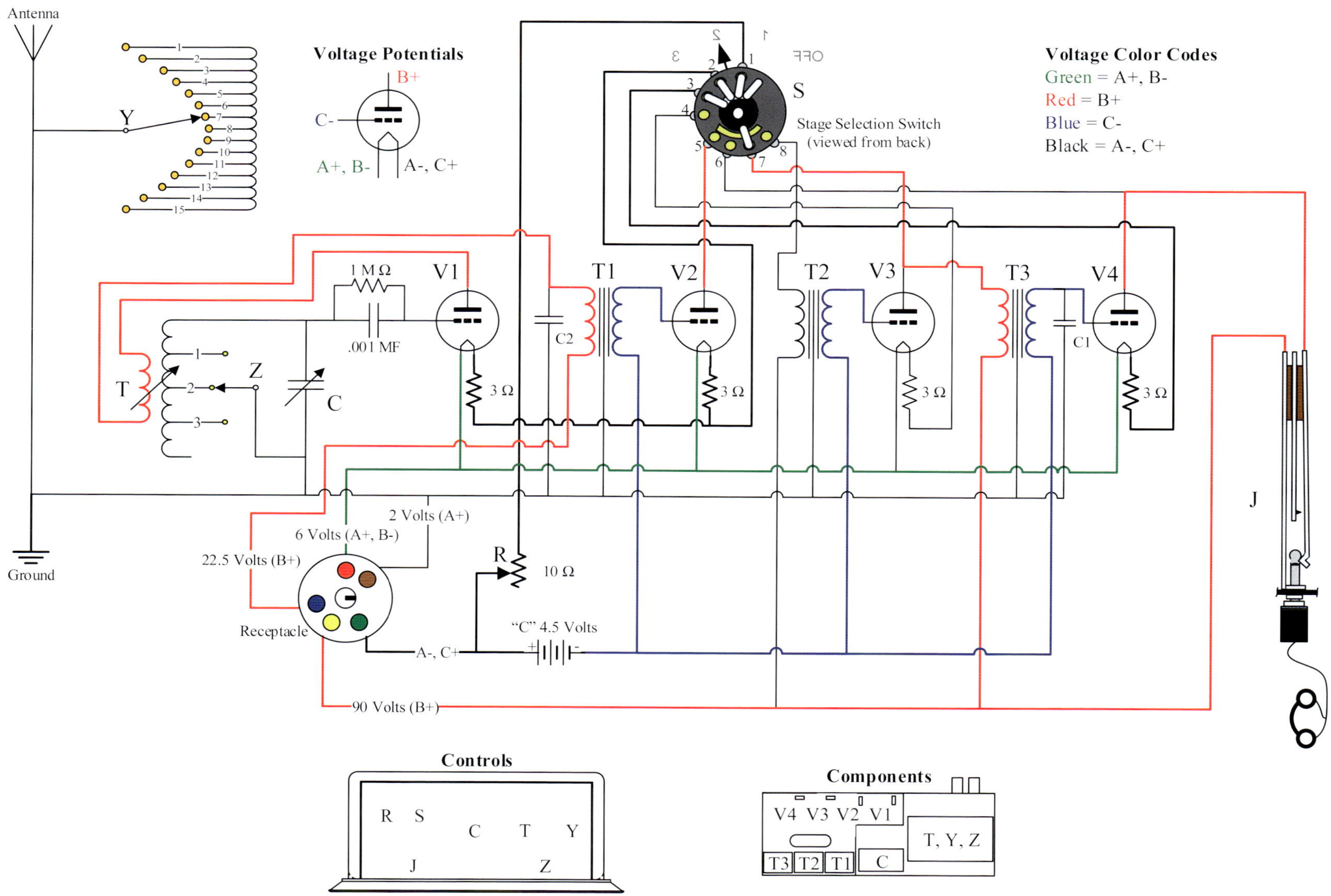

4-R (L) Description of Operation—Switch S in Position 3 (Three AF Amplifier Stages)

When switch S is moved to position 3, the quadruple wiper closes contact points 1, 2, 3 and 4 energizing the filament circuits to V1, V2, V3 and V4. At the same time, the single wiper disconnects contact points 5 and 7, and connects contact points 5 and 8, which connects the plate of V2 to the primary of transformer T2. The radio is now playing with the detector (V1) and three stages (V2, V3, V4) of AF amplification.

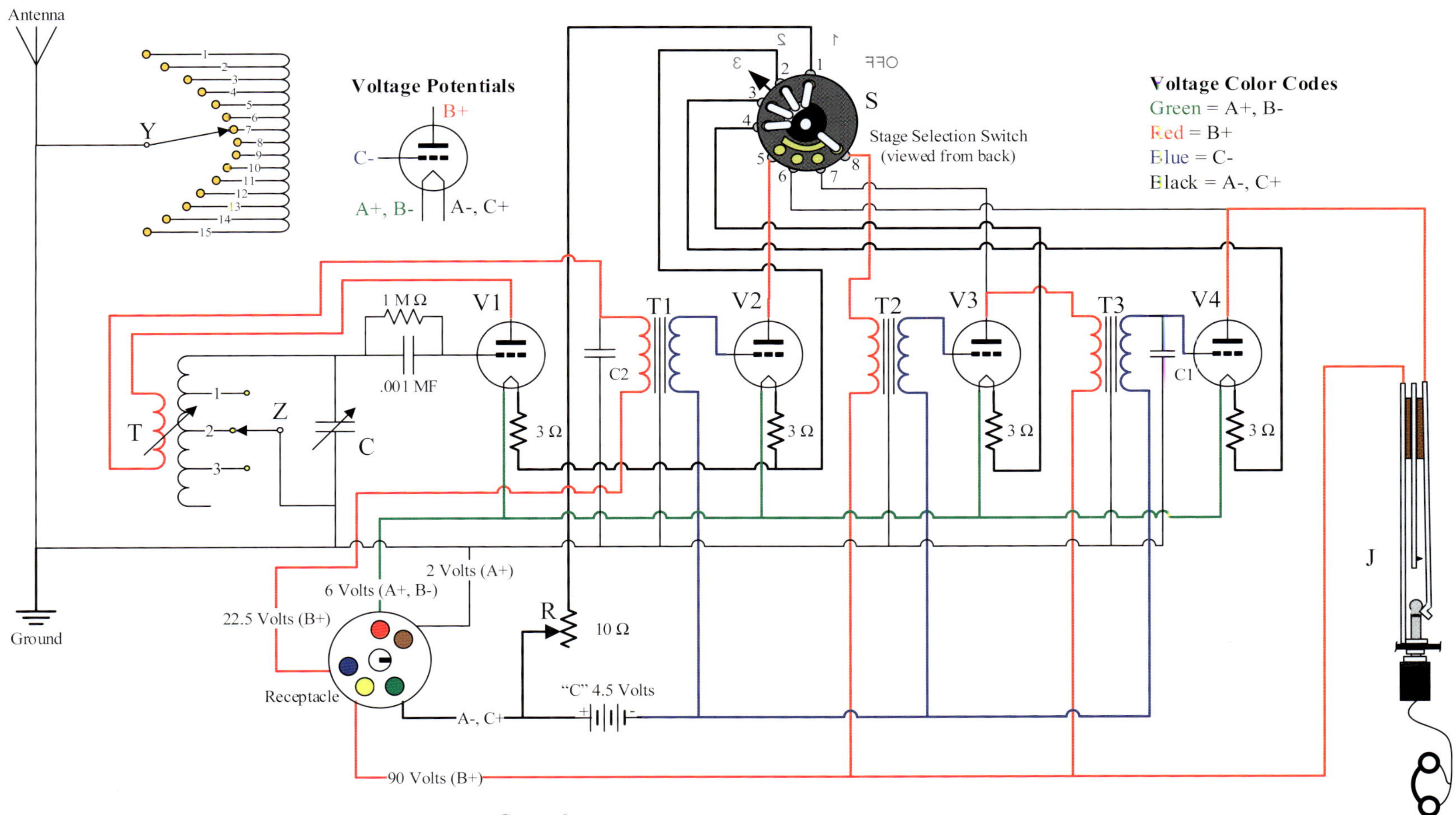

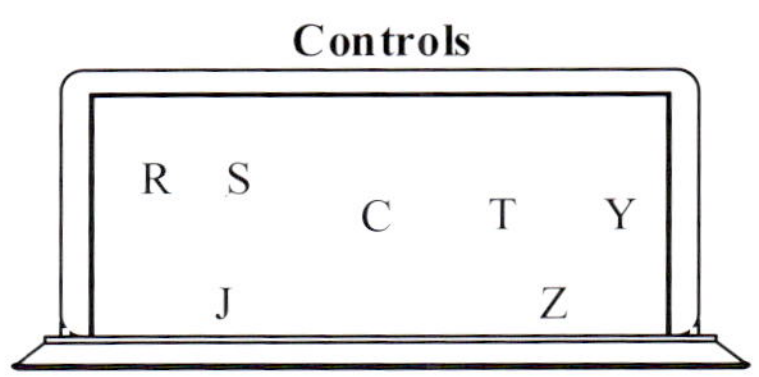

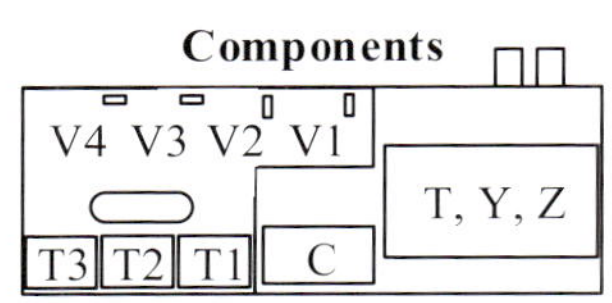

4-R (L) Preparation for Operation

Install the tubes and the "C" battery inside the radio. Connect the antenna and ground to the ANTENNA and GROUND posts. Use the appropriate battery connection diagram, inside the back door of the radio, to make sure the power can be connected properly and agrees with the 4-R (L) Connections diagram below. If the power connections are not correct or otherwise unavailable, utilize the physical and functional descriptions to assist in determining the correct power connections. Hook up the power to the radio. Proceed to 4-R (L) Operation. See Common Radio Attachments (page 8) for additional information related to antennas, loudspeakers and power options.

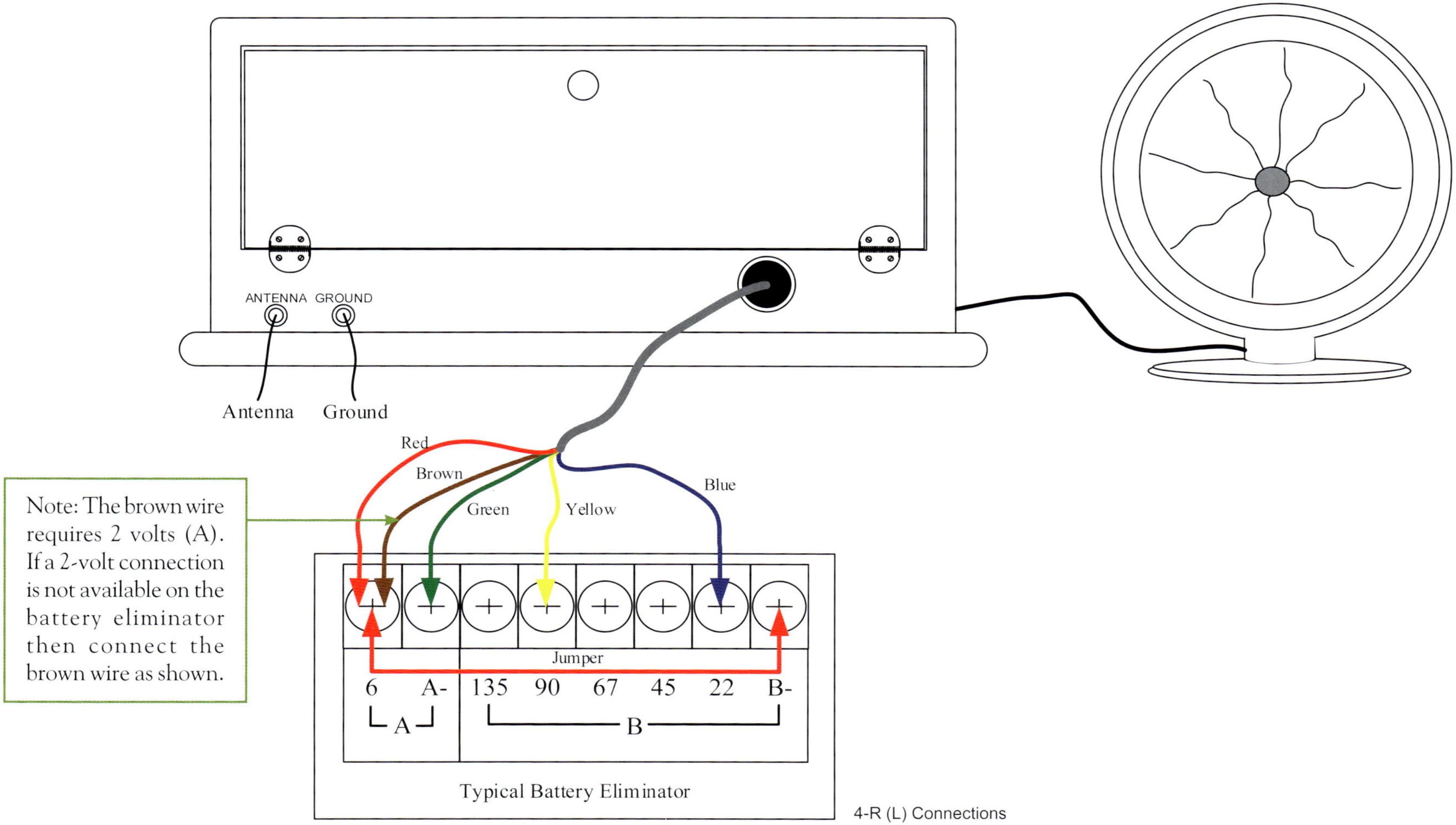

4-R (L) Connections

4-R (L) Operation

Follow the 4-R (L) Preparation for Operation instructions. Set the pointers to the positions shown in the diagram below.

The variable capacitor C selects the stations and the tickler T provides the feedback (regeneration) that increases the volume of the selected signal. Practice using knobs C and T simultaneously.

Insert the headphone plug into jack J. Turn the switch "S" to position 1 or 2.

Gradually move "T" to the right until a hissing sound is heard and hold the pointer to the beginning of that sound. Gradually move "C" to the right until a bird whistle is heard; first high, then descending to low; then a pause will be heard. As you continue to move "C" the whistle is again low, then ascending to high. The station is between these low points or valleys. Move "T" to the beginning of the whistle, while "C" is being moved. If you have C in the valley of the whistle, slightly move "T" to the left to clear the station. For best results operate C and T simultaneously. If an unpleasant howl (oscillations) is heard and the station is lost, turn "T" back to the left until the howl stops.

Adjust the audio level (volume) by setting switch S to position 1, 2 or 3, as appropriate.

Knob Y tunes the antenna circuit. A little manipulation of this switch will enable the operator to obtain improved results.

Knob Z changes the bandwidth of the receiver and is generally left in the position 2. Position 1 decreases the bandwidth and position 3 increases the bandwidth.

Record the control settings for each station in the Station Log (next page). Then, to return to a particular station, set the controls to the settings that were previously recorded in the log for that station.

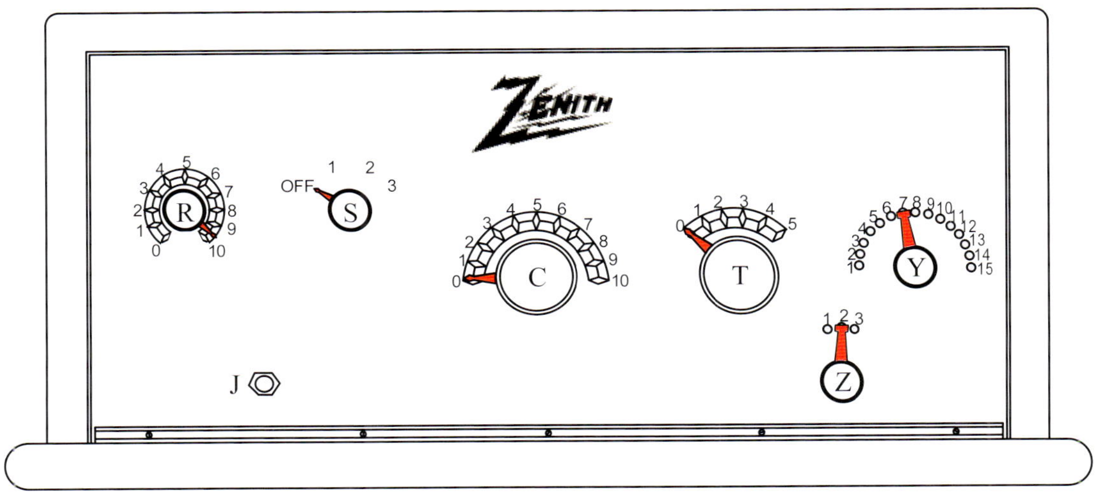

Initial Dial Settings (pointers shown in red for clarity)

4-R (L) Station Log

Name: ____________________________

Date	Time	Station	Frequency	C	T	Y	Z

Chapter 9: Zenith VII

VII General Specifications

Company:	Zenith Radio Corporation
Model:	VII (Super-Zenith)
Year Manufactured:	1925
Manufacture Location:	3620 South Iron Street, Chicago, IL
Serial Number:	3652
Price:	$230
Tube:	UV-201-A (6), two RF amplifiers, one detector, three AF amplifiers
Circuit Type:	Tuned Radio Frequency (TRF)

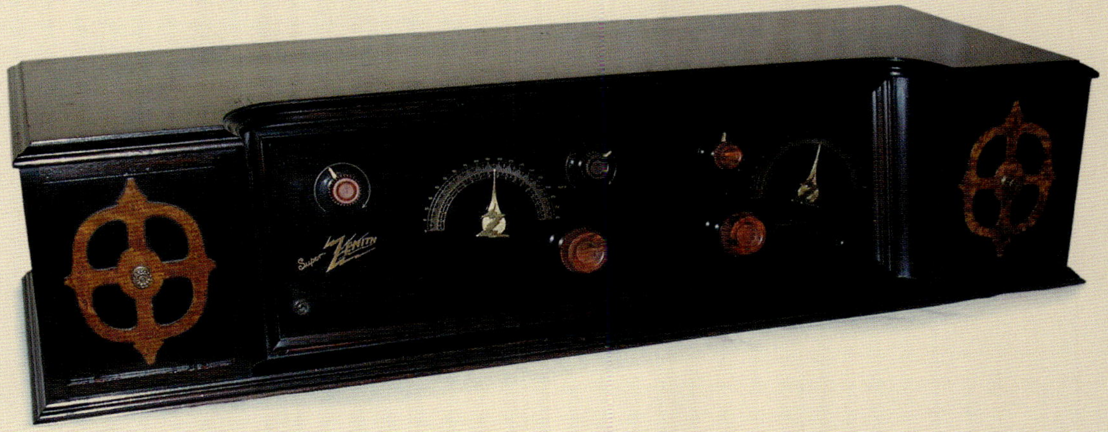

VII General Information

The model VII Super-Zenith was the first radio to be completely manufactured, marketed, and sold by the Zenith Radio Corporation. After its introduction the Chicago Radio Laboratory (CRL) name was dropped.

A major revision to the model VII resulted in the eventual production of two different versions of this set. The first version used a multi-contact switch for selecting from one to three AF amplifier stages to control the volume. The second version employed three rheostats in order to control the volume of the set. Each of these controls performed specific functions, in various parts of the circuit, and thus controlled the gain and volume of the receiver. The Super-Zenith shown in this chapter is the later version with the three rheostats.

Restoration details for the set that was used as the example in this chapter. *This radio was completely restored to like new condition. An adaptor was made and installed for the "D" battery. A new "C" battery was reproduced and installed. The pointers were polished and coated with a protective overcoat. All three vernier pulleys were replaced with plastic pulleys that were provided by the late Mr. Keith Perry. The cabinet was thoroughly cleaned and polished. The back door of this radio is missing. One small knob and pointer were missing and were replaced, however, the red color of the knob is slightly different from the others.*

VII Physical Description Quick Reference Guide

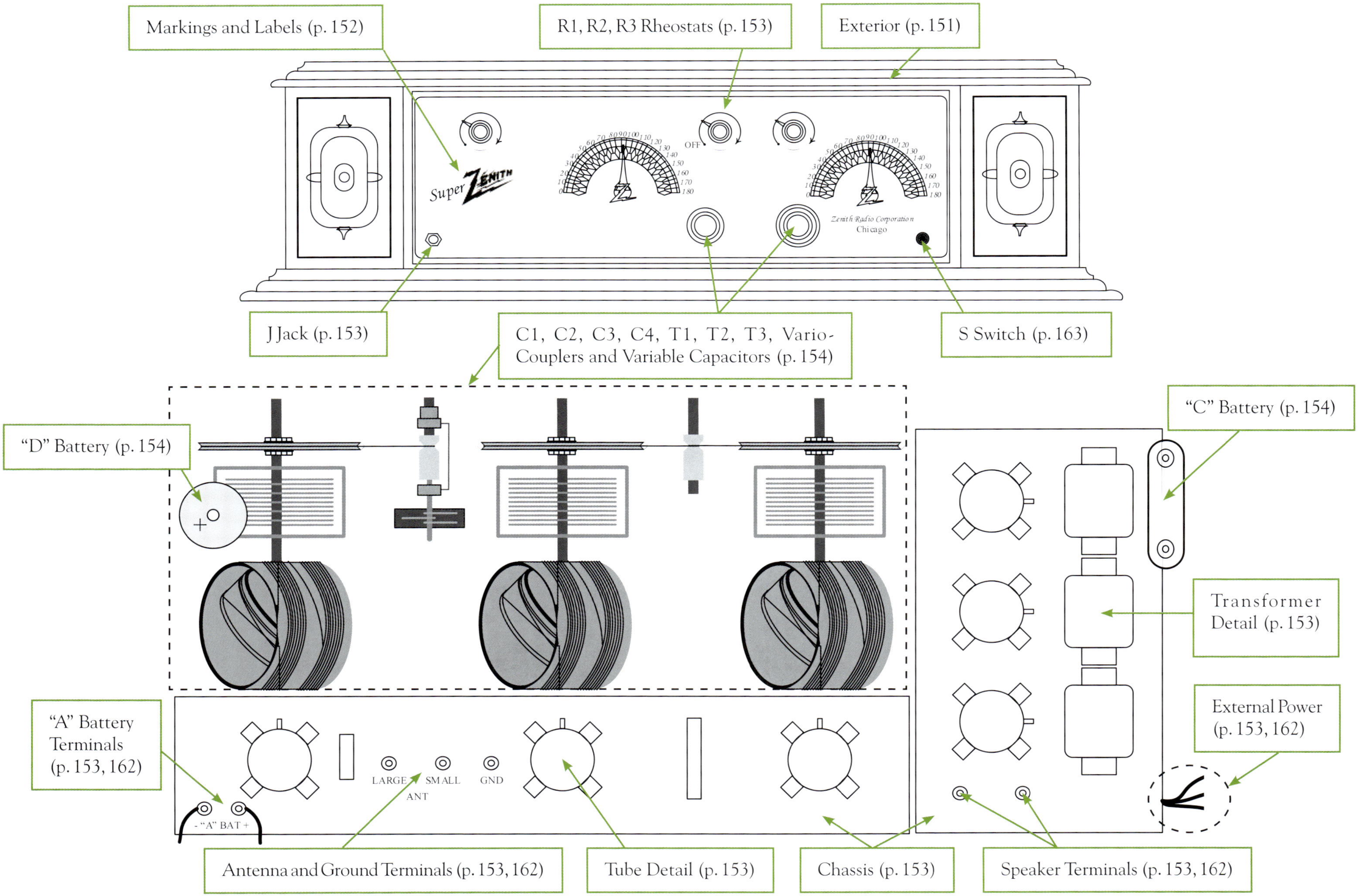

VII Exterior

The cabinet is made of mahogany and stained dark red. It measures 44.75" by 16.25" by 10.75". The cabinet has storage compartments on each end with removable doors. These areas were used for storing batteries. The front panel is brass with a very dark simulated wood finish overcoat.

Front View

Front View
(side doors removed)

Rear View
(back door missing)

VII Markings and Labels

Front View (side doors removed)

Super-Zenith Logo

Zenith Name

Serial Number

Cabinet Maker
(Located underneath cabinet.)

VII Chassis

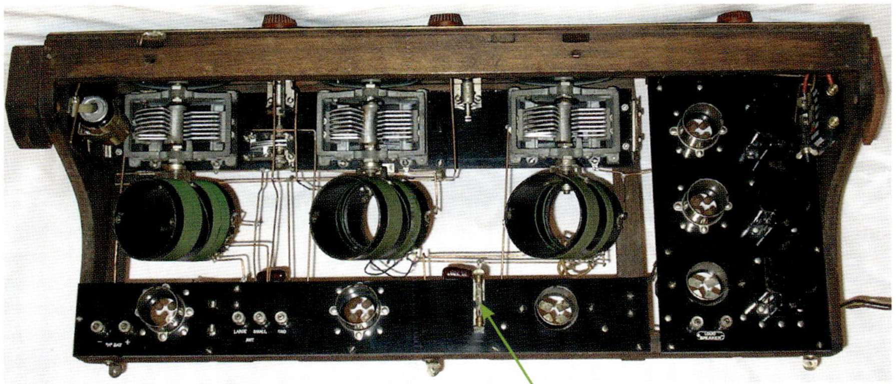

Top View

Component Layout

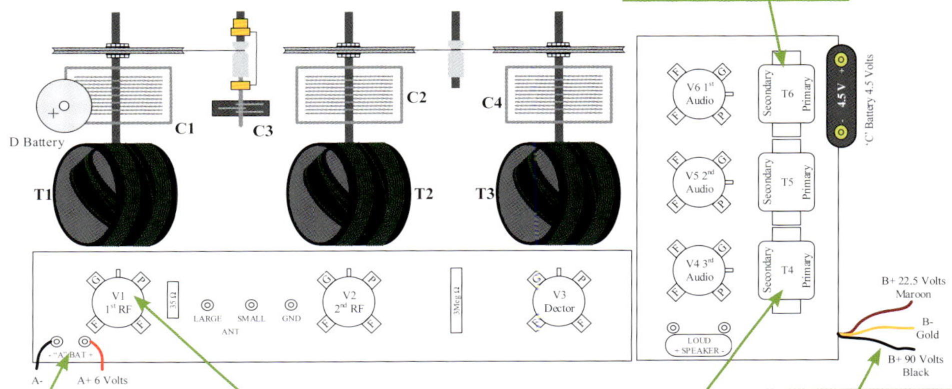

Back View

Underneath View

VII Vernier and Battery Detail

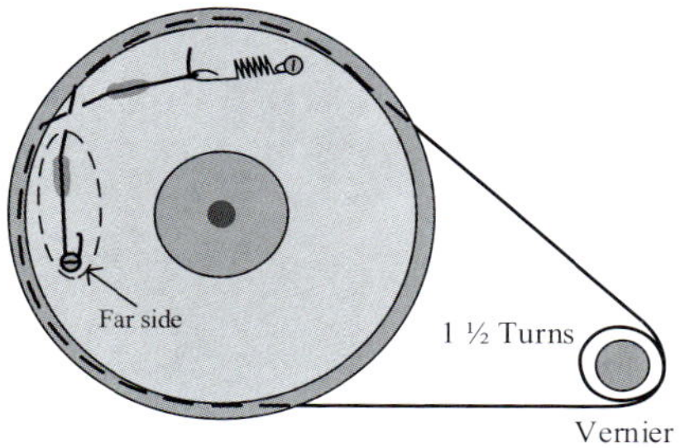

C1 Vernier Dial Detail

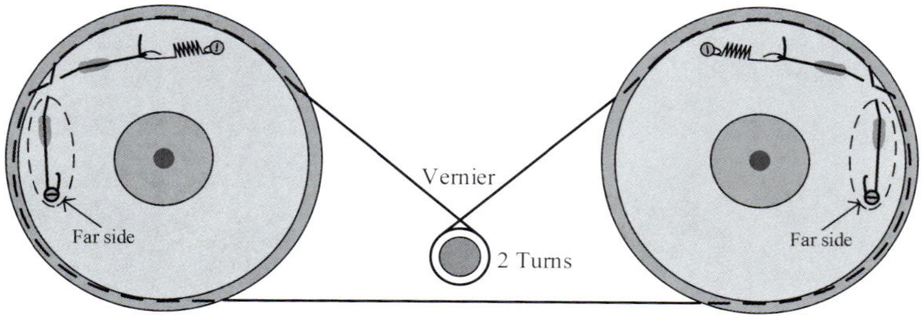

C2-C4 Vernier Dial Detail

"C" Battery

Rear View

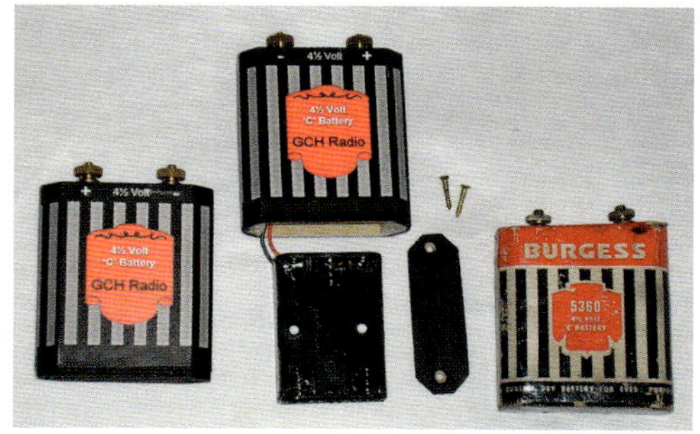

Reproduced and Original "C" Batteries

Installed "D" Battery and Adapter

"D" Battery and Adapter

"D" Battery and Adapter

An adapter is required because the original D battery had a negative case and the modern D battery has an insulated case. See "D" Battery Adapter Modification, page 244.

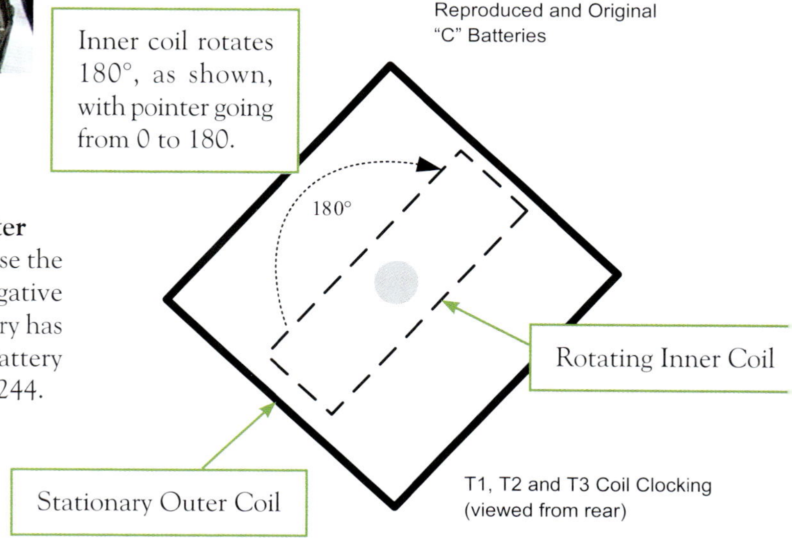

T1, T2 and T3 Coil Clocking (viewed from rear)

VII Functional Description Quick Reference Guide

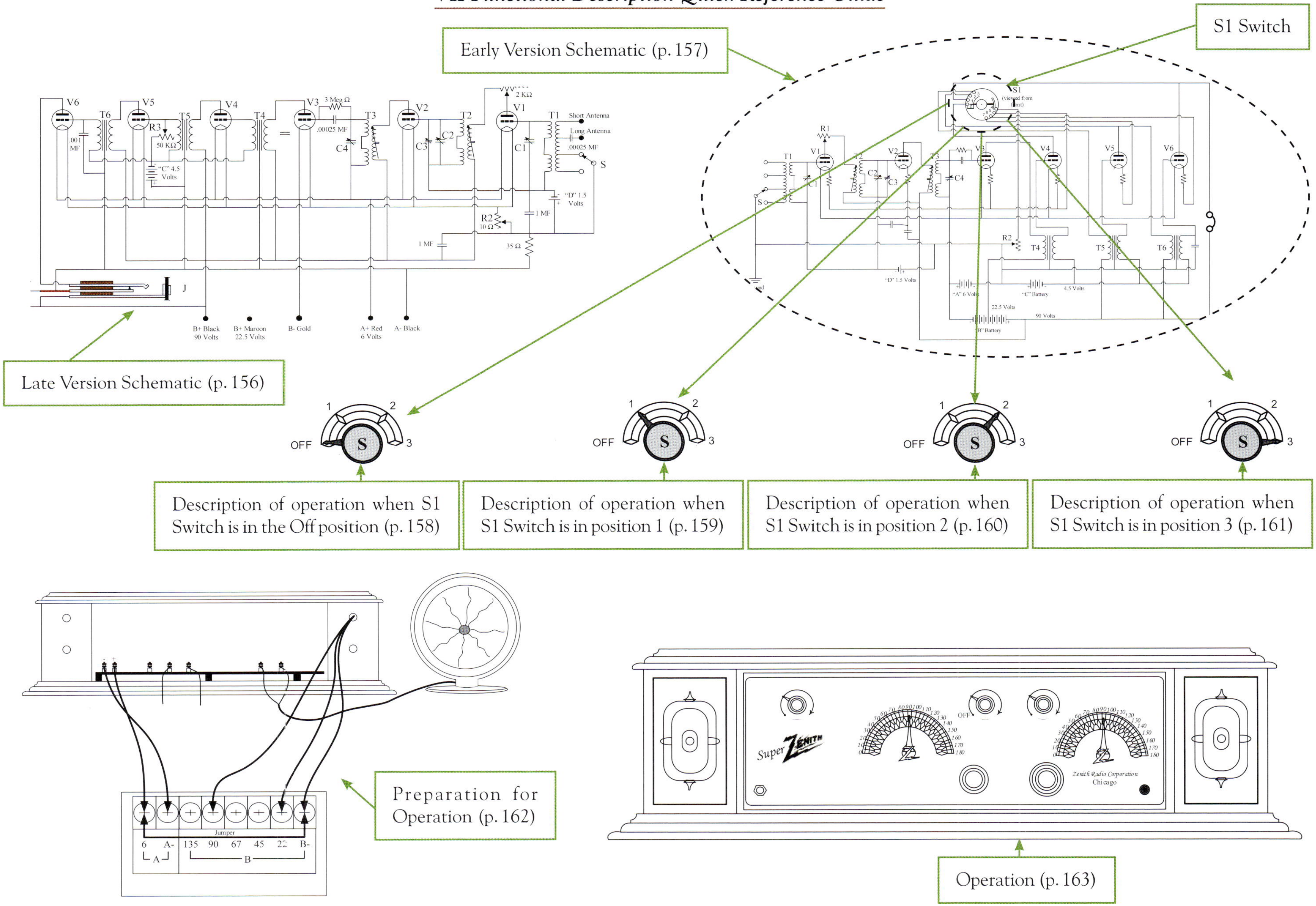

VII Schematic (Late Version)

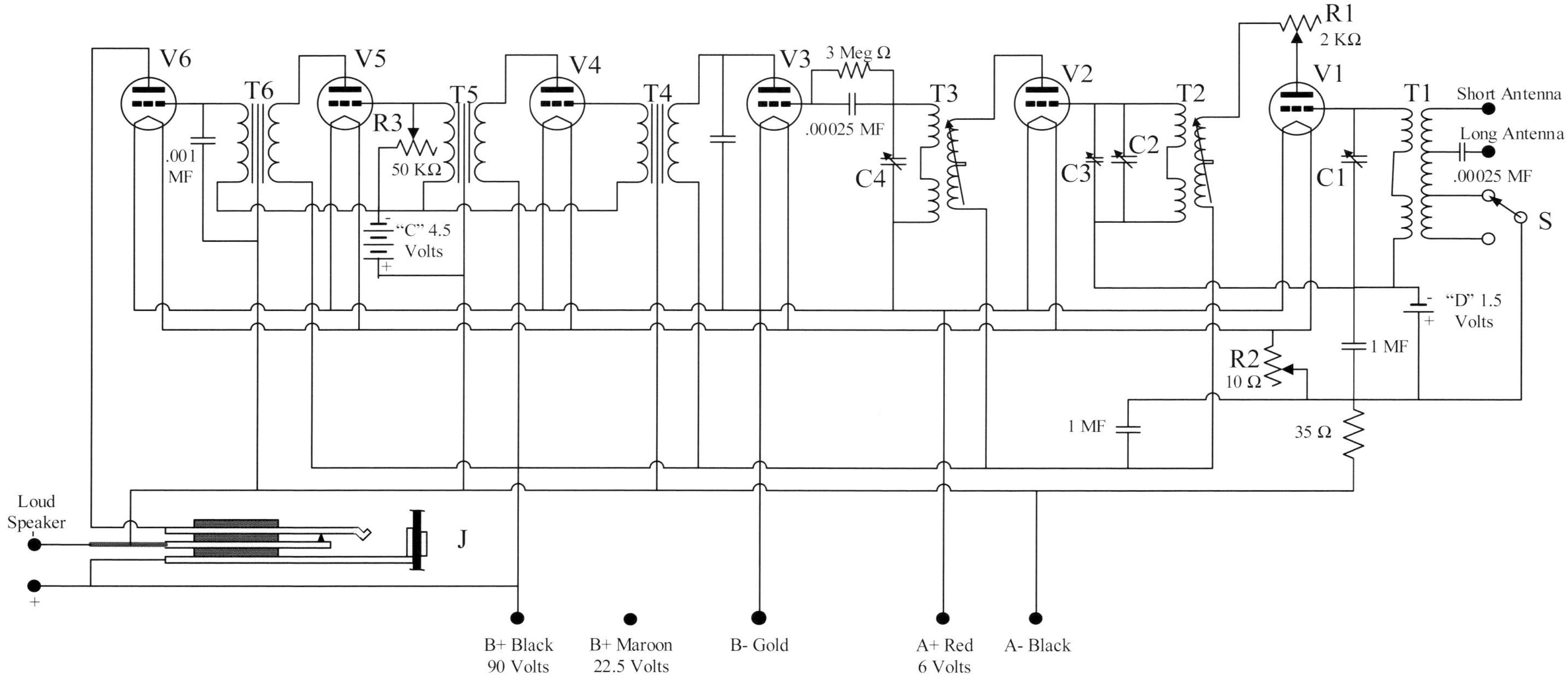

VII Schematic (Early Version)

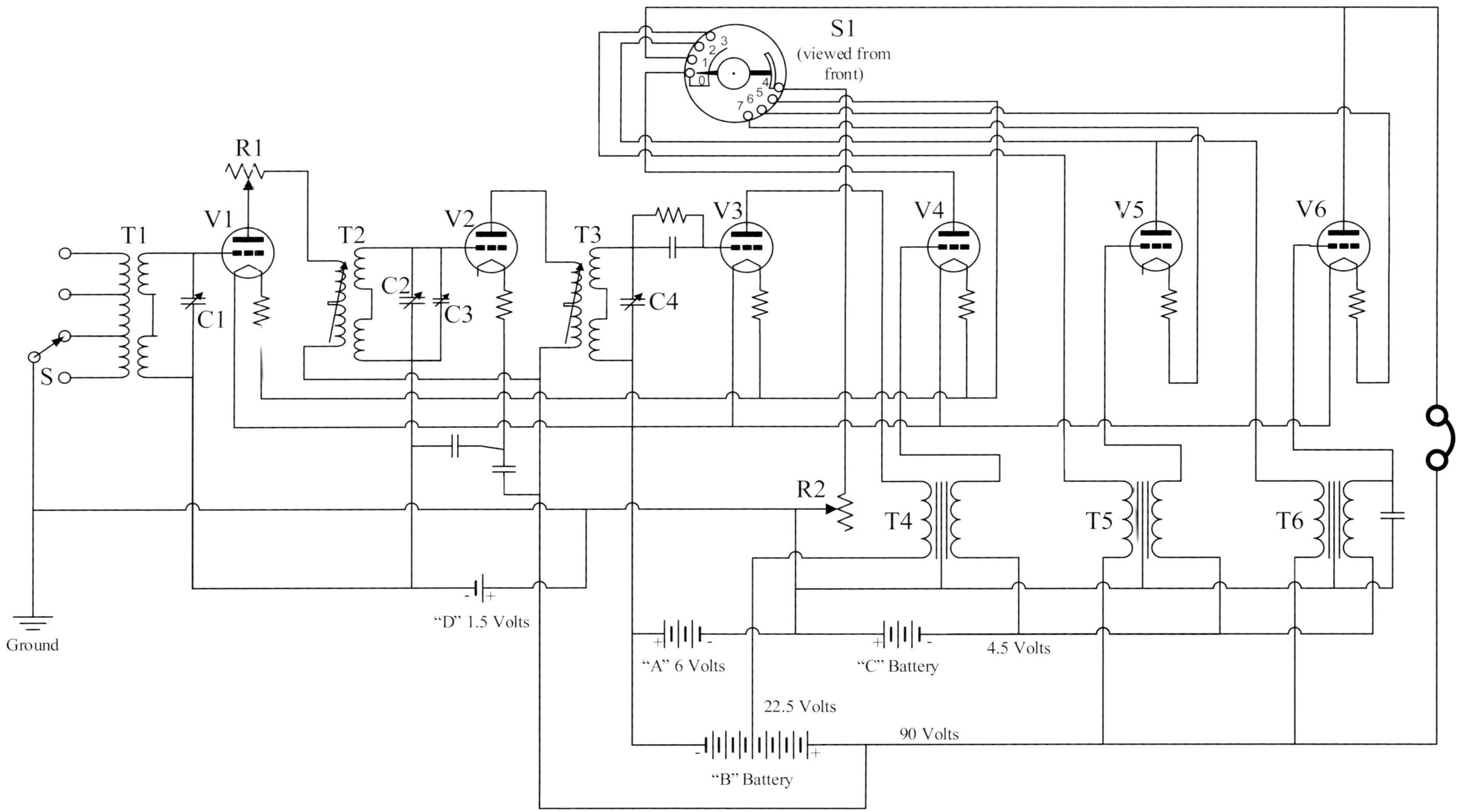

VII (Early Version) Description of Operation—Switch S1 in Off Position

Switch S1 is in the OFF position. The tubes are inactive because the filament circuits are open. The plate voltage for tubes V1, V2, V4, V5, and V6 is 90 volts and V3 is 22.5 volts. A negative grid bias of −1.5 volts is applied to V1 and V2, and a negative bias of −4.5 volts is applied to grids of V4, V5 and V6. This is considered the initial state of the radio for future discussions of circuit operation.

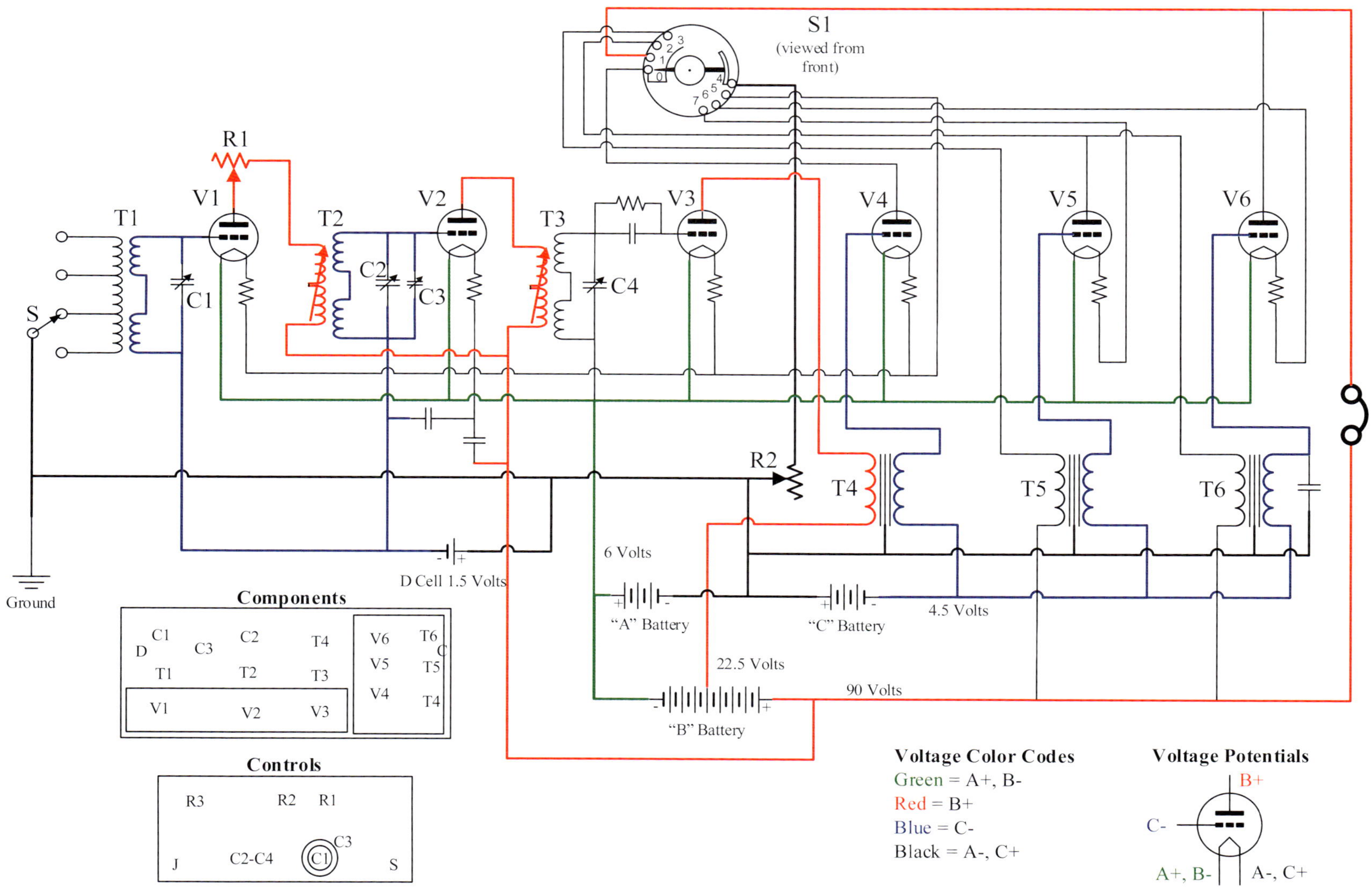

VII (Early Version) Description of Operation—Switch S1 in Position 1

Switch S1 is moved to position 1. Switch S contacts 4 and 5 are shorted, which closes the ground side of the filament circuits and activates tubes V1, V2, V3 and V4. Switch S contacts 0 and 1 are connected, which directs the AF signal from the plate of V4 to the headphones. The radio is now playing with two stages (V1, V2) of RF amplification, one detector stage (V3) and one stage (V4) of AF amplification.

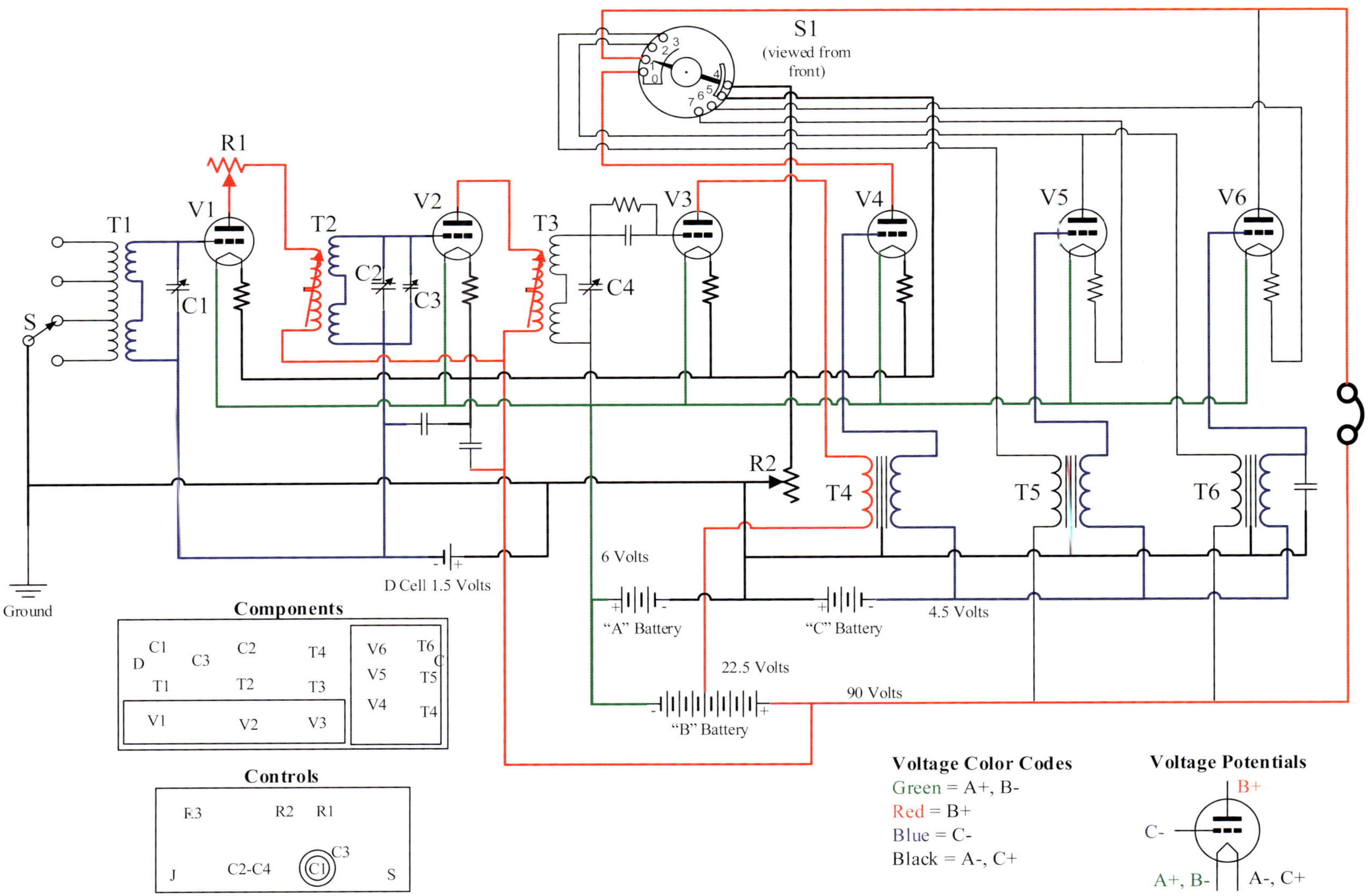

VII (Early Version) Description of Operation—Switch S1 in Position 2

Switch S1 is moved to position 2. Switch S contacts 4, 5 and 6 are shorted, which closes the ground side of the filament circuits and activates tubes V1, V2, V3, V4 and V6. Switch S contacts 0 and 2 are shorted, which directs the AF signal from the plate of V4 to the primary of transformer T6. The secondary of T6 passes the AF signal to the grid of V6. The plate of V6 sends the AF signal to the headphones. The radio is now playing with two stages (V1, V2) of RF amplification, one detector stage (V3) and two stages (V4, V6) of AF amplification.

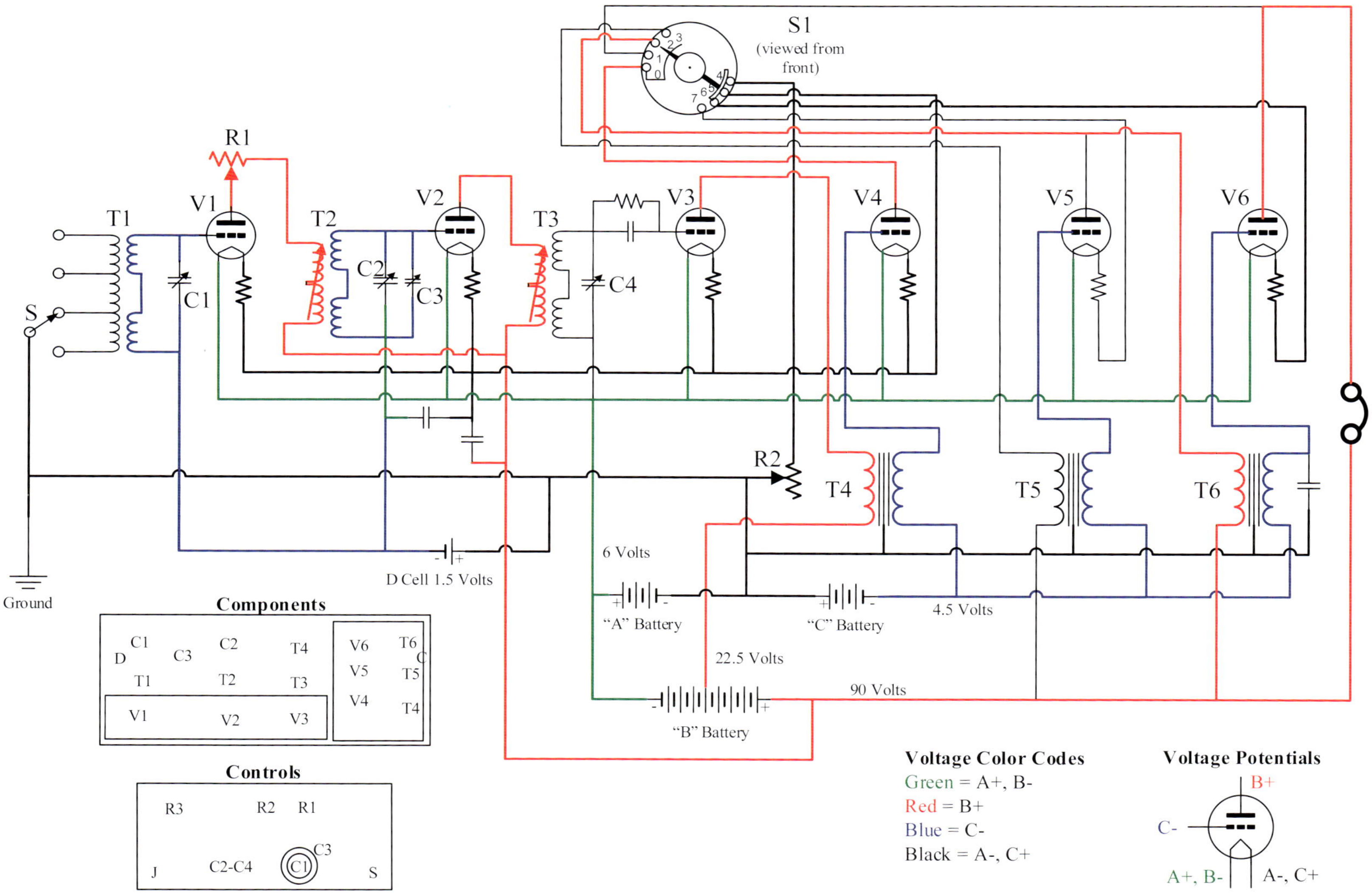

VII (Early Version) Description of Operation—Switch S1 in Position 3

Switch S1 is moved to position 3. Switch S contacts 4, 5, 6 and 7 are connected, which closes the ground side of the filament circuits and activates tubes V1, V2, V3, V4, V5 and V6. Switch S contacts 0 and 3 are connected, which directs the AF signal from the plate of V4 to the primary of transformer T5. The plate of V6 sends the AF signal to the headphones. The radio is now playing with two stages (V1, V2) of RF amplification, one detector stage (V3) and three stages (V4, V5, V6) of AF amplification.

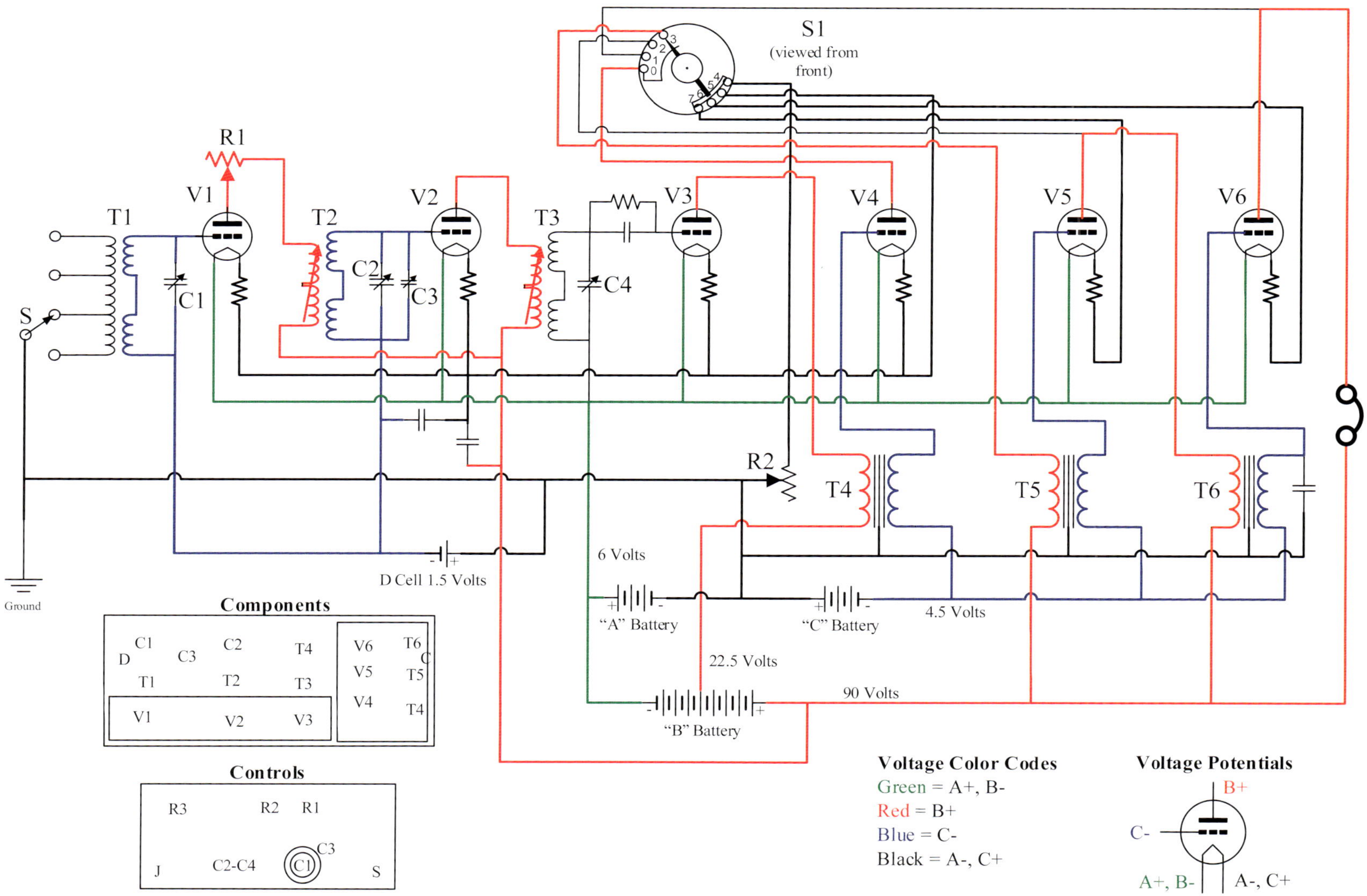

VII Preparation for Operation (Late Version)

Install the tubes, "C" battery and the "D" battery adaptor-assembly inside the radio. Connect the antenna and ground to the appropriate ANTENNA and GROUND terminals. Use the battery connection diagram, inside the back door of the radio, to make sure the power can be connected properly and agrees with the Super-Zenith Connections diagram below. If the power connections are not correct or otherwise unavailable, utilize the physical and functional descriptions to assist in determining the correct power connections. Proceed to VII Operation. See Common Radio Attachments (page 8) for additional information related to antennas, loudspeakers and power options.

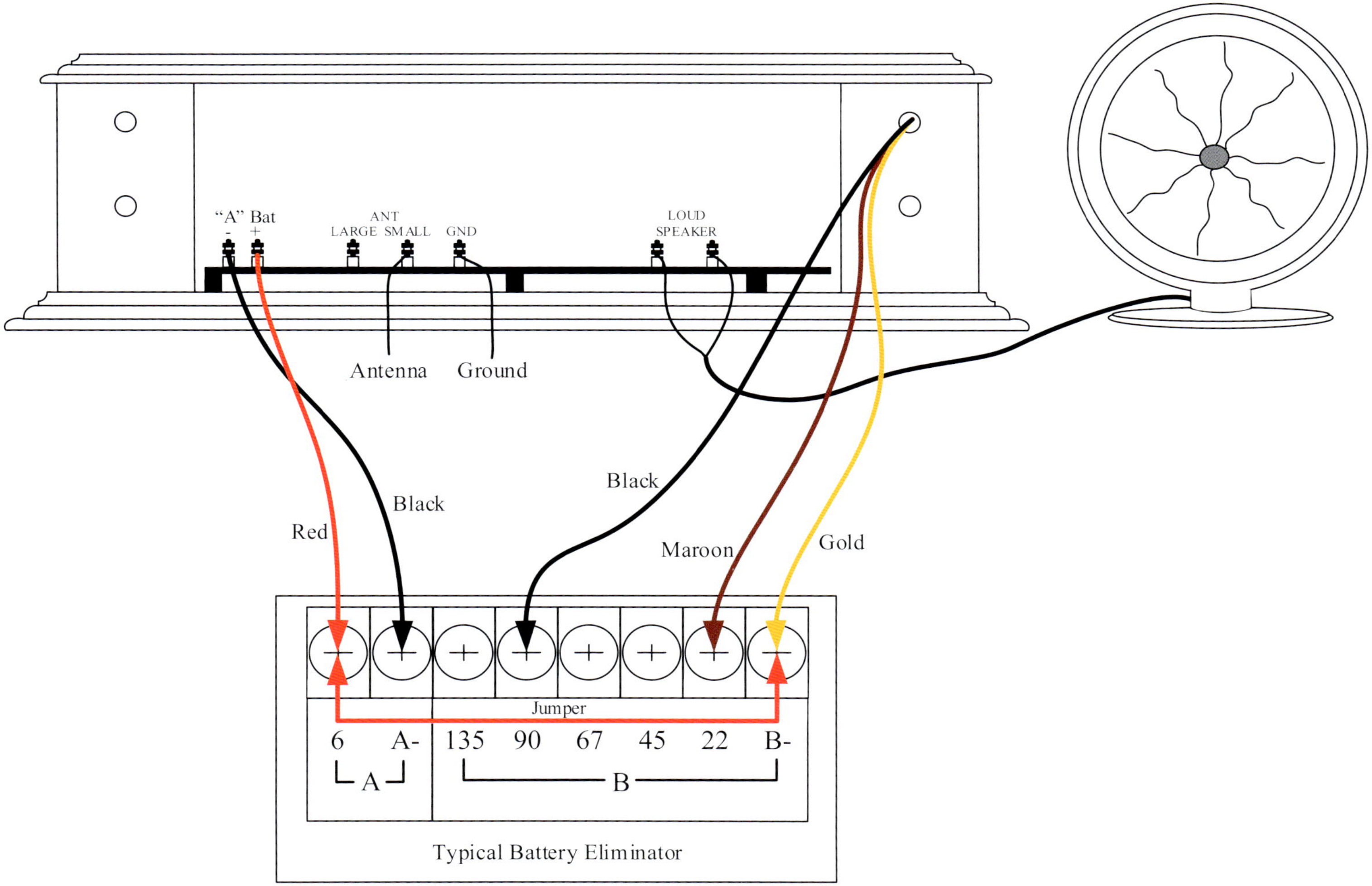

Super-Zenith Connections

VII Operation (Late Version)

Follow the VII Preparation for Operation instructions. Turn "R2" fully clockwise. This sets the voltage on the tube filaments.

Jack J is for allowing the use of headphones for tuning. The method of changing from phones to loudspeaker is entirely automatic; merely removing the phone plug puts the loudspeaker in operation. Inserting the phone plug puts the headphones into operation and disconnects the loudspeaker.

Switch R1 is a compensator. The operator can readily recognize the point when a station is tuned in, by turning the control "R1" completely to the right then bringing it back until a point is reached where the signal is perfectly clear.

Switch S allows greater selectivity when the receiver is being used a short distance from a broadcasting station. To gain this selectivity, the control should be turned to the left. Ordinarily, where no great interference is experienced, the control should be kept in the right-hand position.

The tuning is accomplished by turning "C1" and "C2/C4" simultaneously. These controls will have approximately the same settings, when tuning in a station.

The R3 rheostat is for controlling the volume. Adjust this control appropriately.

The trimmer capacitor C3 is used for fine tuning after the station has been selected.

Record the control settings for each station in the Station Log (next page). Then, to return to a particular station, set the controls to the settings that were previously recorded in the log for that station.

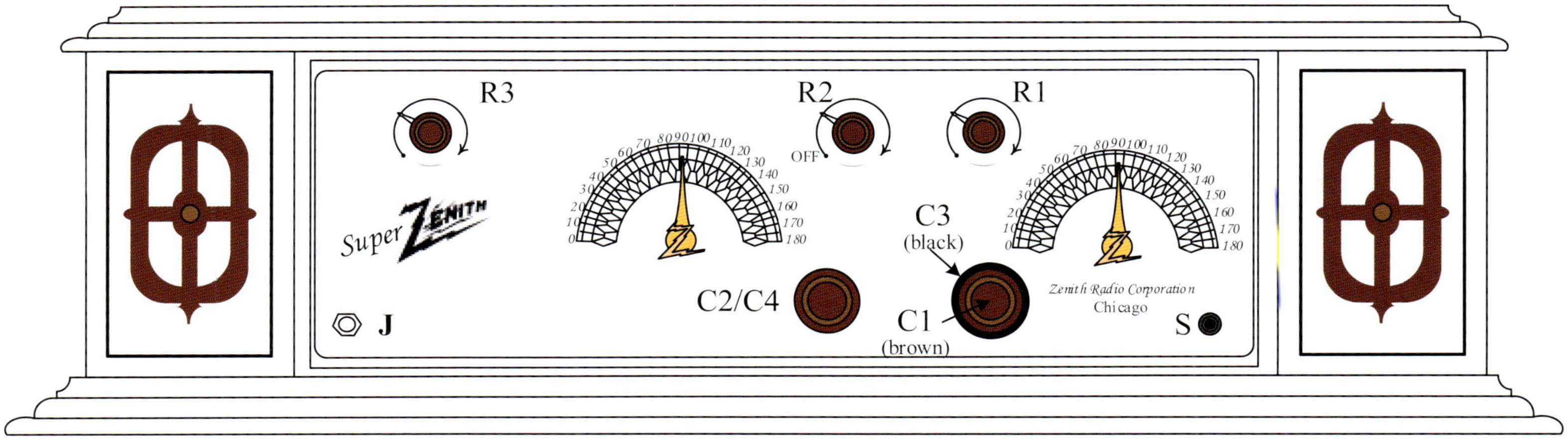

Operation Controls

VII Station Log

Name: ______________________________

Date	Time	Station	Frequency	C1	C2/C4

This page may be copied, scanned or reproduced.

From *The Early Zenith Radios: The Battery Powered Table Sets 1922–1927*, © 2014 by Gilbert M. Hedge with Durell M. Roth. ISBN 978-0-7643-4674-3. Schiffer Publishing Ltd. www.schifferbooks.com

Chapter 10: Zenith 11

11 General Specifications

Company:	Zenith Radio Corporation
Model:	11
Year Manufactured:	1927–1928
Manufacture Location:	3620 South Iron Street, Chicago, IL
Serial Number:	27130
Price:	$110
Tube:	UV-201-A (6), three RF amplifiers, one detector, two AF amplifiers
Circuit Type:	Tuned Radio Frequency (TRF)

11 General Information

This radio was found with the original Zenith Radio Corporation (3220 Iron Street, Chicago, Ill.) guarantee. The bottom portion had been filled out, but was not torn off and mailed in.

Restoration details for the set that was used as the example in this chapter. *There were no major problems noticed with this radio when it was restored. A new power cable was installed to replace the old worn cable. The cabinet is original and only needed polishing with crème polish. This radio was thoroughly cleaned. The only parts that needed replacing were the 2 capacitors, which can be seen in the various pictures.*

11 Physical Description Quick Reference Guide

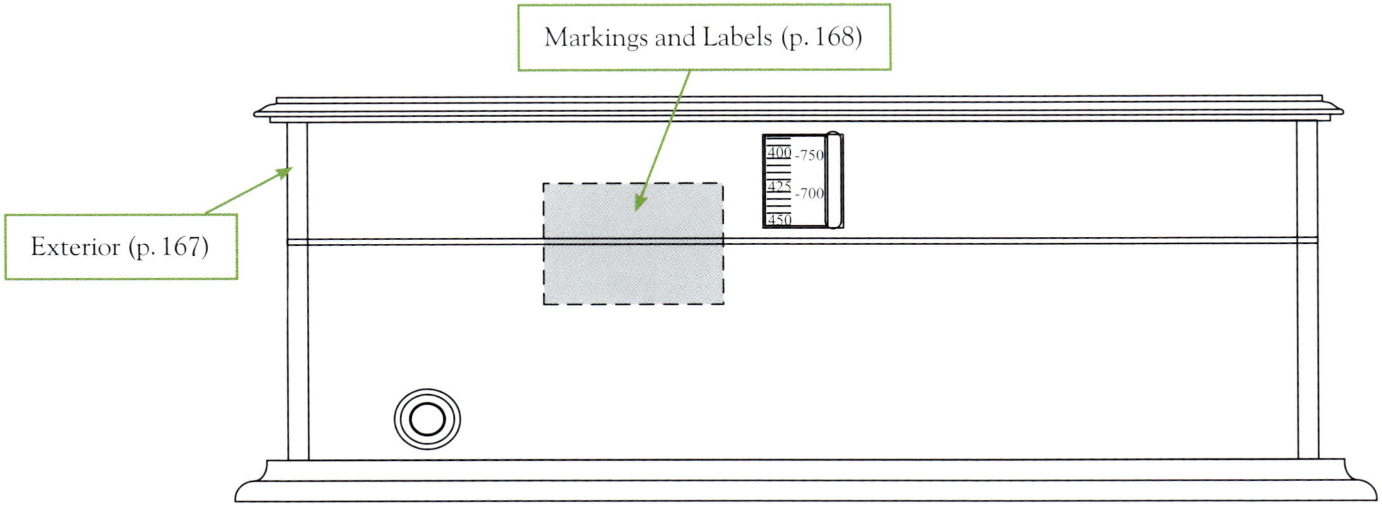

Note: Shaded item is located inside set.

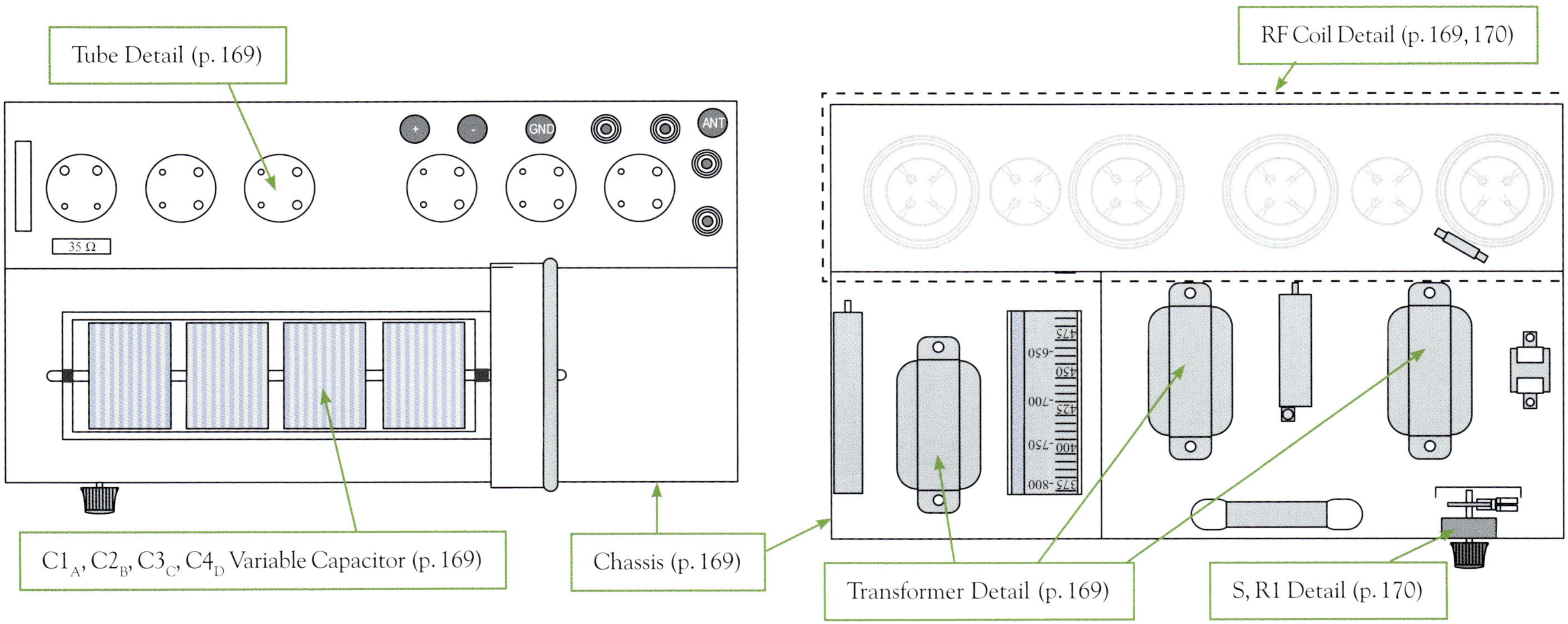

11 Exterior

The cabinet is made of mahogany or walnut veneer and measures 25" by 13.25" by 10.25". Tuning is accomplished with a single thumb wheel.

Front View

Rear View

11 Markings and Labels

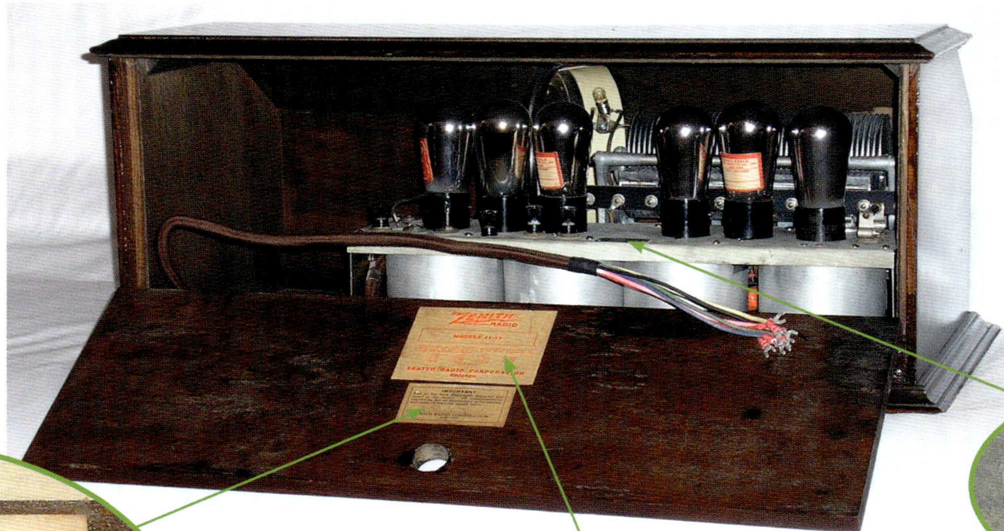

Rear View

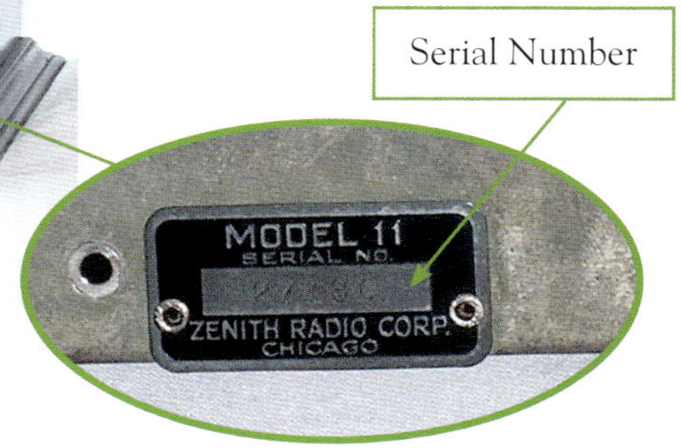

Identification Tag

Inspection and
Test Label

Identification Label

11 Chassis

Top View

Underneath View

11 Component Layout

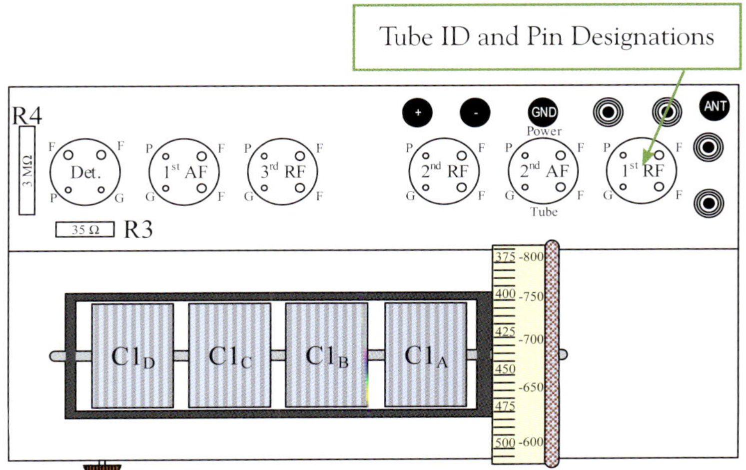

Top View

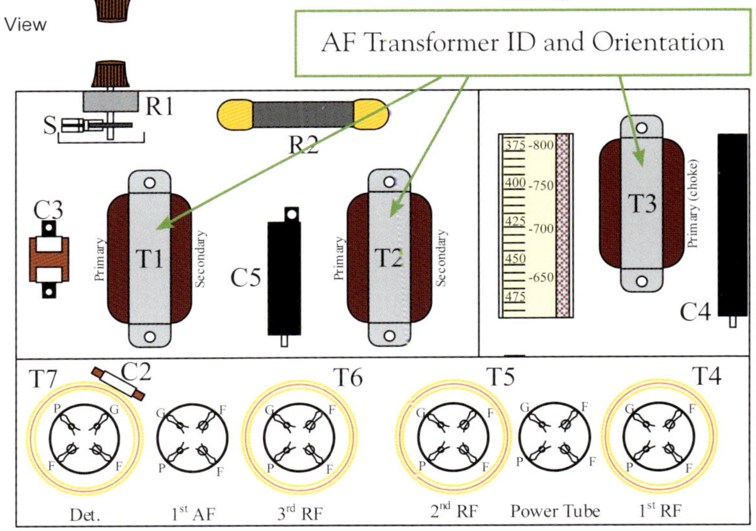

Underneath View (metal shields removed)

Dial Light Replacement

The dial light bulb may be replaced by meshing the tuning capacitor plates entirely together. The bulb will be accessible from the rear of the cabinet.

Use #46, 6.2-volt, miniature screw bulb, available from Antique Electronic Supply at www.tubesandmore.com.

11 Component Detail

RF Coils (metal shields removed)

S Switch, View 1

S Switch, View 2

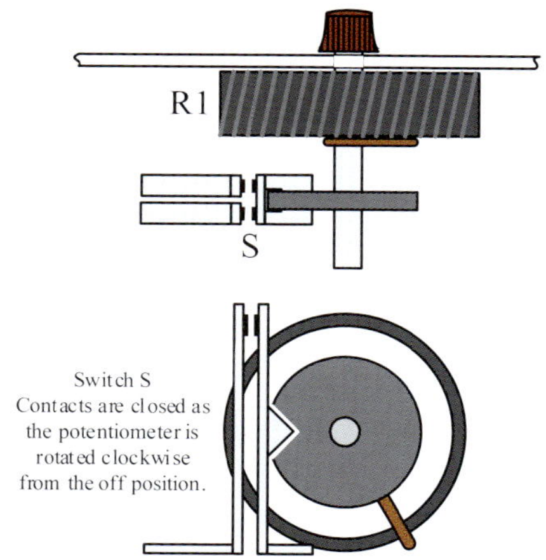

S, R1 Detail (shown in off position)

11 Functional Description Quick Reference Guide

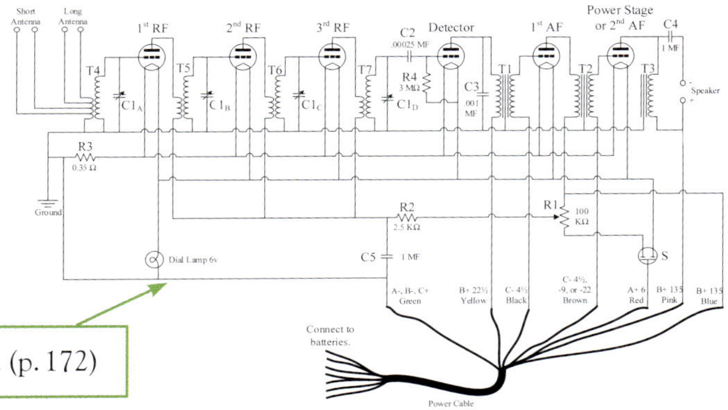

Schematic (p. 172)

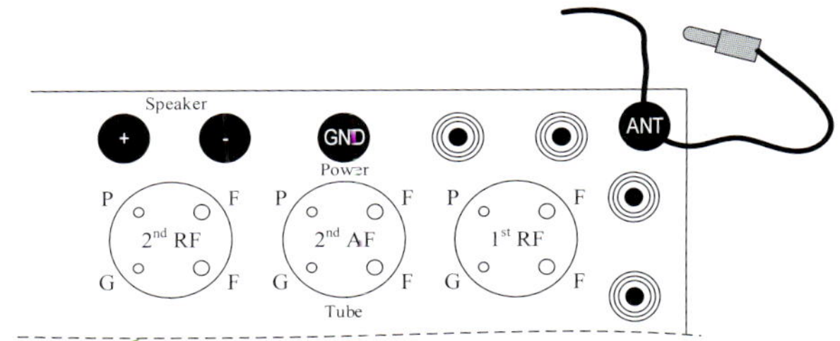

Preparation for Operation (p. 173)

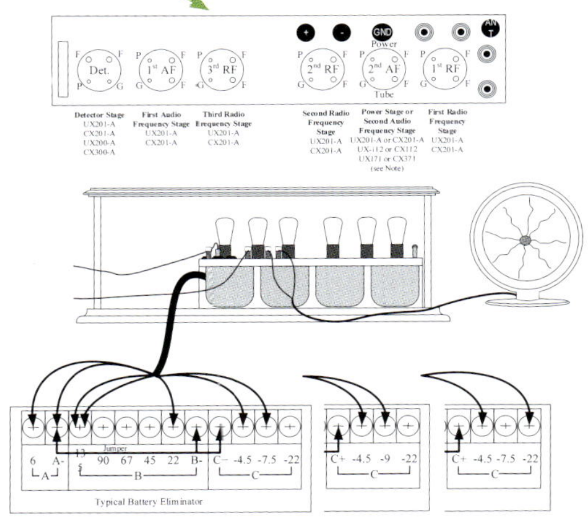

Operation (p. 174)

11 Schematic

The components and circuitry of this radio are similar to the previously presented sets; therefore the description of operation will not be duplicated.

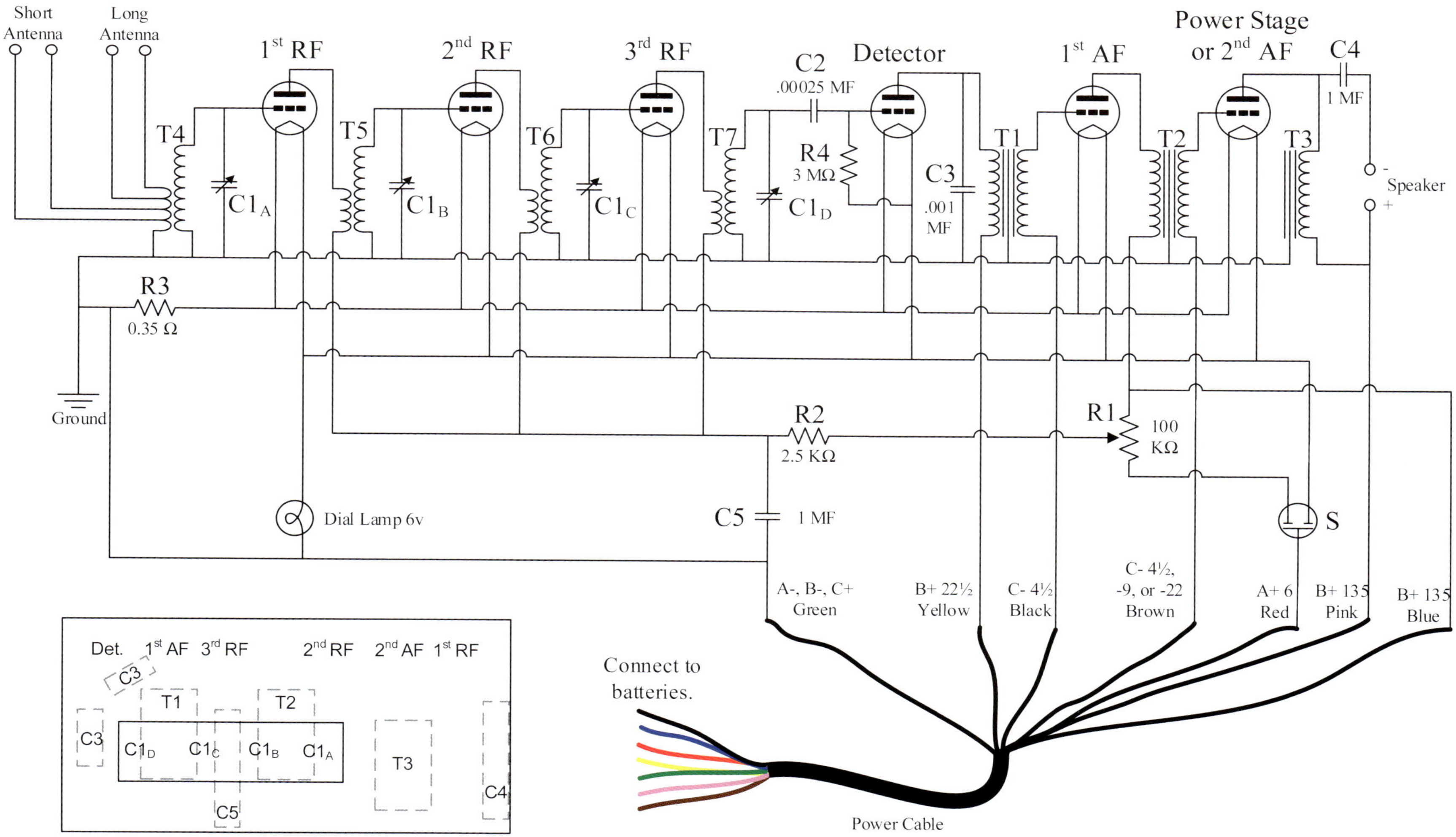

11 Preparation for Operation

Connect the antenna, ground, and loudspeaker. Notice that some of the connections have different options depending on what types of tubes are used. Connect the power as shown in the Zenith Connections diagram or Eliminator Connections diagram below. If the power connections are incorrect for any reason, utilize the physical and functional descriptions to assist in determining the appropriate connections. Proceed to 11 Operation. See Common Radio Attachments (page 8) for additional information related to antennas, loudspeakers and power options.

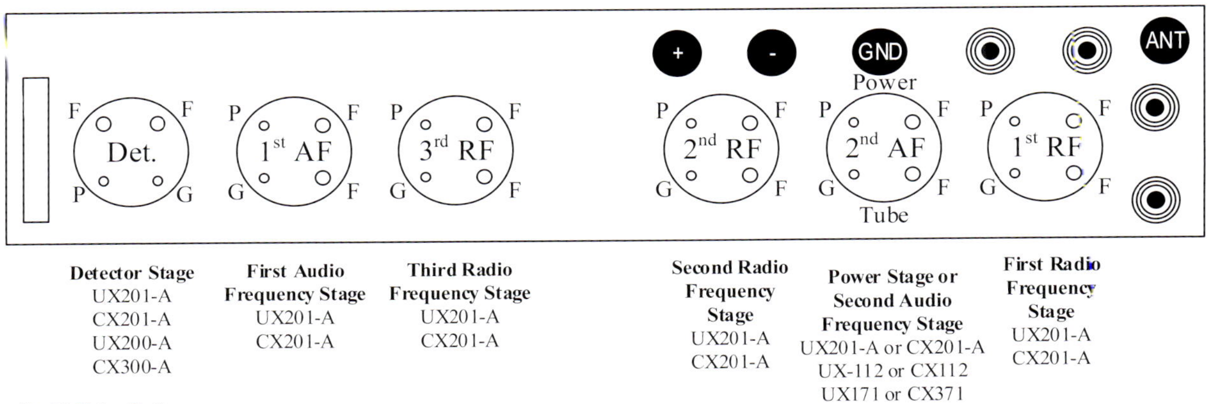

Zenith 11 Tube Options

Power Connections

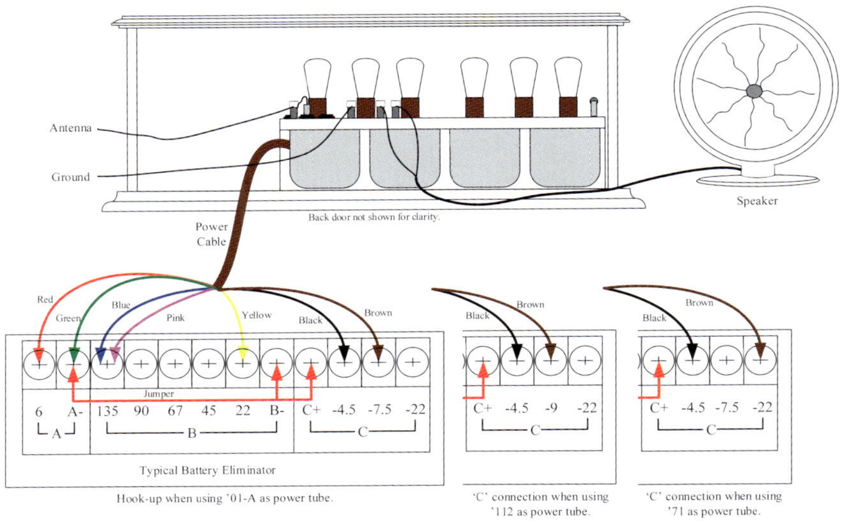

11 Operation

Follow the 11 Preparation for Operation instructions on page 173.

Insert the Antenna Connector Plug into the appropriate jack, which is determined by the length of the antenna, as indicated in the Antenna Connections diagram below. Use trial and error to obtain the best antenna jack.

Tuning is accomplished with the large drum type dial. Fine adjustments are essential to receiving distant or weak stations. The volume control knob not only controls the volume, but also the sensitivity of the receiver. If it is not turned far enough, distant stations will not be received. If it is turned too far, distortion will occur.

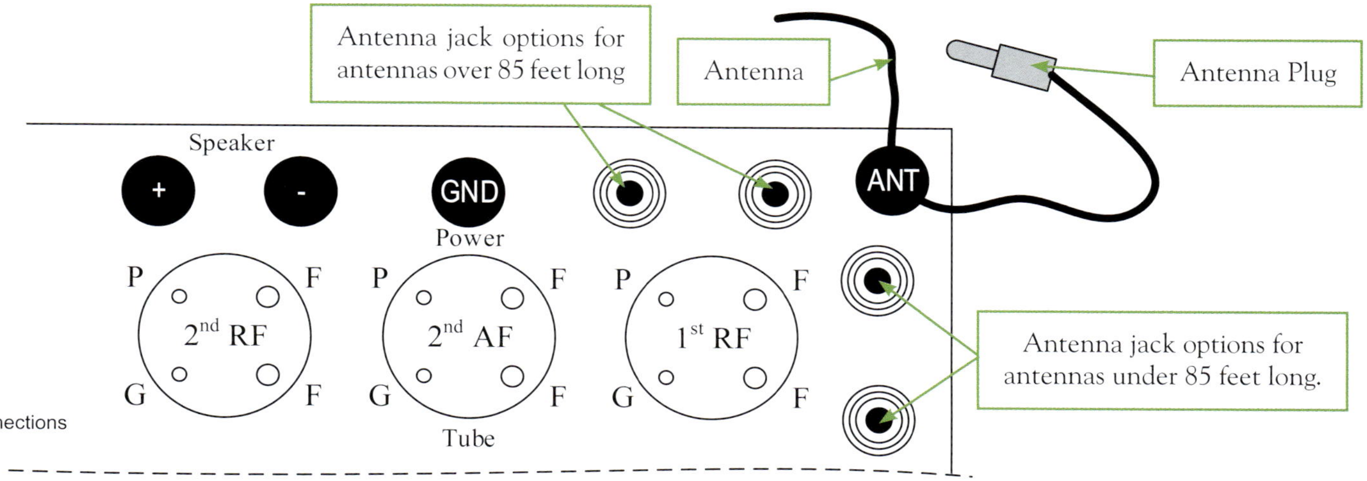

Antenna Connections

Front View

Chapter 11: Zenith 12

12 General Specifications

Company:	Zenith Radio Corporation
Model:	12
Year Manufactured:	1927–1928
Manufacture Location:	3620 South Iron Street, Chicago, IL
Serial Number:	14860
Price:	$100
Tube:	UV-201-A (6), three RF amplifiers, one detector, two AF amplifiers
Circuit Type:	Tuned Radio Frequency (TRF)

12 General Information

The model 12 did not have any markings on the dial. It was intentionally left blank, so that the owner could physically mark the station locations on the dial.

Restoration details for the set that was used as the example in this chapter. *When this radio was acquired the front portion of the sculptured base was missing. A new base was built and attached. The cabinet was striped, stained and lacquered. A new multi-wire power cable was installed. Transformer T2 had an open primary winding and was replaced with one from a parts set. The switch was bent; it was straightened and now works fine.*

12 Physical Description Quick Reference Guide

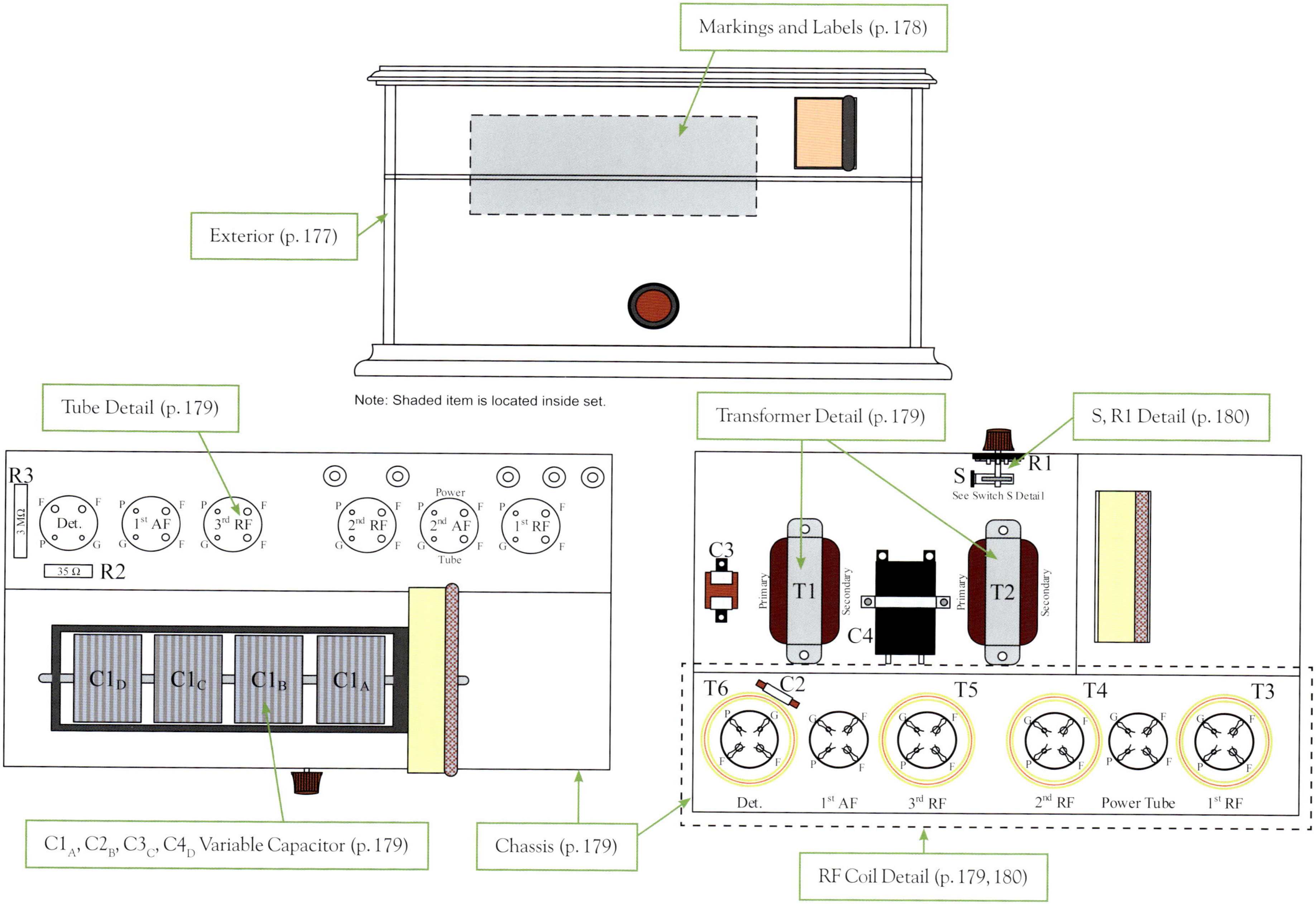

12 Exterior

The radio measures 19" by 13.25" by 10.25". The cabinet is brown mahogany. Tuning is accomplished with a single thumb wheel.

Front View

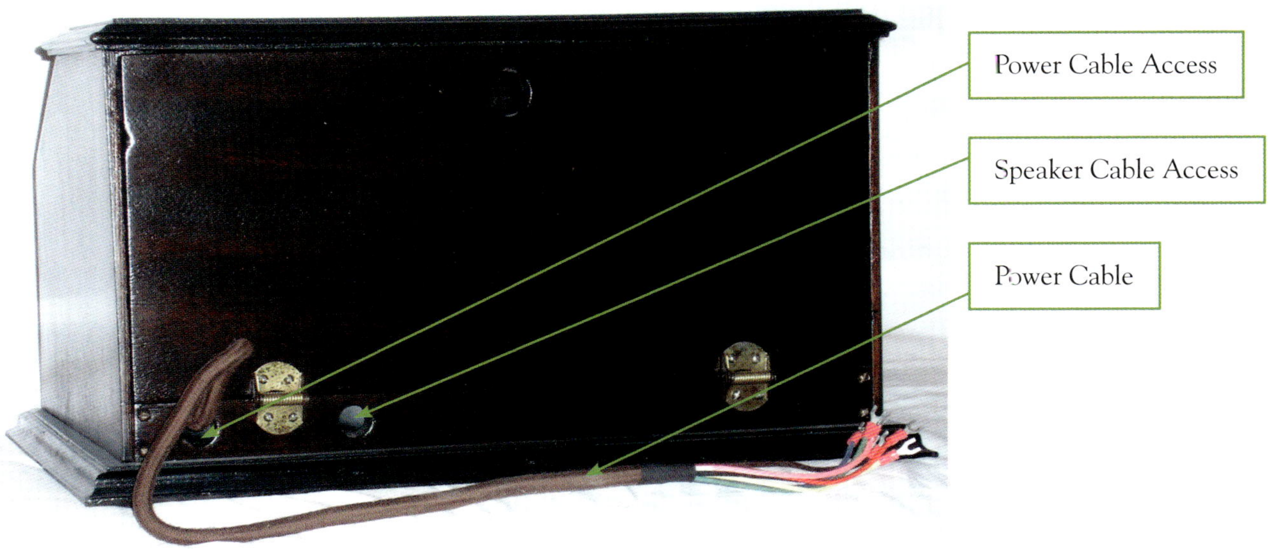

Rear View

12 Markings and Labels

Rear View (door open)

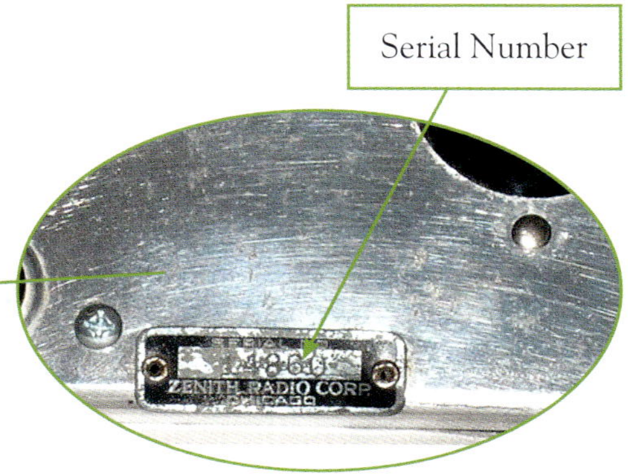

Identification Tag

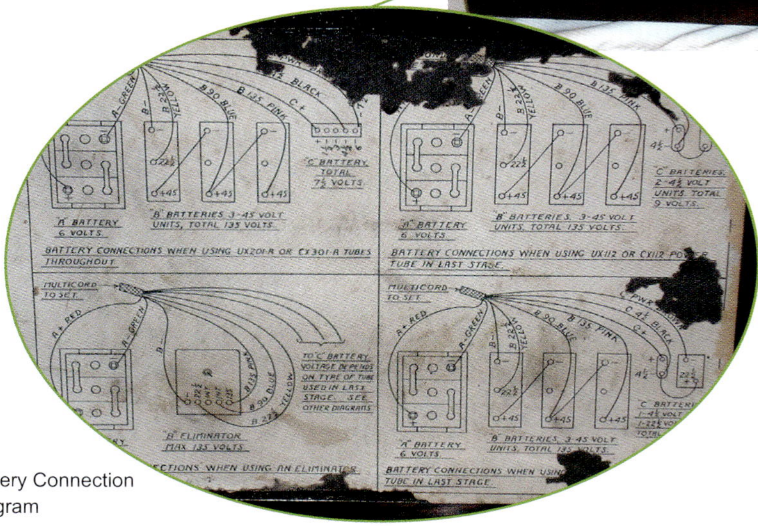

Battery Connection Diagram

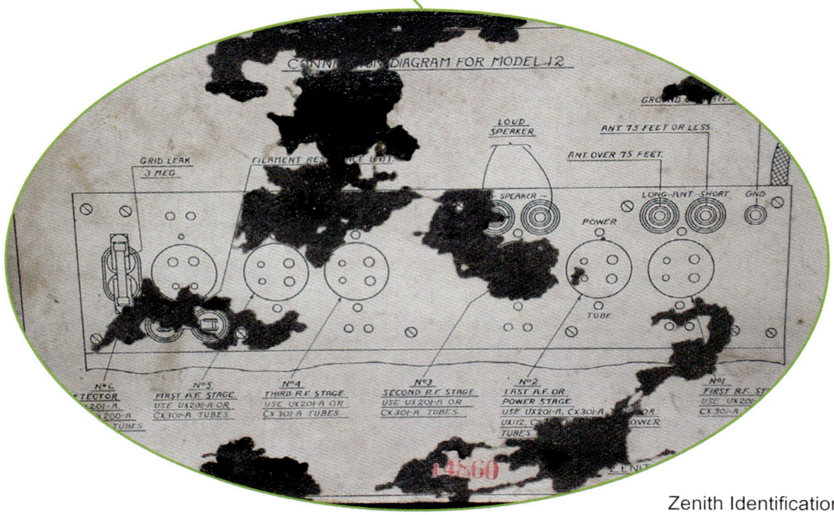

Zenith Identification and Component Layout

12 Chassis

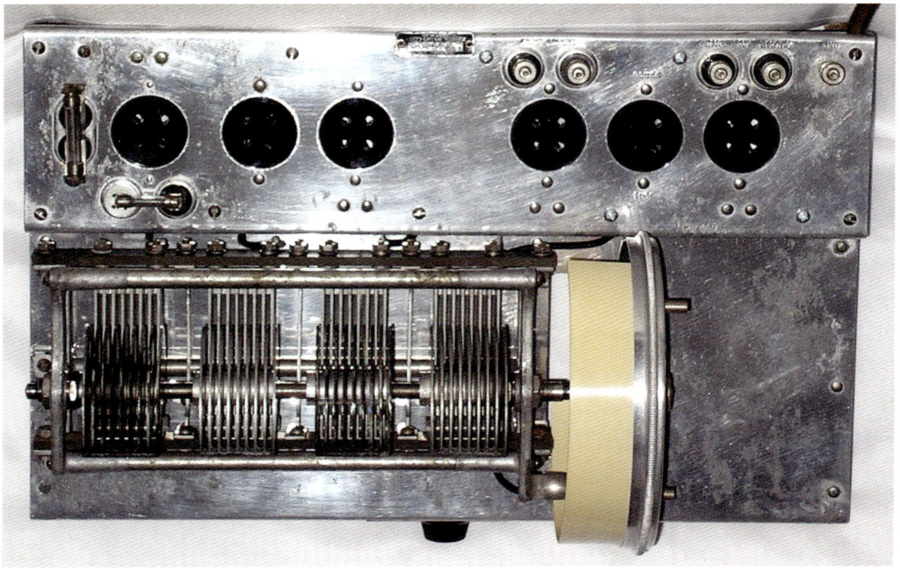

Top View

Underneath View

12 Component Layout

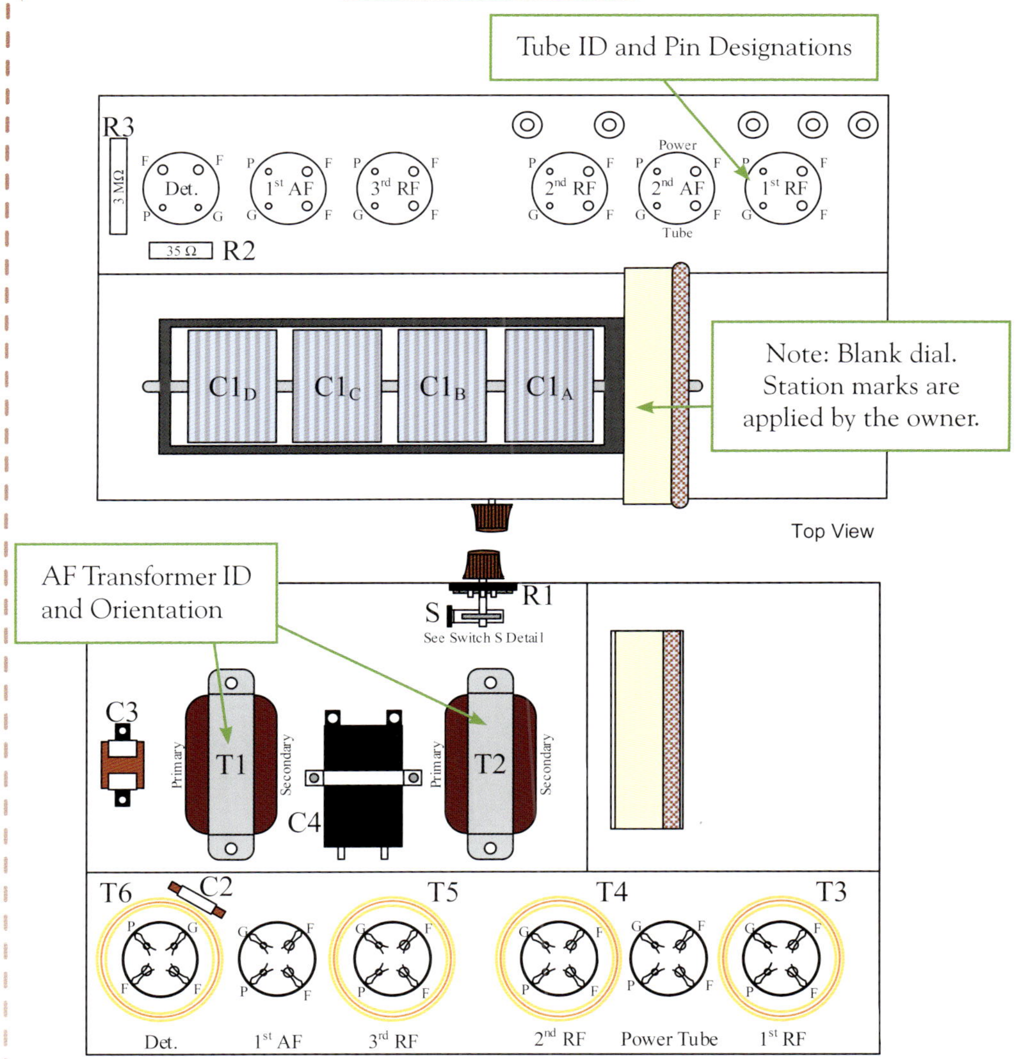

Top View

Underneath View

Dial Light Replacement

The dial light bulb may be replaced by meshing the tuning capacitor plates entirely together. The bulb will be accessible from the rear of the cabinet.

Use #46, 6.2-volt, miniature screw bulb, available from Antique Electronic Supply at www.tubesandmore.com.

12 Component Detail

RF Coils (covers removed)

S Switch and R1

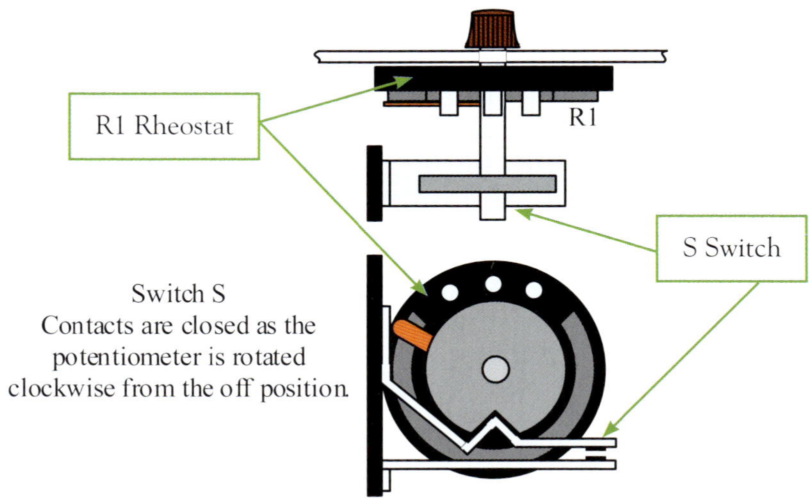

S, R1 Detail
(shown in off position)

12 Functional Description Quick Reference Guide

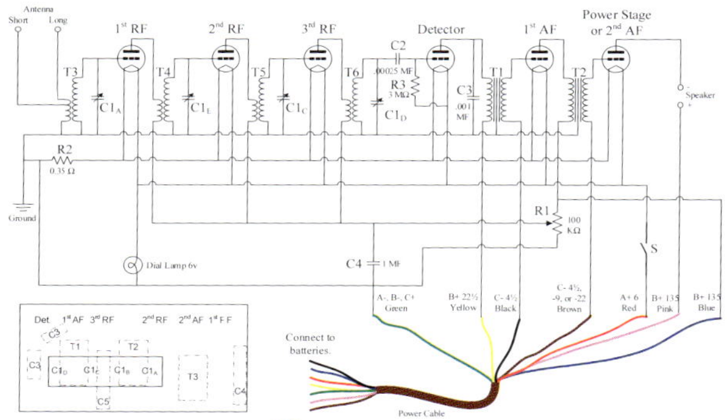

Schematic (p. 182)

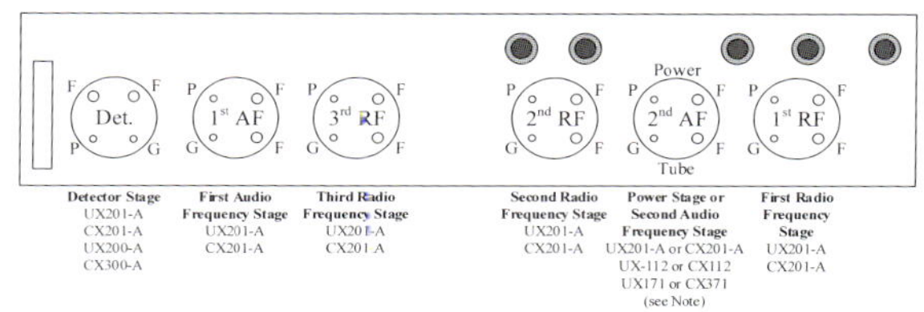

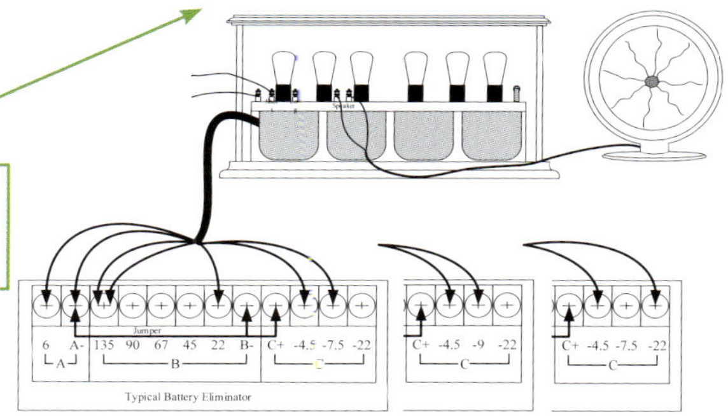

Preparation for Operation (p. 183)

Operation (p. 183)

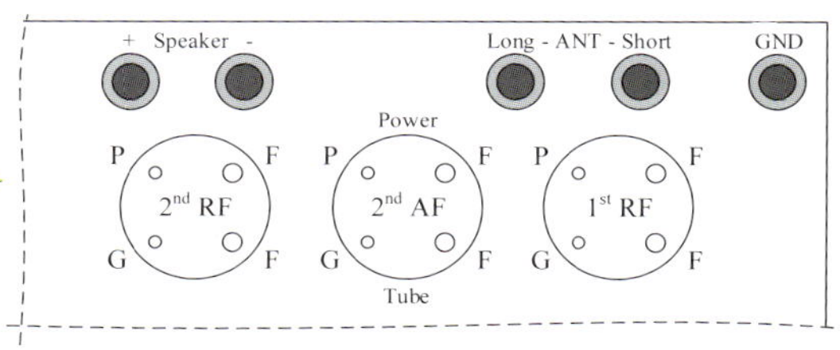

12 Schematic

The components and circuitry of this radio are similar to the previously presented sets; therefore the description of operation will not be duplicated.

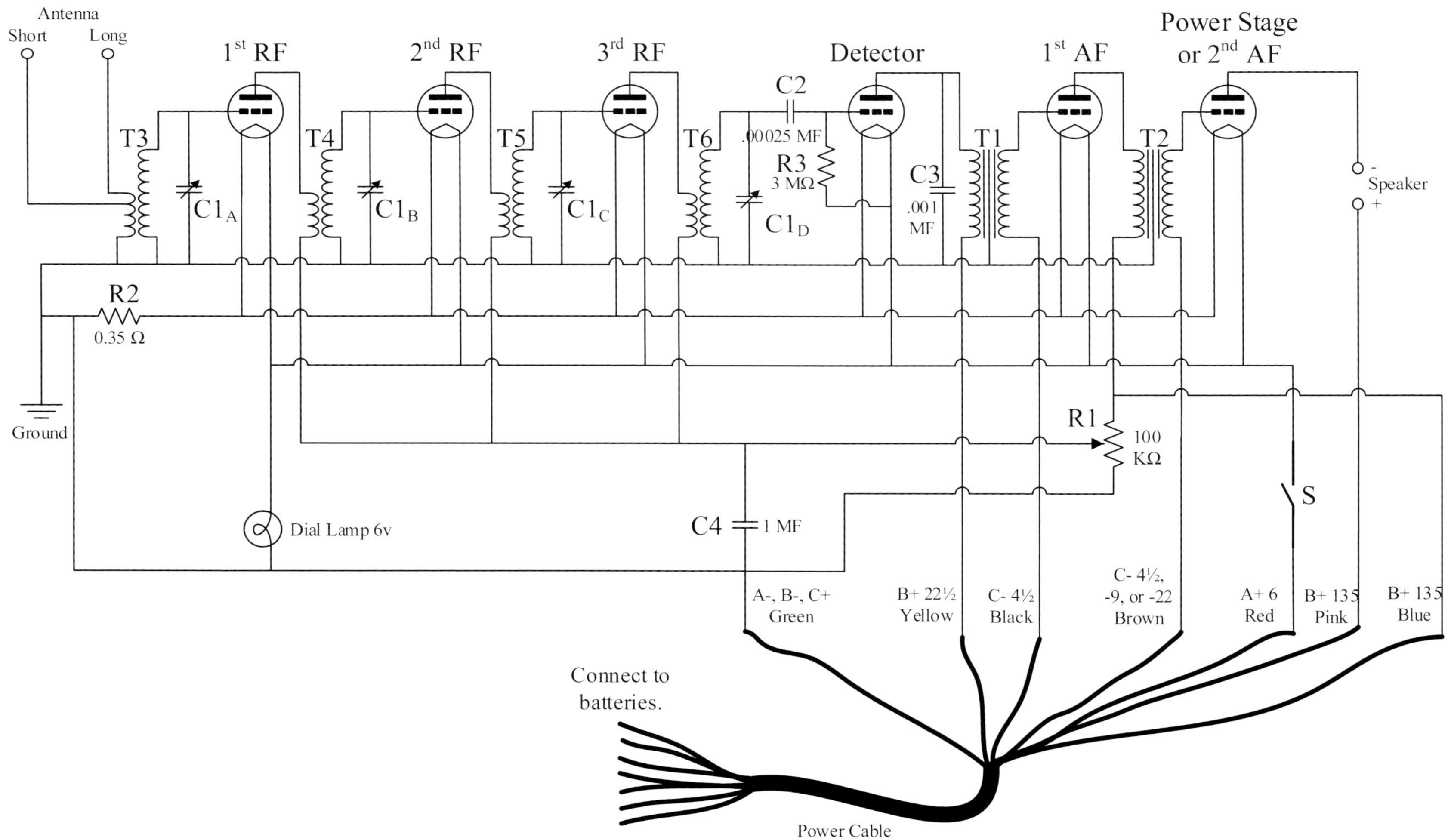

12 Preparation for Operation

Notice that some of the connections have different options depending on what type of tube is used. Connect the power as shown in the Zenith Connections diagram or Eliminator Connections diagram. If the power connections are incorrect for any reason, utilize the physical and functional descriptions to assist in determining the appropriate connections. Proceed to 11 Operation below. See Common Radio Attachments (page 8) for additional information related to antennas, loudspeakers and power options.

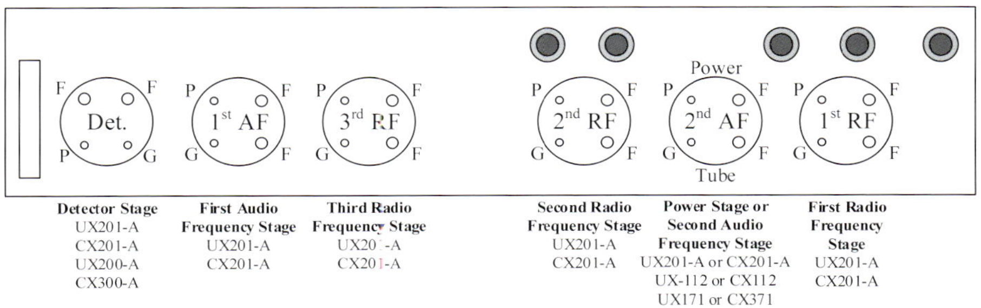

Zenith 12 Tube Options

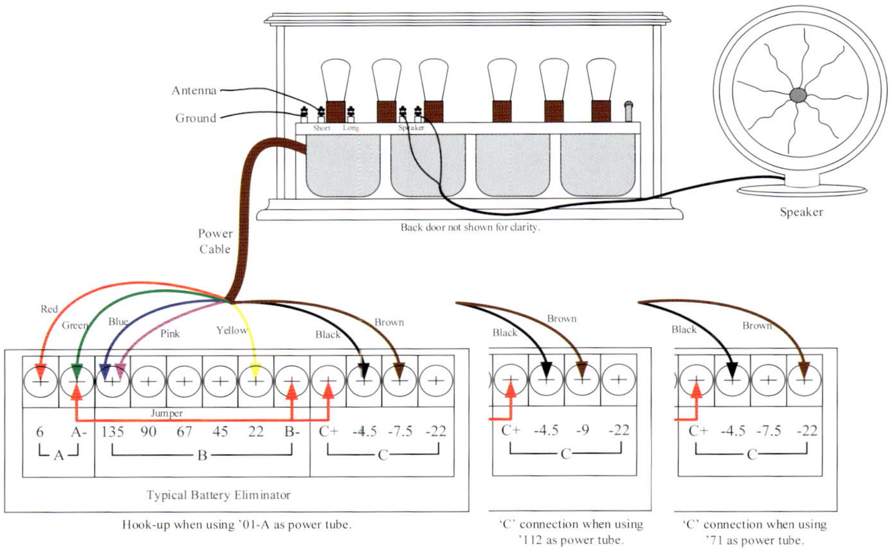

Power Connections

12 Operation

Follow the 12 Preparation for Operation instructions.

Connect the antenna, which is determined by the length of the antenna, as indicated in the Antenna Connections diagram below.

Tuning is accomplished with the large drum type dial. Fine adjustments are essential to receiving distant or weak stations. The volume control knob not only controls the volume, but also the sensitivity of the receiver. If it is not turned far enough, distant stations will not be received. If it is turned too far, distortion will occur.

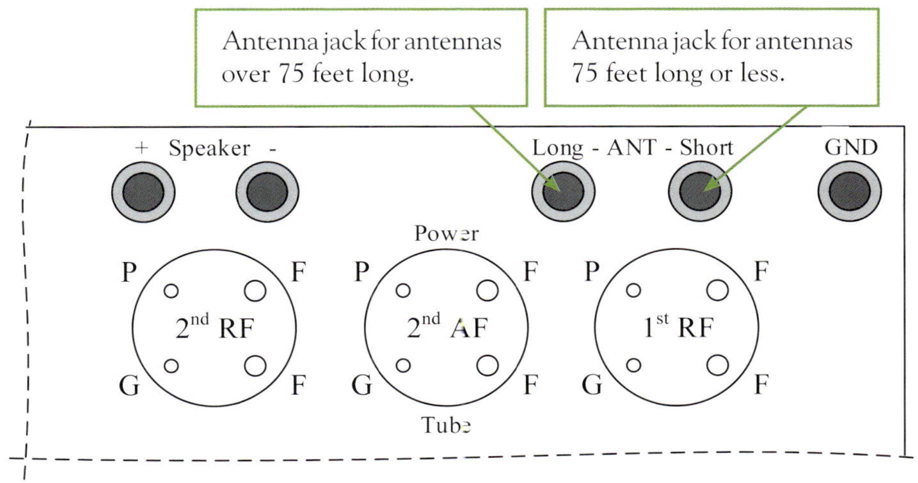

Antenna Connections

Front View

Section II

Useful Details for Understanding Zenith Models and Marketing

Chapter 12: **Notes on the Early Zenith Radios**

The majority of the topics in this chapter are applicable to multiple Zenith radio models or to models that have not been previously mentioned. Much of the information is straightforward: for example, the discussion of 3-R and 4-R versions. Other material is more speculative, e.g., the missing Zenith model numbers. Further research or finds may shed more light on these topics, and allow us more decisive conclusions in the future.

ZENITH RADIO CORPORATION

3620 IRON STREET

CHICAGO, ILL.

Guarantee

We guarantee Zenith Radio Receiver Model 11, Number 26637, to be free from defects in workmanship or material, but should any defect appear in either workmanship or material, within a period of ninety days after the date of original sale of this receiver by a retail dealer, and provided that we are notified within the ninety day period, we agree to repair or replace with new parts any such defect in workmanship or material when the receiver covered by this guarantee is delivered transportation charges prepaid at our factory at 3620 Iron Street, Chicago.

If upon investigation at our factory it appears that the repairs required are due to misuse, negligence, exposure to the elements or to ordinary wear and tear and are in no manner due to defective workmanship or material, the owner will be so advised by letter. The owner may then request the return of the receiver in the condition reported or may authorize its repair at the customary charges, and reshipment with such charges and transportation C. O. D.

ZENITH RADIO CORPORATION ILLINOIS SEAL

This guarantee becomes effective only when the attached coupon is filled in and mailed within ten days after date of purchase to the Zenith Radio Corporation, 3620 Iron Street, Chicago, Illinois.

ZENITH RADIO CORPORATION

By

Original Zenith Guarantee

CRL and Zenith Names

Both Chicago Radio Laboratory (CRL) and Zenith names appear on many of the early Zenith sets because of a patent agreement involving the Armstrong regenerative circuits, entered into by CRL prior to the formation of the Zenith Corporation. This was a serious problem for Zenith in that their equipment designs incorporated many such circuits and it was quickly learned that the patents were not transferable. Too, Major Armstrong had ceased selling licenses for the use of his circuits. Zenith considered possessing an Armstrong regenerative patent as essential to being able to manufacture and sell radios. The problem was resolved by making Zenith the sales agent and having CRL perform the manufacturing activities.

Both names appeared on all subsequent models (1-R, 3-R, 4-R, etc.) until the Super VII was introduced in early 1925, at which time the CRL name was dropped. The only exception to this was the non-regenerative 1924 Super-Portable. The non-regenerative 2-M and 3-M AF amplifiers did not fall under the Armstrong patent agreement; however the CRL name and the Zenith names both appear on these sets for two reasons. First, CRL was the manufacturer and Zenith was the sales agent, and second, the aesthetic design of the 2-M and 3-M remains consistent with the matching 1-R, which did contain the patented Armstrong regenerative circuit.

Models Not Produced

There are two sets that Zenith considered introducing to the broadcast market that were never actually put into production.[1] The first was the 2-R radio, shown at right, and second was the optional 1-M single stage AF amplifier (similar to the 2-M and 3-M) intended to be sold with the 1-R receiver. Zenith may have built prototypes of these two sets during the initial research and development stages, but no known examples exist.

Zenith Model 2-R

Missing Model Numbers

In the first few years of business, Zenith assigned the following model numbers to the radios that they produced: 1 (1-R), 2 (2-R), 3 (3-R), 4 (4-R), 7 (VII), 8 (VIII), 9 (IX), 10 (X). Model numbers were assigned to each successive set, in a fairly consistent manner, as the sets were designed, manufactured and introduced to the public. Notice that there are two model numbers, 5 and 6, missing from the list. There are also two radios that Zenith manufactured, that do not have assigned or identified model numbers: Super-Portable, and Phonograph Set.

The first radio that does not have a Zenith assigned model number is the Super-Portable. The only identifying marks on the Super-Portable are a serial number on the chassis and the letters *ZEV* on the antenna cover of some models. Most likely the letters stand for "*Z*Enith *V*" in that the Super-Portable was manufactured after the introduction of the 3-R and 4-R, but before the Super-Zenith (VII).

The other radio that does not have a Zenith assigned or identified model number is the Phonograph Set. There is no information available that indicates what kinds of markings were actually on the radio. It would make sense that the Phonograph Set would be the model 6 or VI, because it is the only set left without an identified model number and the model 6 is the only number left without a radio. The model VI Phonograph Set also seems fitting, because it was introduced after the 3-R and 4-R, about the same time as the Super-Portable, but before the VII Super-Zenith.

Super-Portable Configurations

There are evidently two different configurations of the Super-Portable radio. One version could have been a prototype that Commander MacMillan tested on his arctic expedition. The other version is in the more recognized configuration such as that shown in the accompanying photo and in *Zenith Radio: The Early Years.*[2] The only indication that the prototype version even exists, is photographs in various antique radio related publications, which are usually attributed to the Zenith Radio Corporation.[3, 4, 5]

A close examination of the prototype picture reveals that the three RF transformers are either missing or obscured, and that two additional batteries (probably "C") are located toward the center of the set between the two AF transformers. The quality of the photograph is poor which may obscure the RF transformers and/or distort other components.

Phonograph Set

The Zenith Phonograph Set is a radio panel that can be inserted into the Victor Phonograph models 215, 400, 405 or 410. It fits into the left side of the phonograph console, where the records would usually be stored.[6] The circuit is regenerative and similar to the 3-R radio, but has only one tap switch and the pointers are similar to the pointers on the Super-Zenith. The panel has holes for the tubes (exposed), so that the user could easily access them, after it had been installed into the phonograph console.

Zenith's initial sales network for its radios consisted of phonograph dealers across the country, so it was a logical step for Zenith to provide a radio panel that could be installed into the phonograph sets. This way the dealer could sell radios, phonographs or radio-phonograph combinations. The dealer would usually sell and then install the Phonograph Set into the Victor Phonograph console. Other manufacturers produced similar radio panels for insertion into various phonographs.

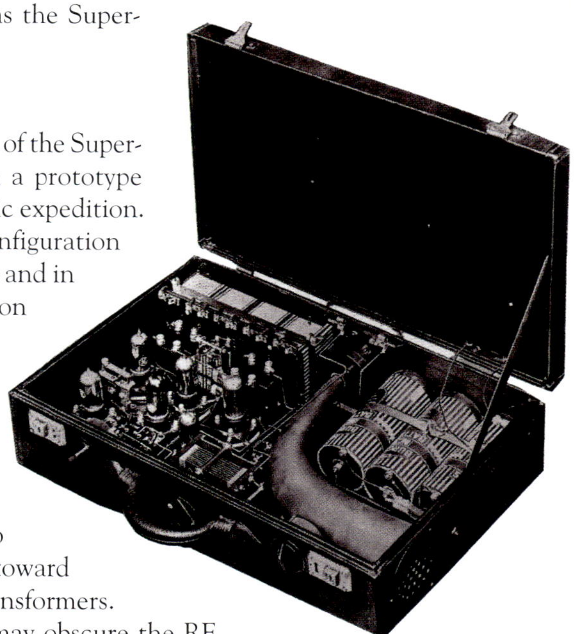

Notice the difference in the layout between the prototype version on the left and the usual Super-Portable version on the right.

Zenith Phonograph Set[7]

Victor Phonograph Model VV-405S

3-R and 4-R Versions

Zenith introduced the 3-R and 4-R in 1923 and for some reason, did not introduce a new radio model the next year; they just revised the early versions and continued selling them throughout 1925, along with the Super-Zenith VII. As compared to the rest of the industry, that was a long time to be selling the same radio.

The revision of the late version consisted of using the same basic regenerative circuit with updated voltages to the plates of the tubes. The power cable was also changed from 7 to 5 wires. The stage selection switch was upgraded from the lobed camshaft switch to a multi-contact rotary switch and the order of audio frequency amplifier stages selected was also changed. The peep hole was removed from the front panel. The late version of the 3-R and 4-R does not play any better than the early version.

3-R and 4-R Differences

The 4-R, which sold for $95, was a more economical version of the 3-R, which sold for $165. The 4-R cabinet had a boxier look with a black panel with exposed tap switches. It lacked the X tap switch, the vernier dials and the speaker connections at the rear of the set that were included on the 3-R. The 3-R had the more ornate cabinet with the slanted front, faux wood grain panel, and hidden tap switches. The testing that I carried out resulted in the 3-Rs and 4-Rs performing pretty much the same, except for one particular 4-R (E) that seemed to outperform all the rest.

3-R and 4-R Power Cables

Zenith utilized a power cable, with a Jones Multi-Plug on one end, for connecting the batteries to the 3-R and 4-R radios. There were two different versions of these cables, a 7 wire cable and a 5 wire cable. The 7 wire cable is used with the early version of the 3-R (E) and 4-R (E) radios. The 5 wire cable is used with the later version of the 3-R (L) and 4-R (L) radios. The difference between the two cables is that the 7 wire cable has an extra pink and black wire. All other wires are the same for both cables. The 7 wire cable can be used with either version (early or late) of the 3-R or 4-R radios. The 5 wire cable can only be used with the late version of the 3-R (L) and 4-R (L) radios.

Special Note: Some 3-R and 4-R early version cables did not have the brown wire and will only work with those radios. For more information review the description for your radio.

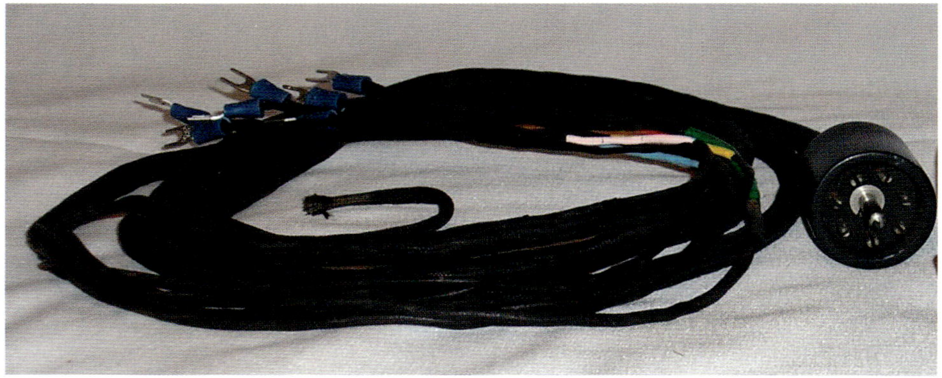

7 Wire Cable with Jones Multi-Plug

Left is the 5 wire multi-plug and right is the 7 wire multi-plug.

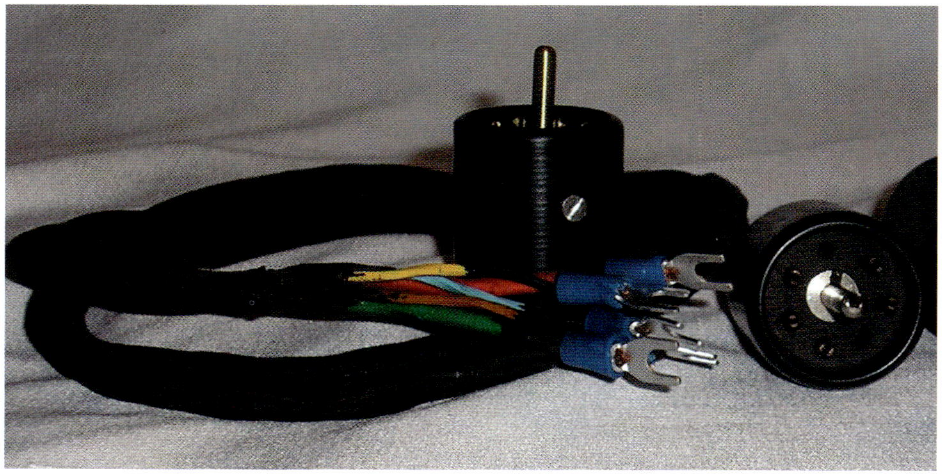

5 Wire Cable with reproduced multi-plug (similar to the Jones Multi-Plug)

Other radio manufacturers (Kennedy, Magnavox, etc.) utilized power cables, with the Jones Multi-Plug connector, for connecting the batteries to their radios. These battery cables produced by other manufacturers are probably wired or color coded differently. Before using any power cable, always make sure the correct voltage is applied to the appropriate wire, pin, connector, etc., or you run the risk of damaging the radio, its internal components, or even creating a safety hazard.

Chapter 13: **Successes and Failures**

As a result of collecting and restoring these early Zenith radios for many years, I have realized numerous facts regarding the design and production of their equipment. Some of the more esoteric observations are included here.

First, Zenith Stage Selection Development explains the three different successive designs that Zenith developed, in multiple radio models that allowed the user to control the volume of the radio, prior to the invention of rheostat volume control.

Second, Zenith Portable Radio Broadcast Station describes the station, which contained an early version of the 3-R radio.

Next, The Zenith Pot Metal Problem explains the use of pot metal which caused the deterioration of many of the internal parts in several of the early Zenith radios.

Last, the section Failure of the Zenith Super-Portable investigates what the problem may have been with the Super-Portable radio, a problem that resulted in Zenith halting its production.[1]

Zenith Radio Corporation, 3620 Iron Street, Chicago, Illinois (Zenith Salesman's Handbook, circa 1927)

Zenith Stage Selection Development

Many of the radios that were produced in the early 1920s utilized some sort of stage selection control for increasing or decreasing the volume; rheostat volume control had not been invented. One stage is one AF amplifier tube. The detector is not considered as an AF amplifier stage, although it does provide some AF amplification. The volume is raised by selecting more stages or lowered by selecting fewer stages. The total amount of amplification that is available depends on how many stages there are. Over the first few years of producing radios, Zenith developed several different designs for stage selection and each new design was introduced with a new radio model.

The first stage selection method that Zenith used to control the volume was the plug and jack method, which was incorporated into the 1-R radio and optional 2-M and 3-M AF amplifiers. The 2-M provided two stages of AF amplification and the 3-M provided three stages of AF amplification. When the operator wanted to change the volume, the headphones or loudspeaker plug was pulled out of one jack and inserted into another jack. The physical action of inserting the plug into the jack would make the internal jack switches open and close, which would automatically route the filament voltage and plate voltage to the appropriate circuits.

The second stage selection method that Zenith designed was the lobed camshaft mechanism, which was introduced with the early version of the 3-R and 4-R. As the switch is rotated, the lobes on the camshaft would open and close the switches,

Plug and Jack Method (2-M and 1-R)

which would automatically route the filament voltage and plate voltage to the appropriate circuits. The camshaft switch mechanism was a significant improvement over the physical plug and jack method.

The third method that Zenith used to control the volume was the multi-contact rotary switch, which was introduced in the late version of the 3-R and 4-R. As the switch is rotated, the multi-contacts would route the filament and plate voltage to the appropriate circuits. The first Super-Zenith VII radios that were produced included this stage selection method, but shortly after introduction, the Super-Zenith was revised to incorporate the new rheostat volume control. The first version of the Super-Zenith VII was the last Zenith radio to use stage selection to control the volume.

Lobed Camshaft (4-R E)

Multi-Contact Rotary Switch (3-R L)

Model of the portable radio broadcast station WJAZ ("Special Edition" 1925 Kenworth, Die-Cast Metal Vehicle, Ertl Collectibles, 1995)

Zenith Portable Radio Broadcasting Station

Zenith designed and built a portable radio broadcasting station in late 1924 or early 1925 that was mounted on the bed of a Federal truck.[2] The portable transmitting station was completely self-contained, including a 100-watt generator. The Zenith onsite staff only had to install the folding antenna. The original purpose of the portable transmitting station was to help to locate a site for a new Zenith radio station around the Mt. Prospect, Illinois area.[3] Originally the portable transmitting station had the call sign WJAZ, which was changed to WSAX in August 1925.[4] The new Zenith radio station at Mt. Prospect, Illinois retained the call sign WJAZ.

The portable radio broadcast station also included an early version of the 3-R that was used for checking the transmitter modulation and maintaining communication, if desired.[5]

Once that task of locating the new site was completed, the portable transmitting truck was designated as a promotional vehicle for Zenith and started touring the nation promoting Zenith Radio Corporation and broadcasting in general.[6] Zenith obtained much publicity as a direct result of this nationwide tour.

The *Houston Chronicle* of Friday, October 23, 1925, states that the WSAX portable radio broadcast station broadcasted, on Thursday, from Houston, Texas. The article also noted that the crew included R. H. G. Mathews, chief transmitting engineer, who was in charge of the station and who also had directed the construction of the portable radio broadcast station.

Rear views showing the Zenith equipment that was used in the portable radio broadcast station (*Radio News*, January 1925, pg. 1149).

The Zenith Pot Metal Problem

Each radio has to be restored based on its own merits. Occasionally the collector acquires a "farm fresh" radio that does not need a restoration or may only require a partial restoration...maybe just replace the capacitors, clean the cabinet and apply a coat of good furniture polish. Typically that is not the case with the 1-R, 2-M, 3-M, 3-R, 4-R or Super VII Zenith radios; they just about always require an extensive restoration, due to deterioration of the parts. The reason is that in the first few years of producing radios, the Zenith Radio Corporation had various radio parts made from cheap low-grade alloy referred to as pot metal.

Pot metal refers to an inexpensive low-melting-point alloy that was unstable over time and tended to bend, distort, crack, shatter and pit with age.

This was not a problem that the Zenith Radio Corporation would have had to deal with because it takes decades for the pot metal to deteriorate. Only the collectors of Zenith radios have to confront this extensive problem.

The parts that could be affected by this problem are the gears in the 1-R, the pulleys in the 3-R and Super VII, the pointers in the 1-R, 2-M, 3-M, 3-R, 4-R, the tuning condenser shaft in the 1-R, 3-R, and some pointer shafts on 1-R, 2-M, 3-M, 3-R, and 4-R. For some reason, occasionally the parts of a particular radio are not affected. The effects of pot metal are so severe that only on rare occasions can they be repaired.

For those parts that are affected, the best solution is to get another good part from a spare parts set, one that was unaffected, or to reproduce the parts yourself. Because this pot metal problem was so extensive and affected so many of the early Zenith radios, the reproduction of all these parts will be covered in Chapter 15, Zenith Radio Restoration Guide and Tips, page 223.

Damaged Parts Removed from a 1-R

Damaged 3-R Pulleys

Damaged Pointers and Shafts

Super-Portable before restoration

Failure of the Zenith Super-Portable

When Zenith introduced the Super-Portable (Companion) radio in early 1924, it sold for $224. It was entirely self-contained in a traveling case, described as "smaller than the average suitcase," and it included tubes, batteries, horn speaker and built in antenna. The controls were located outside so that it could be played in a closed or open configuration, making it truly portable. Zenith marketed the Super-Portable radio as allowing the user to take it on camping trips, to the beach, or to enjoy it in the home. The company maintains that the 1924 Super-Portable radio was the "world's first portable radio."[7]

There was a problem with this radio that resulted in fairly low production numbers. There seems to be some confusion on exactly how many were produced. Evidently most were either returned to the manufacturer by the purchaser or they were recalled by Zenith. One report stated there were 1,000 produced and 600 of them were returned.[8] Another report was that 6,000 were produced and they all came back at least twice.[9] There's even a folklore story that Zenith President Eugene McDonald publicized the destruction of the returned radios by destroying them in a huge bonfire.[10] Judging from these reports and the fact that there are not very many in existence, the problem was obviously a very serious one.

To my knowledge, there is no information available that would indicate the nature of the problem. Did it not play well? Do the batteries run down quickly? Did it just quit playing? What exactly was the problem?

I was fortunate enough to have obtained a 1924 Zenith Super-Portable radio, in good condition. Moreover, it was complete—it even had original Zenith UV-99 tubes—so I decided to restore it, to the point of reproducing the "A", "B" and "C" batteries. The only serious problem I found during restoration involved the first and second Acme–2 RF transformers that were open. There was a note inside the first transformer that stated "Repaired both #2 coils 3-28-1940." I used substitute transformers, but they didn't work very well. Five used Acme transformers were purchased and they were all open. The only option left was to rewind them using 30-gauge lacquer-coated wire as a substitute for the original 40-gauge cloth-covered wire. All other parts in the radio remain original except for the "D" battery which required an adapter, since the original "D" battery had a negative case.

With all problems solved the radio began to play. I theorized that the batteries might have been the original problem because of a statement in an article about this radio that read "…the Zenith did not work very well and its batteries ran down." As a result of this information, I decided to perform some tests to see if I could determine if they were indeed the problem, and also determine how long they would last.[11]

My test results are presented in the Test Activity Log table. The tests were conducted by recording the A voltage, the B voltage and the start time. The radio was then turned on and frequently monitored, and I would occasionally change stations. When I turned off the radio the end time was recorded along with the A voltage and the B voltage again. This was continued until the radio quit playing.

Acme R-3, Acme R-2 and Acme R-2 RF Transformers

Re-wound Acme R-2 Transformer

I used the 1929 Trav-Ler Model T Portable radio as a baseline for how long the A batteries should last (Trav-Ler suggests that the A batteries should last for 75 hours). The Zenith portable actually outperformed the Trav-Ler, playing for over 80 hours on the same set of batteries, despite the fact that it had six 99 tubes and the Trav-Ler only had five 99 tubes. The Trav-Ler is also five years younger. Was the Zenith Super-Portable radio ahead of its time?

No adjustments were made for using the "D" cells, 9-volt, and AAA batteries in the reproduced "A", "B" and "C" batteries. The test data confirms that battery consumption of the Zenith Super-Portable radio was not a problem.

The performance of the Super-Portable radio over the test period was very satisfactory. The volume of the horn speaker was always acceptable and could be heard across the room. The tuning required a little patience as the single knob tuning is a little misleading, since the rheostat knobs have a great deal of effect on the volume of the station being tuned. If the rheostats are turned to low the station cannot be heard, and if they are turned to high a hissing noise is heard (feedback). With a little tinkering, many stations, both strong and weak, could be heard. The antenna is located in the lid of the case and most of the tests were performed with the case in the open position; however some were performed in the closed position. The radio seemed to perform slightly better in the open position. The radio was also tested with headphones that resulted in excellent performance as expected. During the testing there were never any problems or performance issues that would suggest that the radio should be returned to the manufacturer.

Test Activity Log

Day	Start A Volts	End A Volts	Start B Volts	End B Volts	Start Play Time	End Play Time	Total Time	Cumulative Play Time
1	4.61	4.2	91	85.2	3:00 PM	9:00 PM	6	6
2	4.38	4.2	87.2	84.7	3:00 PM	5:00 PM	2	8
2	4.27	4.1	85.4	82.9	5:40 PM	10:00 PM	5.3	13.3
4	4.34	4.03	85.9	81.9	8:30 AM	1:20 PM	4.9	18.2
4	4.17	3.97	83.3	80.9	3:45 PM	9:05 PM	5.3	23.5
5	4.2	3.96	83.2	80.6	3:40 PM	7:40 PM	3.6	27.1
6	4.17	3.96	82.8	80.4	7:45 PM	9:45 PM	2	29.1
7	4.12	3.89	82.1	79.6	7:45 AM	11:45 AM	4.2	33.3
7	4.04	3.88	81.2	79.3	2:45 PM	5:30 PM	3.3	36.6
7	3.99	3.86	80.6	79	7:30 PM	9:30 PM	2	38.6
8	4.04	3.8	81	77.9	8:30 AM	4:15 PM	7.7	46.3
15	4.06	3.9	81.3	79.1	7:45 AM	8:45 AM	1	47.3
15	3.99	3.64	80.1	74.9	10:00 AM	9:00 PM	11	58.3
16	3.88	3.6	77.1	73.5	7:00 AM	1:30 PM	5.5	63.8
16	3.72	3.49	74.8	71.9	3:00 PM	9:00 PM	6	69.8
18	3.76	3.31	73.9	70.1	2:45 PM	9:45 PM	6	75.8
19	3.6	3.32	72	69.9	6:15 AM	7:30 AM	1.3	77.1
19	3.52	3.23	71.3	69.1	9:15 AM	11:15 AM	2	79.1
19	3.49	2.74	70.9	67.3	2:35 PM	8:00 PM	4.6	83.7

Restored 1924 Zenith Super-Portable Radio

After restoration, testing and many hours of operating this particular radio, I was only able to identify one catastrophic problem that would have caused this device to be returned to the manufacturer. That problem was the failure of the Acme transformers. I now find myself speculating as to what was wrong with these transformers; was there a problem in design, materials (wire, solder, flux), or workmanship?

Since only a single radio was tested it isn't reasonable to attribute all those returns to the Acme transformers. It might prove enlightening if other Zenith Super-Portable radio owners could check the Acme transformers or even perform similar testing.

The 1924 Zenith Super-Portable radio was one of the first truly portable radios manufactured for the general public. When it was introduced it would have been considered high-tech, incorporating the new low voltage UV-199 tubes and single dial tuning as well as being fully self-contained. Based on the performance of this radio I believe the Zenith Super-Portable radio might have been a real success story for Zenith if engineering had not made the decision to use the Acme transformers, which likely resulted in its failure, early death and obscurity.

Chapter 14: Promotional Material

Zenith created various forms of promotional material that was focused on the general public, and other types of material that was focused on providing product-related information to the distributors, dealers and the sales force.

Publicity image examples of the radios are presented here, along with various examples that show the radios in use. Several of these include famous people.

Also included are several examples of pamphlets that Zenith would have provided to the dealers. The salesman would then give the brochures to prospective customers, with hopes of enticing them to purchase a radio. Each of these brochures had blank spaces, so that the local dealer could apply their contact information.

One example of Zenith's promotional material to its distributors and dealers was the Zenith Salesman's Handbook that was provided, free of charge, to educate the sales force about the history of Zenith, the line of radios that Zenith offered, and how to sell Zenith radios. The Salesman's Handbook is reprinted here in its entirety, and includes some interesting descriptions and information about the early Zenith radios.

Zenith also created a series of Zenith Trade Letters that were periodically sent to each distributor and dealer to inform them about new models that were available, problems with previous models, and the availability of various advertising materials (mats, signs, brochures, etc.) for local advertising. Several of these Trade Letters, which also include information about the early Zenith radios, are reprinted here.

Zenith Promotional Material

Publicity Examples

R. H. G. Mathews amateur radio station 9ZN. Notice the 1-R and 2-M on the table and a picture of a 2-R on the wall, just below the clock (*QST*, February 1923).

Zenith equipment used by the MacMillan Arctic expedition. Notice what looks like a 1-R and 2-M in a single cabinet, at bottom right (*QST*, July 1923).

Arctic explorer Commander MacMillan listening to a 1-R (*QST*, July 1923).

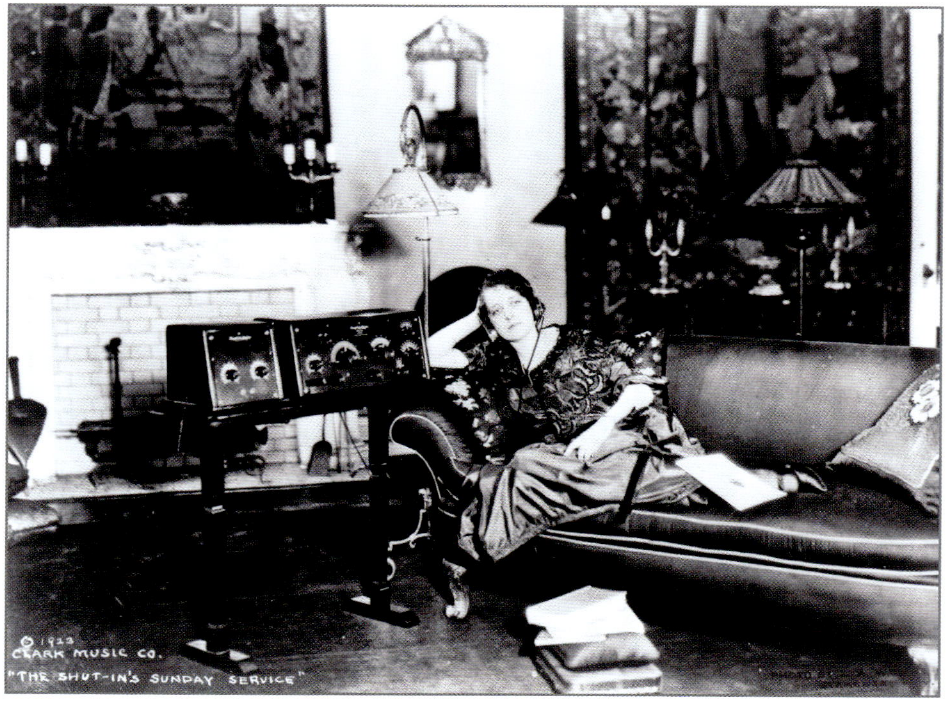

Unidentified lady listening to a 1-R and 2-M radio. Probably a publicity photograph (Clark Music Co., 1923).

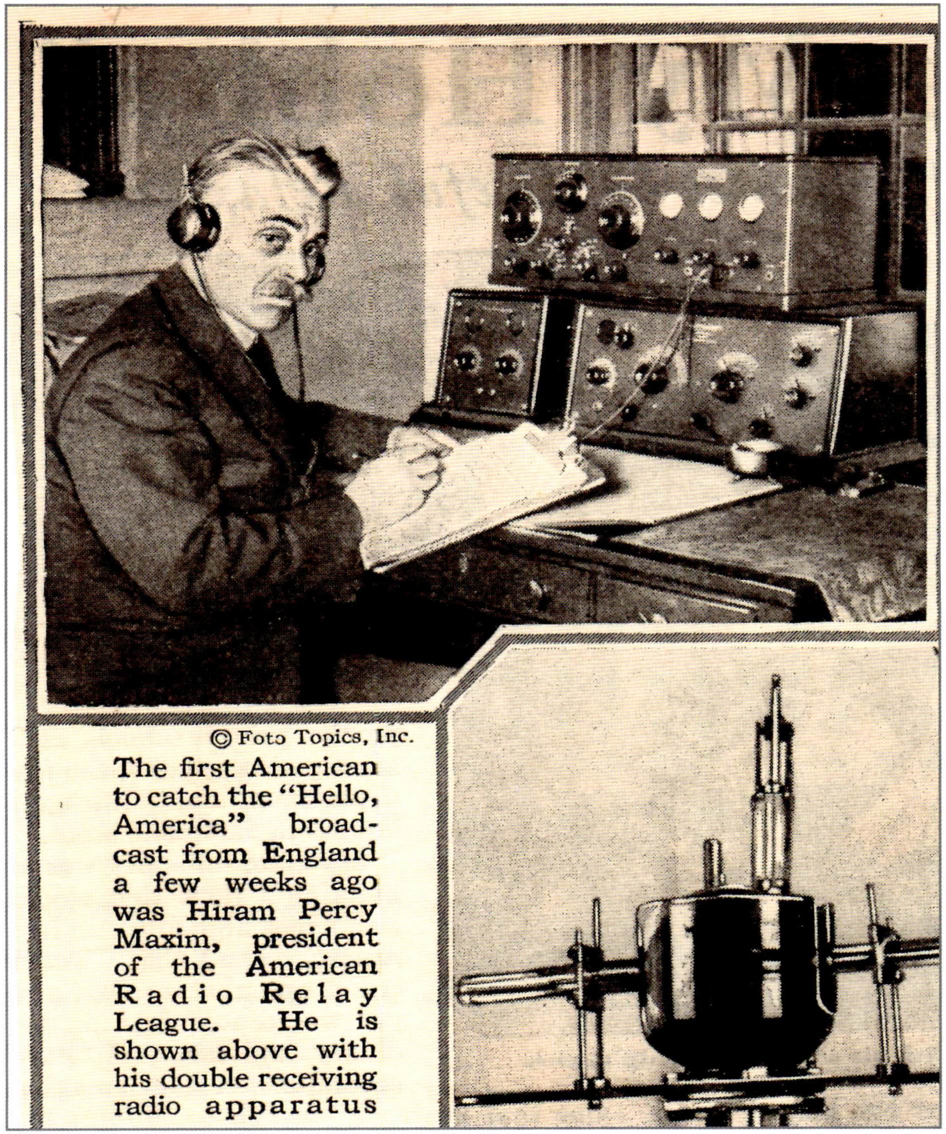

Hiram Percy Maxim, president of the ARRL, with a 1-R and 2-M (*Popular Science*, April 1924).

1-R Advertisement. Notice that the radio shown is a 2-R (*QST*, November 1922).

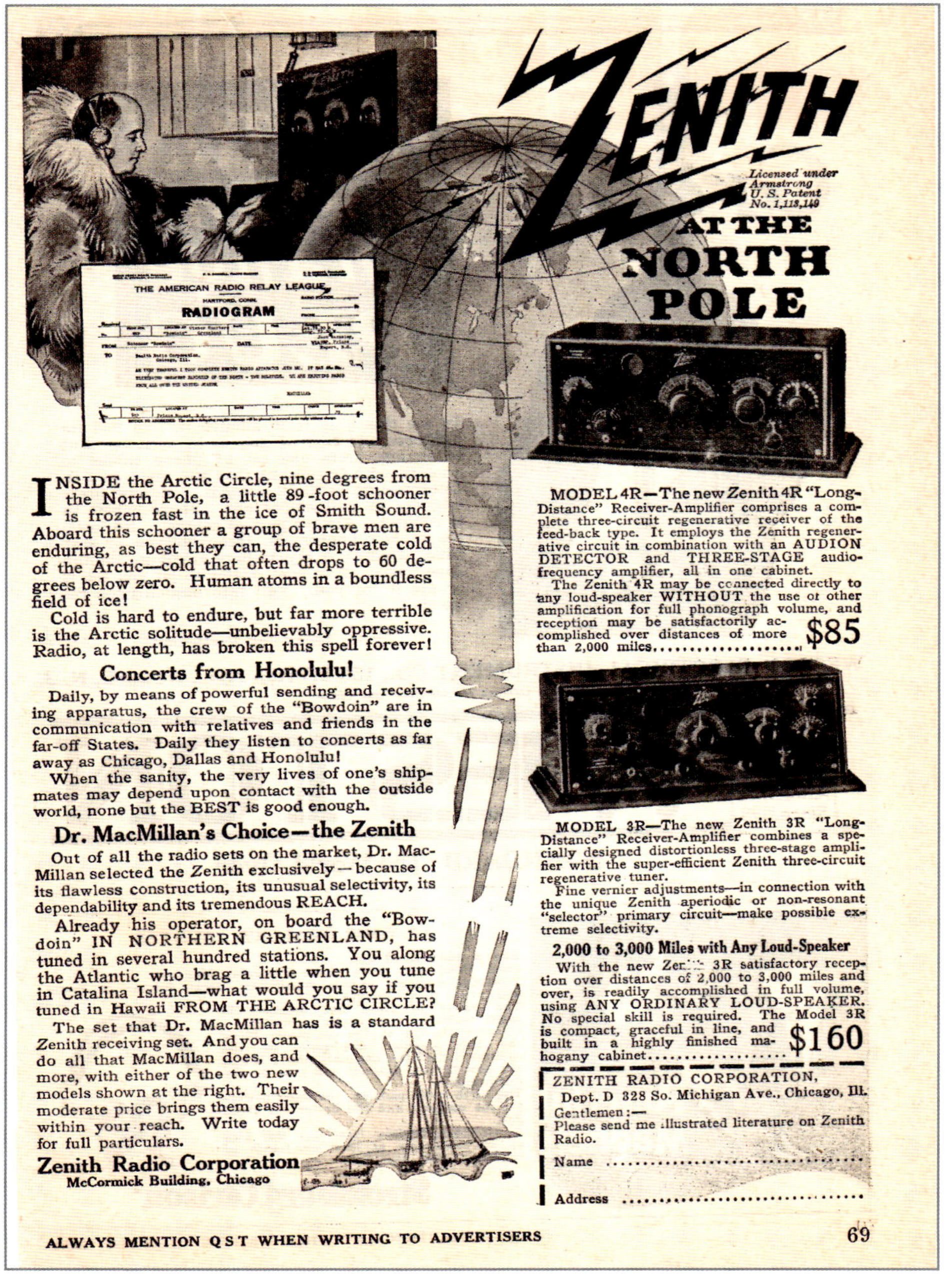

3-R and 4-R Advertisement (*QST*, January 1924)

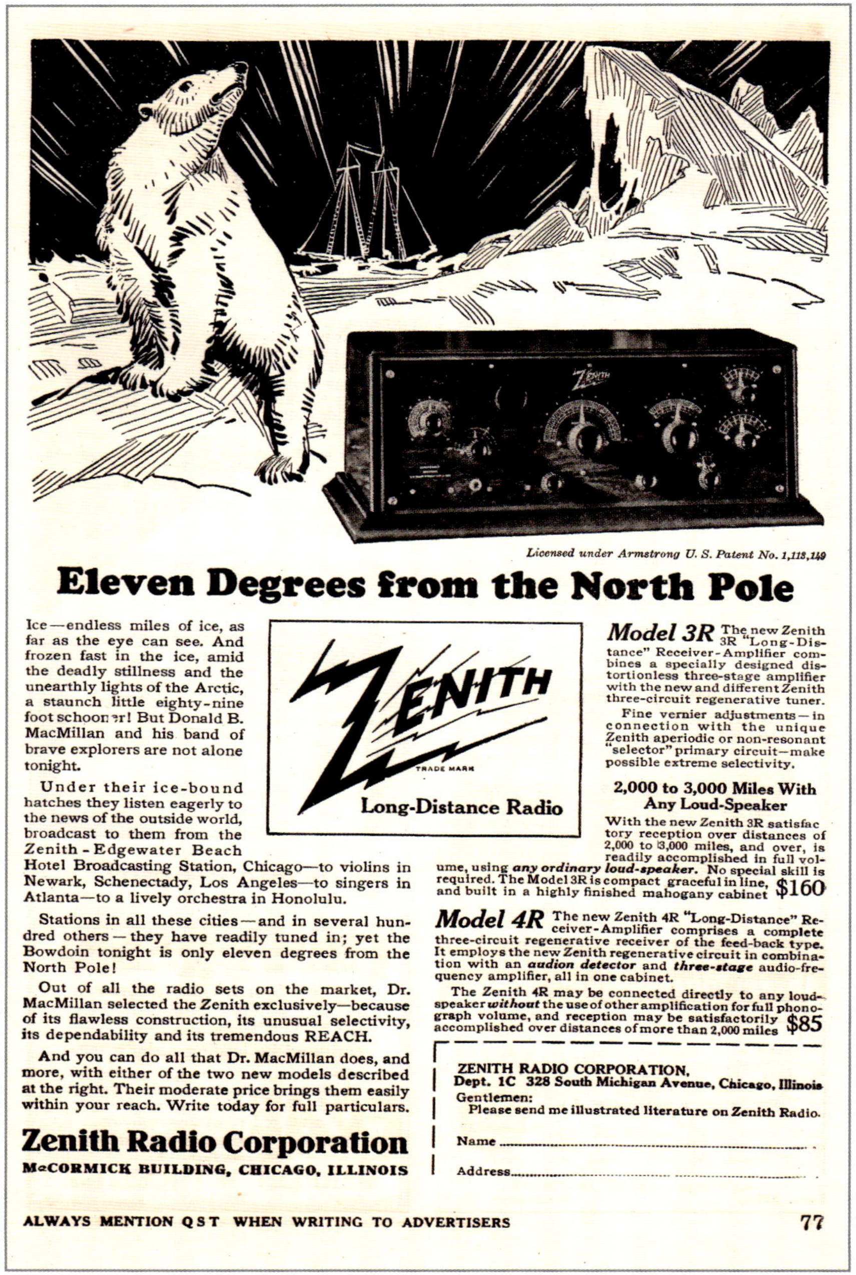

3-R and 4-R Advertisement (*QST*, March 1924)

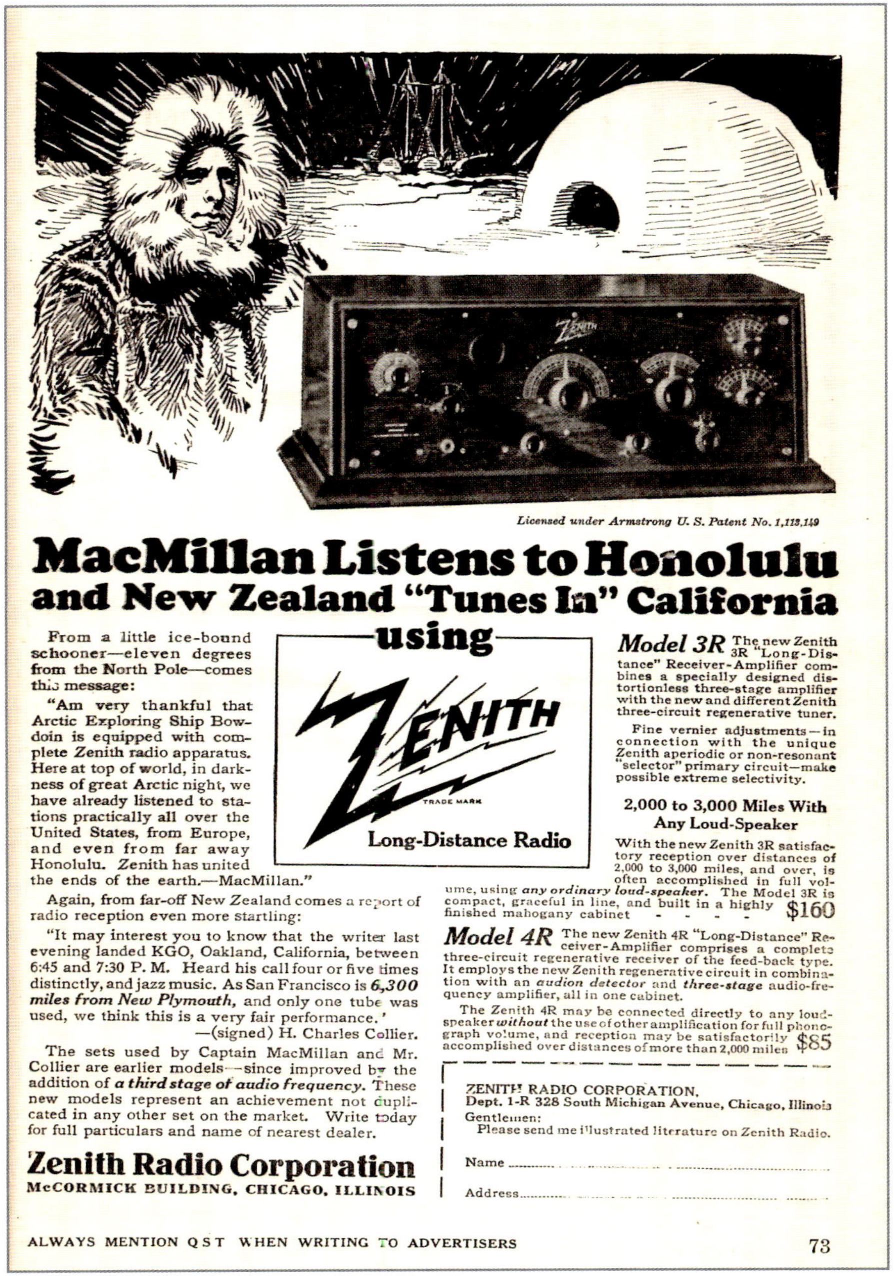

Licensed under Armstrong U. S. Patent No. 1,113,149

MacMillan Listens to Honolulu and New Zealand "Tunes In" California

using

ZENITH
TRADE MARK
Long-Distance Radio

From a little ice-bound schooner—eleven degrees from the North Pole—comes this message:

"Am very thankful that Arctic Exploring Ship Bowdoin is equipped with complete Zenith radio apparatus. Here at top of world, in darkness of great Arctic night, we have already listened to stations practically all over the United States, from Europe, and even from far away Honolulu. Zenith has united the ends of the earth.—MacMillan."

Again, from far-off New Zealand comes a report of radio reception even more startling:

"It may interest you to know that the writer last evening landed KGO, Oakland, California, between 6:45 and 7:30 P. M. Heard his call four or five times distinctly, and jazz music. As San Francisco is ***6,300 miles from New Plymouth,*** and only one tube was used, we think this is a very fair performance.'
—(signed) H. Charles Collier.

The sets used by Captain MacMillan and Mr. Collier are earlier models—since improved by the addition of ***a third stage of audio frequency.*** These new models represent an achievement not duplicated in any other set on the market. Write today for full particulars and name of nearest dealer.

Zenith Radio Corporation
McCORMICK BUILDING, CHICAGO, ILLINOIS

Model 3R The new Zenith 3R "Long-Distance" Receiver-Amplifier combines a specially designed distortionless three-stage amplifier with the new and different Zenith three-circuit regenerative tuner.

Fine vernier adjustments—in connection with the unique Zenith aperiodic or non-resonant "selector" primary circuit—make possible extreme selectivity.

2,000 to 3,000 Miles With Any Loud-Speaker

With the new Zenith 3R satisfactory reception over distances of 2,000 to 3,000 miles, and over, is often accomplished in full volume, using ***any ordinary loud-speaker.*** The Model 3R is compact, graceful in line, and built in a highly finished mahogany cabinet - - - - $160

Model 4R The new Zenith 4R "Long-Distance" Receiver-Amplifier comprises a complete three-circuit regenerative receiver of the feed-back type. It employs the new Zenith regenerative circuit in combination with an ***audion detector*** and ***three-stage*** audio-frequency amplifier, all in one cabinet.

The Zenith 4R may be connected directly to any loud-speaker ***without*** the use of other amplification for full phonograph volume, and reception may be satisfactorily accomplished over distances of more than 2,000 miles $85

ZENITH RADIO CORPORATION,
Dept. 1-R 328 South Michigan Avenue, Chicago, Illinois
Gentlemen:
Please send me illustrated literature on Zenith Radio.

Name ....

Address ....

ALWAYS MENTION QST WHEN WRITING TO ADVERTISERS 73

3-R and 4-R Advertisement (*QST*, June 1924)

Zenith brochure, circa 1924, 9.25" x 6".

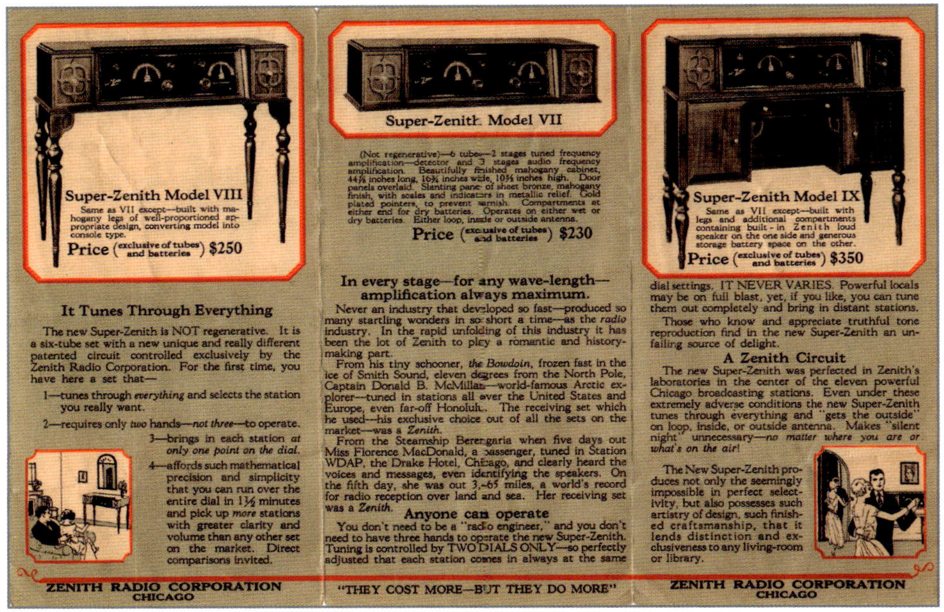

Reverse Side. Reprinted with permission of Zenith Electronics LLC.

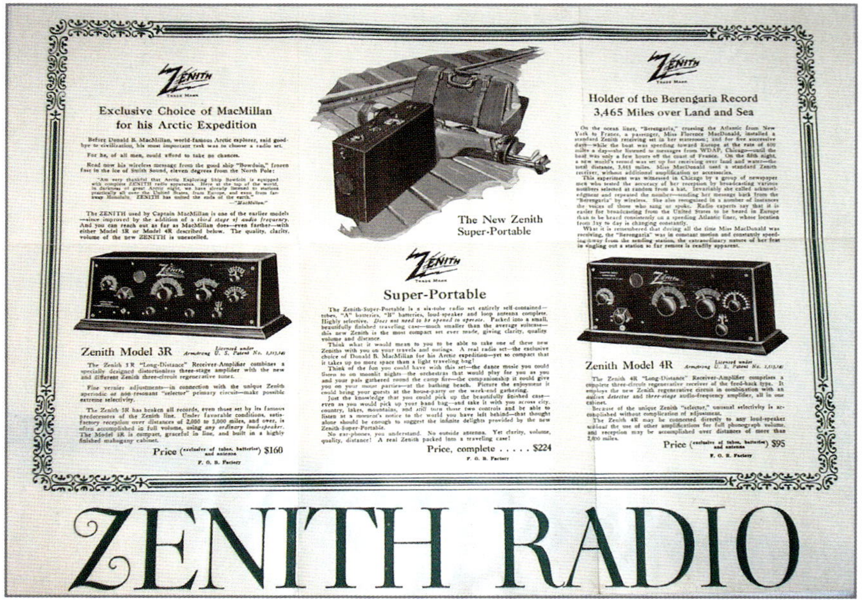

Zenith brochure, circa 1924, 14.5" x 11".

Reverse Side. Reprinted with permission of Zenith Electronics LLC.

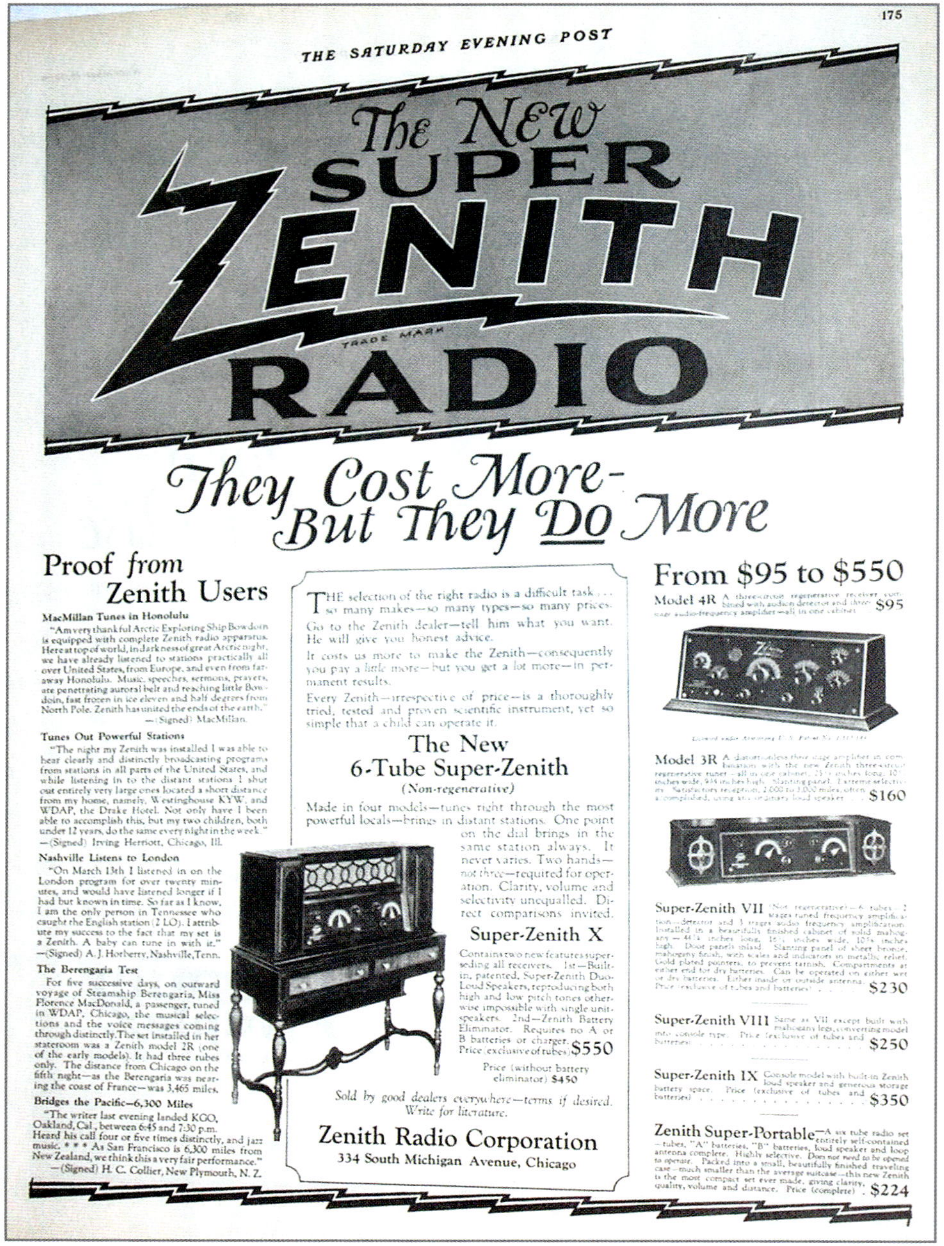

Notice the Zenith Super-Portable, bottom right. By the time this full-page advertisement was distributed, Zenith had already ceased production of the Super-Portable radio, due to performance issues (*Saturday Evening Post,* November 15, 1924, pg. 175).

Zenith 11 Advertisement (*Radio News,* January 1928)

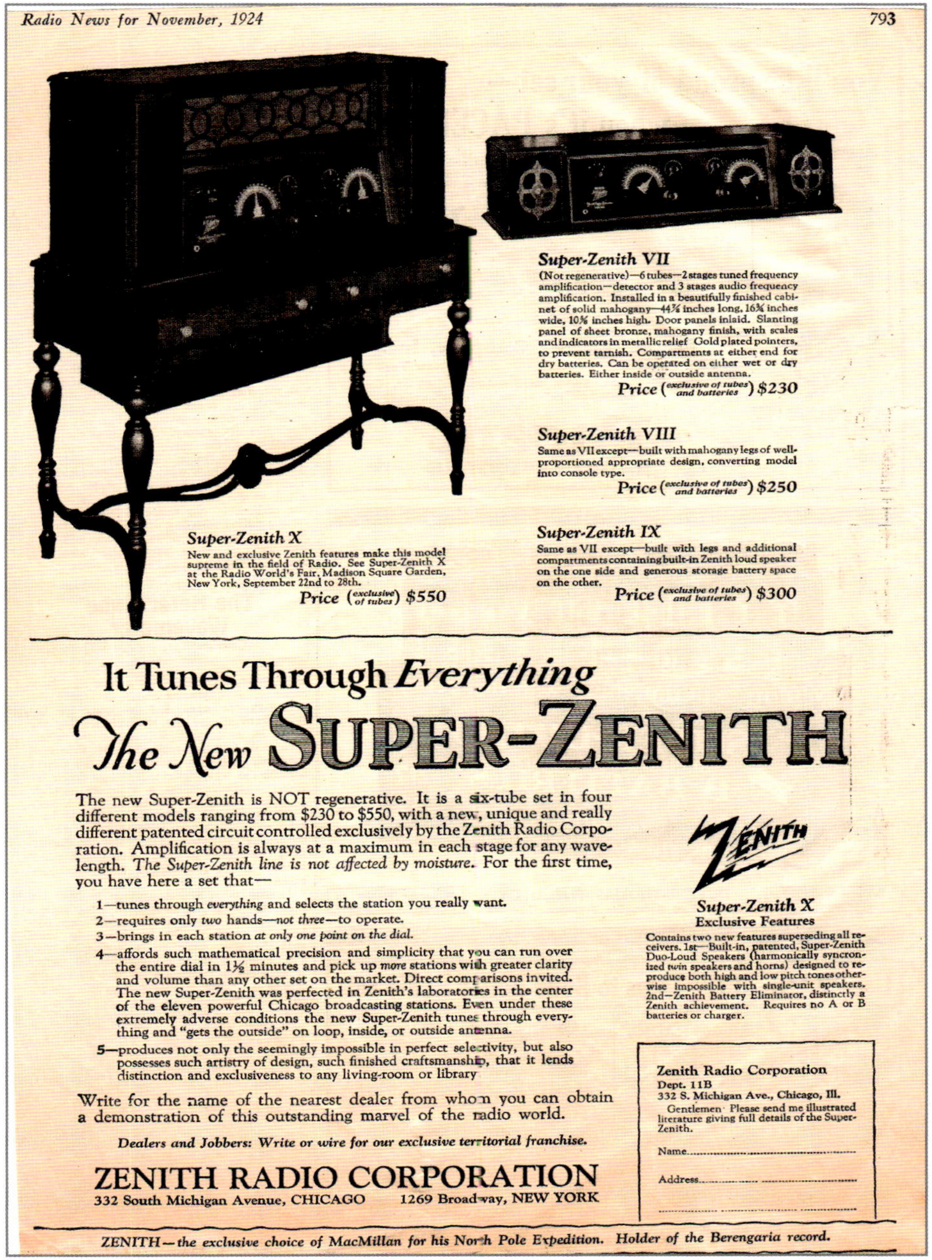

Super-Zenith Advertisement (*Radio News,* November 1924)

Zenith brochure, circa 1924, 13" x 5.5".

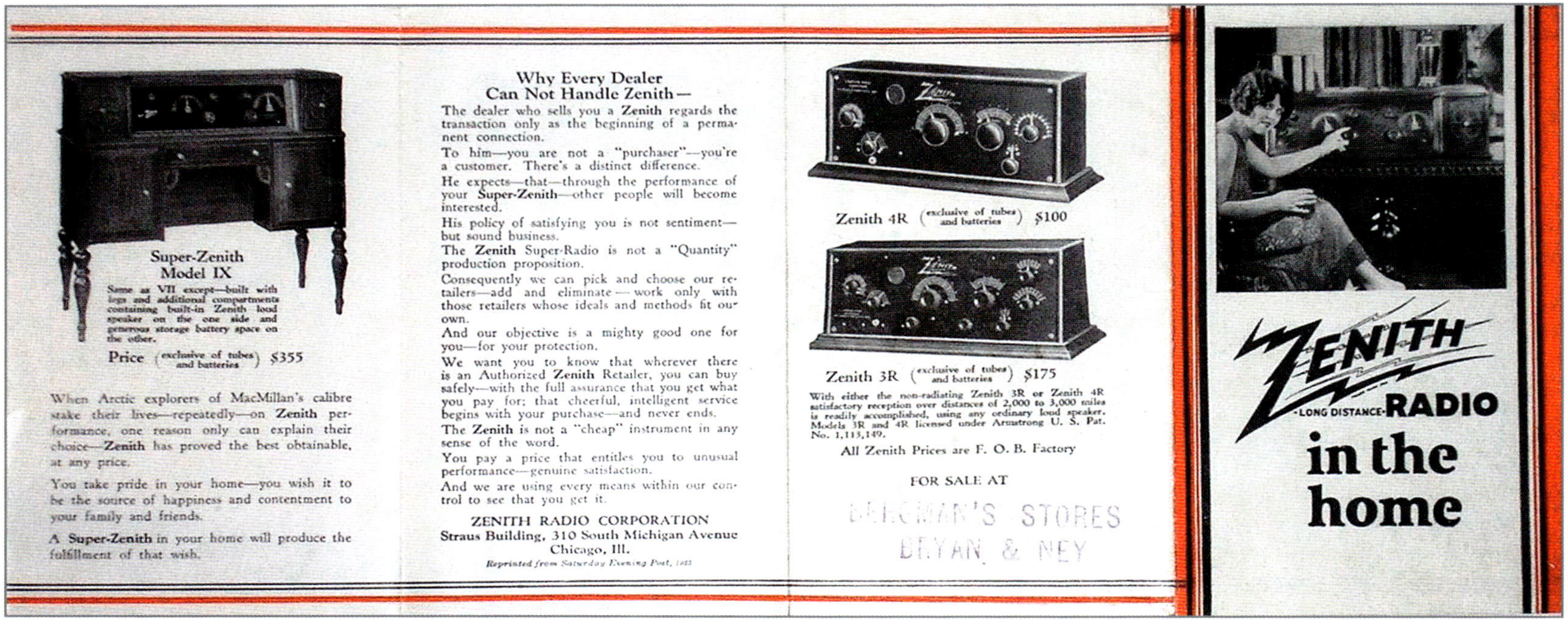

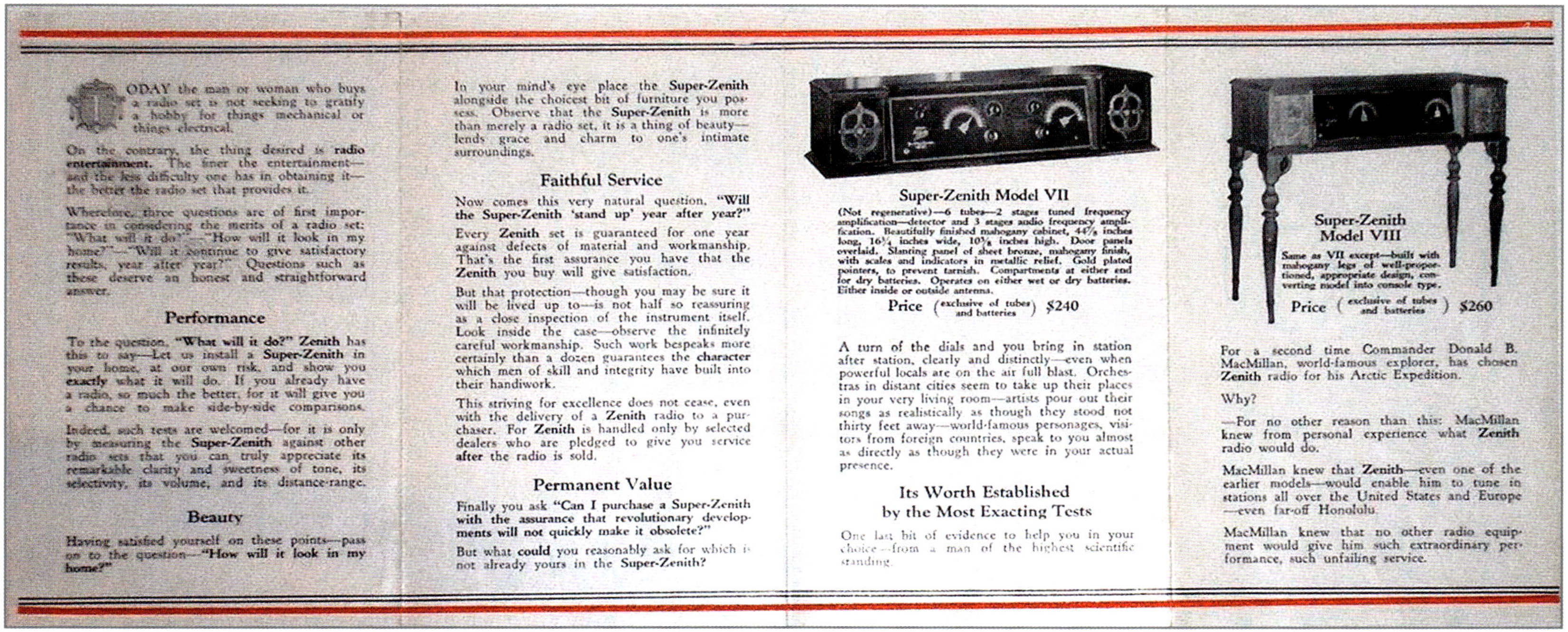

Reverse Side. Reprinted with permission of Zenith Electronics LLC.

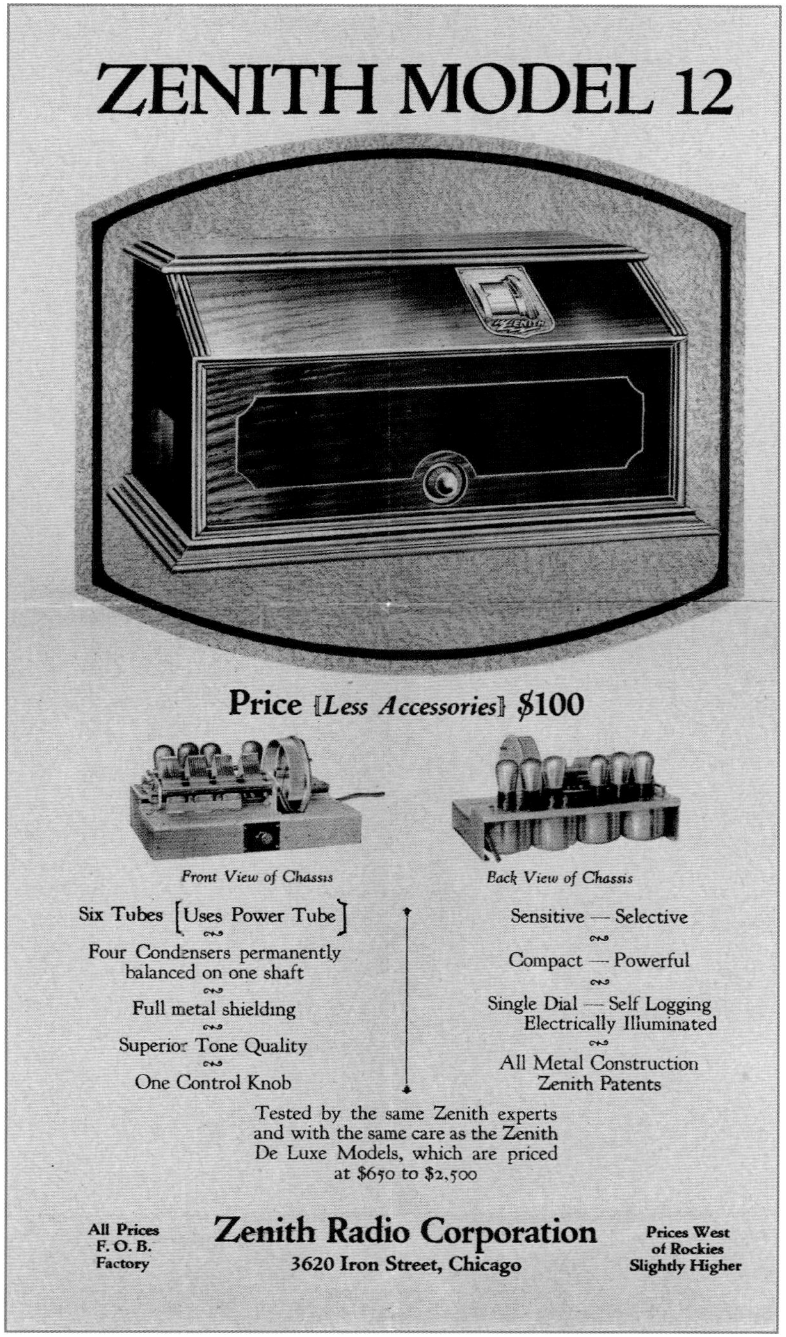

Zenith 12 brochure, circa 1927, 11.25" x 6.5".

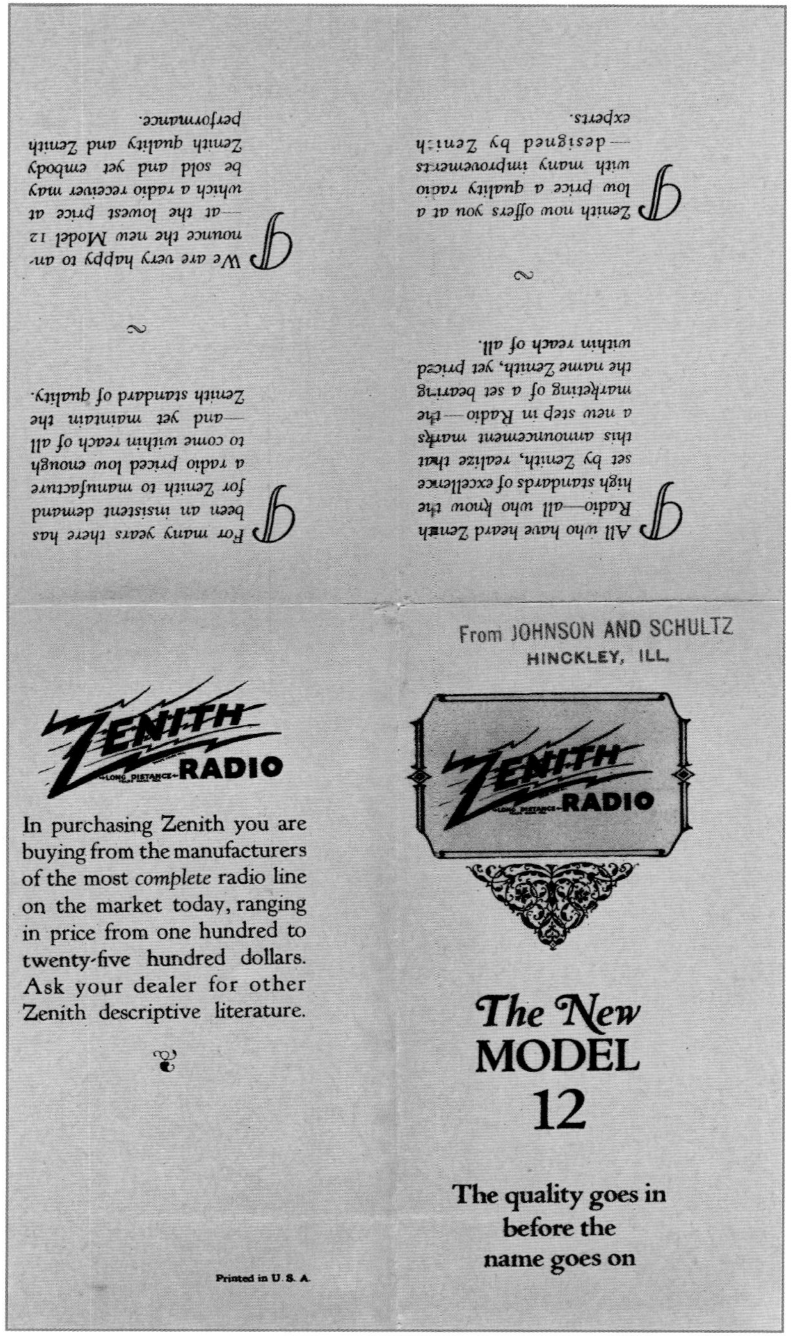

Right: Reverse Side. Reprinted with permission of Zenith Electronics LLC.

Zenith 1927–1928 Salesman's Handbook

(Reprinted with permission of Zenith Electronics LLC)

1927-1928

Salesman's Handbook

ZENITH
TRADE MARK REG.
→LONG DISTANCE← RADIO
TRADE MARK REG.

THIS HANDBOOK IS NOT FOR GENERAL DISTRIBUTION—It is for the exclusive use of salesmen selling Zenith Radio. Its purpose is to show salesmen how to present Zenith Radio intelligently to intending purchasers. A study of that which is inside these covers will make it possible for a salesman to make more money selling Zenith. This handbook should be placed in a substantial binder by the salesman. The edition is limited.

NOTICE TO DEALERS

Insist upon your salesmen studying this handbook. It will make better and more successful salesmen.

ZENITH RADIO CORPORATION

3620 IRON STREET CHICAGO, U. S. A.

INTRODUCTION

THIS booklet is designed to help you sell Zenith radio. If you will study the contents carefully, review it from time to time, it will be a big source of inspiration, will fortify you for your work, and will help you more than anything else to sell Zenith sets.

Salesmen like to sell Zenith Radio because it literally lives up to the Zenith slogan, "It costs more, but it does more." There is a certain thrill that goes with the selling of a product which is in a class by itself. Zenith salesmen can throw themselves heart and soul into the work of drawing public attention to the exceptional merits of Zenith radio.

No other set is wrought like it. All through its various stages of making, from the selection of the raw materials, down to the finishing touches on the cabinet, there is the one big objective of high quality regardless of cost. The remarkable tone of Zenith is the answer to all the painstaking care in its manufacture and reflects the great wealth of experience and the successful record of the builder whose achievements in the field of radio go back to the very beginning of the industry.

This booklet tells the history of Zenith radio; where it is made; how it is made; and the reasons for its excellence. It also describes the different styles. So that the reader does not need to rely entirely upon his imagination, the narrative is sprinkled with a profusion of illustrations. And all of this is in a compact and usable form. It is information that Zenith salesmen want to know. It is information that every purchaser of Zenith should know.

The truth needs no coloring to impress the reader or hearer of the fact that Zenith radio is made under exceptional auspices and that there is every good cause for the remarkable Zenith success.

ZENITH HISTORY

As far back as 1915, the first work was done upon Zenith Radio. The two men who started in business in those early days are still active in the Zenith organization—one is our chief engineer. From a meagre beginning, the business has developed under the influence of high ideals and principles, so that today it stands without a peer in the high quality radio field.

From the very beginning, Zenith radio became identified in the public mind with quality of the highest order. Nothing has been spared in energy, time, and expense, to make the Zenith set as good as could be. Many of the first sets are still in use and the owners are reluctant to part with them. The same quality that makes Zenith radio stand out today, characterized Zenith radio sets of yore.

Its outstanding quality has always put Zenith in the way of unusual opportunities. One of the early successes directly due to its fine reputation, and which gave the name Zenith much additional renown, was the famous Berengaria radio test. That put Zenith Radio in the public eye in an especial manner. There was something Lindberghian about the episode, for it was the first demonstration of how a fast moving ship bound for France picked up broadcasting in Chicago for four days hand-running. A Zenith radio set on board the Berengaria turned the trick. Newspaper men, ship officials, and a large number of people having no connection with the Zenith organization, conducted the test.

In the field of broadcasting the Zenith Radio Corporation won considerable fame. Our station contributed much to the stirring history of broadcasting in its early days. There has been no more enterprising and certainly no more successful station than W.J.A.Z. Its popularity is attested to in part by the receipt on one occasion of over 150,000 letters from the radio audience. These included responses from every state in the union, from Canada, Mexico, many countries of South America, and from a large number of ships at sea. Sixteen girls were engaged to open and answer that mail. Photographs of that huge cargo of mail perpetu[illegible]

2

On another occasion over 50,000 paid telegrams sent by appreciative listeners evidenced again the dominant position of W.J.A.Z. No other station received such great public approval.

The name Zenith is inseparably connected with the advent of radio into the North Pole region. Commander Donald B. MacMillan selected Zenith for contact with the rest of the world while he was hemmed in by snow and ice fields during his long arctic sojourn. The story of how he received news almost daily; how, from time to time, he heard the voices of his relatives addressing him over the radio; how the terrible loneliness of North Pole life was finally conquered, adds a brilliant page to the romance of radio in its first development. Zenith radio proved equal to every call upon it by Commander MacMillan, and as a result it has been an indispensable part of his equipment on every North Pole journey since. Millions of radio listeners pleasantly remember the programs given at W.J.A.Z. in MacMillan's honor every Wednesday evening. Only MacMillan and his crew know how much those programs helped to brighten the long arctic nights.

Of course, the Berengaria and MacMillan occasions could not have been except for the exceedingly high quality of Zenith radio. That quality has stood every test so far. That high quality is responsible for the Zenith Radio Corporation enjoying the distinction of building more high grade radio sets than any other manufacturer.

The Zenith Radio Corporation attracted public attention far and wide when it entered the courts in the year 1926 to prove that the old, antiquated law of 1912 was out of date and valueless, in so far as present day broadcasting is known. Something had to be done to force Congress to pass new legislation which would control the growing interference between stations. Zenith won its case with the government and immediately following the decision of the Attorney General that the old 1912 law was ineffective, Congress bestirred itself and prepared to pass the type of legislation which was so much desired. Officials of the Zenith Radio Corporation took a leading part in helping to frame and pass the new law. If it had not been for the action of the Zenith Radio Corporation in proving that the old law was ineffective, the United States would not have the Federal Radio Commission today, nor the law backing up the Commission.

The Zenith Radio Corporation has pledged itself in writing to do everything within its power to see that the constitutionality of the new radio law is preserved and to use its best influences to prevent anyone from attempting to break down the new law. Furthermore, it is on record as intending to cooperate with other interests in seeing that any defects in the new law are cured at an early date thru amendments to be introduced in the next session of Congress.

ZENITH POLICY

Enduring business is the superstructure reared upon the groundwork of sound policies. Zenith policies toward the distributor, dealer, and consumer, are in harmony with the most advanced theories and principles. The more important are worthy of mention:

1. **We franchise Zenith distributors and dealers.** One of our fundamental policies provides for the sale of Zenith by a limited, yet carefully selected number of wholesale distributors and dealers. The Zenith franchise by its very nature can appeal only to high class merchants. The highest grade radio set should be handled only by distributors and dealers who can sell radio in a high class way. We consider this phase of merchandising so important that our distributors and dealers are all under contract. We originated this system in radio.

2. **We do not believe in making cheap radios.** We do not believe that it is a sound policy to build a radio set to retail at less than $100. It has yet to be proven that the dealer can make a real profit on a radio set selling under $100. There is as much time consumed in the sale of the $50.00 set, as one twice or thrice that price, but it is an unsatisfactory transaction, from the point of view of the dealer and the consumer. The dealer's profit is so small as not to justify the selling effort and the customer has a set that cannot be a source of genuine satisfaction. A transaction involving the sale of a radio set at $100 or more is much better for all concerned. It provides a living profit for the seller and a worth while radio set for the buyer.

3. **We believe in a limited output of Zenith sets.** The demand for Zenith has always been considerably greater than our ability to care for, but we would rather have an under- than an over-supply. None of our distributors and dealers are confronted with the necessity of selling Zenith sets at a loss. Zenith customers do not therefore find that their radio sets are suddenly shrunk in value because of an advertised bargain sale. Zenith values have always been maintained at highest level, principally for the reason that there have never been enough Zeniths made to keep step with the demand.

4. **New Models.** Refusing to build more radio sets than the market can naturally consume has a still greater advantage than explained above. Under our policy, when the radio season ends, it is possible immediately to bring out new models and refinements representing the latest developments in the art. If Zenith radio sets were over-produced, we would, like so many manufacturers, be prevented from marketing improved constructions until the old stock had been worked off. In connection with our alert laboratory, this manufacturing policy explains why Zenith has always led the other manufacturers of radio.

ZENITH PATENT ALLIANCES

On March 11th, 1927, we became the first licensee of the Radio Corporation of America, the American Telephone & Telegraph Company, the General Electric Company, and the Westinghouse Company, under more than one-hundred radio patents. This alliance was welcomed on both sides, for it made common cause between two of the most outstanding pioneers in the radio field—ourselves and the above mentioned associated companies.

This patent license is of especial interest to distributors, dealers, and consumers, because it enables us to employ in the making of Zenith Radio, the best known devices in the radio art. The Zenith laboratory is thereby unhampered in its efforts to give the public the result of its exceptional experience and ability.

Except for this license, no radio manufacturer can build a radio set without infringing on one or more of the patents contemplated by the above alliance. Infringement is subject to severe penalty and in many cases that penalty is tantamount to being put out of business. In our opinion, it is not safe for any distributor or dealer to sell the radio sets of a manufacturer who is not licensed under the fundamental patents owned and controlled by the Radio Corporation of America, and its associates.

There are only a limited number of licensees. The policy of the holders of these patents is to license only substantial manufacturers. It is therefore a high compliment to the Zenith Radio Corporation to be selected as the first licensee. Thus its outstanding position in the radio field, and the great strength of its financial status, were duly acknowledged.

Through foreign representatives and advertising in foreign countries, we are in constant touch with European and other foreign inventors and manufacturers. Late developments in the radio art in any country are sooner or later made available to us.

ZENITH PERMANENCY

Of great interest to the wholesale distributor, dealer, and consumer, is the permanency of the maker of a radio set. As in the case of automobiles, there is nothing so dead as the set made by a defunct radio manufacturer. Why, therefore, should a wholesale distributor or dealer build up sales for a product, the future of which is in doubt?

Zenith Radio Corporation enjoys all of the attributes of permanency. It is one of the financially strong institutions of the industry. Its management is conservative and its policies constructive and far sighted. As this paragraph is written, the Zenith Radio Corporation owes less than $10,000. It has no preferred stock, bonds, mortgages, liens, bank loans, private loans, or borrowings of any kind. Its assets are well over a million dollars.

Naturally suppliers of materials and parts are anxious to have their products incorporated in Zenith Radio Sets, because of our preëminent position in the industry. They recognize that considerable prestige falls to the lot of the seller of materials or parts of Zenith Radio. In consequence we are solicited by the best producers. This coupled with the fact that we pay spot cash for our purchases, gives us the choice of the world's markets.

We have no desire to do the largest radio business. Our sole idea is to keep the Zenith banner at the head of the radio manufacturers' procession, as it has been from the very beginning. Quality and not quantity is our watchword. There is no greater factor in causing substantial growth than a good word carried from a satisfied owner of a Zenith set to a prospective buyer. Every Zenith set is a standing advertisement for us. Zenith owners are enthusiastic advocates of Zenith supremacy. Zenith wholesale distributors and dealers derive much gratification in the sale of Zenith sets because of the permanency of Zenith.

ZENITH FIRST

Leadership is not attained and sustained excepting through achievement. Here are a few of the many developments in which Zenith has been first. Some of these developments are now common property — some are not. They are recounted here merely to give the salesman a better idea of Zenith progressiveness.

First: Slanting panel.
First: 2 variable condensers on one knob.
First: All metal chassis.
First: 3 variable condensers using only 2 knobs.
First: Three stages audio amplification.
First: Workable electric set.
First: Artistic appearance in 1921 (Models 1R and 2M).
First: Multiple loud speakers in one set.
First: 10-tube set commercially.
First: Illuminated dials.
First: Licensee under fundamental patents.
First: Short wave commercial receivers.
First: Radio manufacturer licensed by A. T. & T. for transmitter.
First: Railroad set.
First: Independent manufacturer to build broadcast transmitter.
First: Receivers with reception range from 150 to 550 meters.
First: Three different circuits produced at one time.
First: To adopt present standard line of discounts.
First: No lost motion vernier.
First: Continuous volume control from zero to full volume.
First: Portable broadcasting station.
First: Low loss construction.
First: Common "B" battery for all tubes (1918).
First: 6 gang variable condensers on one shaft.
First: Licensee to use R. C. A. AC tubes.
First: DeLuxe period cabinets.
First: Receiver working within 11 degrees of the North Pole (MacMillan).
First: Complete line from $100 to $2500.
First: Exclusive contracts for wholesale distributors and dealers.

5

ZENITH ACTIVITY EXTENDS OUTSIDE HOME RECEIVERS

Zenith salesmen will do well to mention this episode as a further exemplification of Zenith prestige.

One of the successes of which we feel especially proud, is the first development of radio telephony on freight trains. The first test made of two-way telephone communication, as an aid to operation of freight trains, was conducted by the New York Central Railroad in cooperation with us in the summer of 1926. Those tests ushered in a new era in freight train operation and demonstrated clearly that successful radio telephone service can be maintained between the conductor in the rear, and the engineer in the front, of freight trains regardless of length. Formerly the only means of communication between the front and rear of trains, was by hand, lamp, or whistle signals.

The radio equipment placed on both the engine and the caboose consisted of special receiving-transmitting sets made by us.

ZENITH DEALER FRANCHISES

The value of any radio franchise is measured by the opportunity it affords the dealer to make money. The dealer cannot make a profit handling a cheap radio, because it costs him as much to sell that kind, as it does a higher priced one. Usually the cheap radio set requires a lot of servicing. The radio dealer cannot make a profit handling a line of radio that is sold by too many dealers in a territory. He cannot make a profit when every now and then he is face to face with cut-price competition on his own line. He cannot make a profit when he is forced at the hands of either the manufacturer or the wholesale distributor to buy more merchandise than he can conveniently sell, for that means eventually closing out a lot of stock at ruinous prices. He cannot make a profit when his line of radio is not well advertised or well known, because the cost of selling in such case is prohibitive. He cannot make a profit when the radio sets, because of their indifferent or inferior quality, do not help him to sell more sets.

The Zenith dealer franchise presents the bright side of the above picture in affording excellent money-making opportunities, for the following reasons:

1. Only a limited number of Zenith dealers are appointed in every section.

2. Zenith dealers are carefully selected. Our desire is to have only good business men of sound reputation and of ample facilities, selling Zenith sets.

3. Zenith radio sets are so well known as not to require an expensive selling campaign on the part of the dealer.

4. The high quality of Zenith makes it easy to sell against competition on the basis of actual performance. Zenith dealers welcome the opportunity of a side by side competitive test.

5. The prices of Zenith radio sets insure a living profit.

6. Zenith sets make friends so readily that every sale is quickly followed by an endless chain of others.

7. The call upon the dealer to service Zenith sets is comparatively small.

8. Zenith sets are always as good as dry wheat in the bin, and do not need to be sacrificed by the dealer at special prices.

9. The Zenith license puts the Zenith dealer in the position of perfect freedom from liability of suit because of any patent infringement on our part.

10. The Zenith dealer has the pleasure of knowing that the franchise will be increasingly valuable from year to year.

11. The Zenith dealer needs entertain no doubt about the permanency of the Zenith line.

6

ZENITH SLOGANS

1. **"Costs More—Does More"**

   A true description of our products. We don't make the cheapest radio. We could, but we don't. Our lowest price is $100. In our opinion, no radio set can be listed at a lower price and yet give the owner the quality of reception which he has a right to expect.

2. **"The Quality goes in before the Name goes on"**

   A sermon in a line. Zenith has stood for quality in radio since its start. Every new development coming into our hands is carefully studied from the standpoint of quality before serious consideration is given to its adoption. This slogan is the keynote of our laboratory and production departments.

3. **"There is Profit in Zenith"**

   The foundation of our success. Ask any Zenith wholesale distributor or dealer about Zenith profit. Their answer will be "Zenith is the most dependable profit-maker in my line." Zenith is not hawked around. It is sought after by leading distributors and dealers everywhere. The reason for this popularity is shown in the title.

4. **"Known the World Over"**

   Zenith is internationally advertised and internationally known. Its reputation for quality and dependability goes far beyond the United States. The buyers in foreign countries with their natural leaning toward fine, careful workmanship, and precision manufacture, are among the most enthusiastic users of Zenith.

THE ZENITH 1927-1928 LINE

Never before have we presented such a complete line for our trade and never before have our competitors been faced with such formidable competition.

Our retail list prices run from $100 to $2500. Included are battery sets; batteryless sets; floor models; table models; antennae sets; loop sets; sets which operate without either loop or antennae; three different circuits; 6-tube sets; 8-tube sets; 10-tube sets and hand-made period furniture models of rare design.

At the Chicago Trade Show held in the Stevens Hotel the week of June 12, 1927, Zenith was the central exhibit. No other radio line presented equal sales possibilities. No other radio line was as much sought after. There is hardly a distributing territory which did not have several strong wholesale distributors vieing with each other for the Zenith line. Never has Zenith been in such a strong position and never before has a radio manufacturer had the opportunity to pick and choose his representatives.

7

ZENITH HOLDS WORLD'S RECORD

(Printed through the Courtesy of the New York Telegram)

Saturday, June 18, 1927. THE NEW YORK TELEGRAM RADIO 3

World's 'DX' Record for Hoboken Listener

Gets JOAK, Japan, 7,270 Air Miles Distant

By STUART ROGERS.

HAIL the Lindbergh of the radio—Arthur Wald!

Wald on April 15 completed the longest non-stop radio "DX" flight in the officially verified records of broadcast reception—Hoboken, N. J., to Tokyo, Japan, 7,270 miles in 14 minutes and 37 seconds.

This feat, while by no means comparable or to be mentioned in the same breath with the courageous flight of the immortal airman, holds many of the same elements of adventure, drama and scientific progress.

Wald, first to span the continent and the Pacific Ocean in one leap, tuned "alone." This apparent flippancy on the writer's part—the very suggestion of a parallel in "loneliness" where there can never be a parallel for lone-handed courage and audacity—has its relative place in Wald's less heroic but none the less sensational accomplishment.

...was because Wald tuned "alone" that ...ared not tell a soul of his record-break... performance until official verification ... received from Japan.

Officially Proved.

...e treasured postcard has at last arrived ...now the story can be told.

...ere are still other minor similarities in ... feats of the immortal Lindbergh and the very fortunate Wald.

Wald hopped off on his record-breaking radio journey in almost the same nonchalant, carefree manner as did the young Lochinvar from the West when he soared away from Roosevelt Field. Like Lindbergh, his destination was unknown. He knew only that he was out to get a record if it took all morning.

So, alone in the stillness of early morning, without a witness, in a house made lonesome by the absence of wife and child, he turned on the switch that sent him on his way to worlds and stations unknown. The time was exactly 2:41 o'clock in the morning of April 15. A clear night and the complete absence of static provided ideal conditions for the jump.

The receiver, an undoctored stock model

"My wife and child had been away for two days, so as not to be home while the painting was being done, as we figured it would be uncomfortable for the baby. I live above my place of business and possess the keys to the premises, which are protected against burglary by an electric Central Office system.

"I retired about 10:30 o'clock that evening. While still asleep some one came and knocked at my door. Inquiring who it was, I found that there were two men from the A. D. T. Company which protects our premises. It was 2:30 o'clock. The men stated that there was a break in the alarm, and as I possessed the keys I naturally had to re-open the store to allow them to inspect the premises. There happened to be a slight break in the wiring, so I was assured that everything was O. K.

Picks Up "Chinese."

"When I returned upstairs I went back to bed, but could not fall asleep again. I got up and started to fool with my receiving set, which is a six-tube Super-Zenith, model 7. The set had been repaired by my radio man here in Hoboken, from whom I

and in between them the announcer spoke in English and Japanese, each time giving the call letters and the location—Tokyo.

Listened for Forty Minutes.

"I held on to the station for exactly forty minutes without the slightest fading. While I was listening I carefully wrote down everything I heard and sent my report away, hoping against hope that it would be verified.

"I did not dare tell persons that I had received JOAK—Tokyo, Japan—for fear they would ridicule my story and accuse me of all sorts of trickery. Now I have the verification and I take great pride in telling every one about it.

"I am sending you my entire verification book of stamps, many more of which are due to arrive shortly, as very recently I have again tuned in Mexico City, Los Angeles and Vancouver, British Columbia.

"Other verifications in which I take great pride are HHK, Port au Prince, Hayti; PWX, Havana, Cuba; CZE, Mexico City; KPO, San Francisco; KFI and KHJ, Los Angeles; KGO, Oakland, Cal.; WFAA, Dallas, Texas, and KGW, Portland, Ore.

Post Card

STATION CALL

Power—1kw Antenna Input

"JOAK, Kochirawa Tokyo Chuwo Hoso-kyoku de arimasu" (JOAK, This is Tokyo Central Broadcasting Station)

W.L.—375 Fs.—800kc

Dear Sir:—

We wish to verify with pleasure your report of April 15, 1927 regarding your reception of our program dated April 14th...... It affords good reference to us in checking the extent of our service abroad.

Thanking you for the report, and trusting to hear from you again.

Very truly yours,

TOKYO CENTRAL BROADCASTING BUREAU

Mr. Arthur Wald
Pineles & Welger, Inc.
702 Washington St.
Hoboken, N.J.

U. S. A.

Arthur Wald,
Radio pilot on the Hoboken-Tokyo non-stop "DX" flight.

light than hearing a concert from a powerful station in the Middle West after midnight.

Other Stations "Logged."

Mr. Wald's complete log follows:—

WADC, Akron, Ohio; WCCO, Minneapolis, Minn.; WABC, Richmond Hill, N. Y.; WSB, Atlanta, Ga.; WBAI, Baltimore, Md.; WORD, WSWS, Batavia, Ill.; WEMC, Berrien Springs, Mich.; WEEI, Boston, Mass.;

FOR LONG DISTANCE RECEPTION

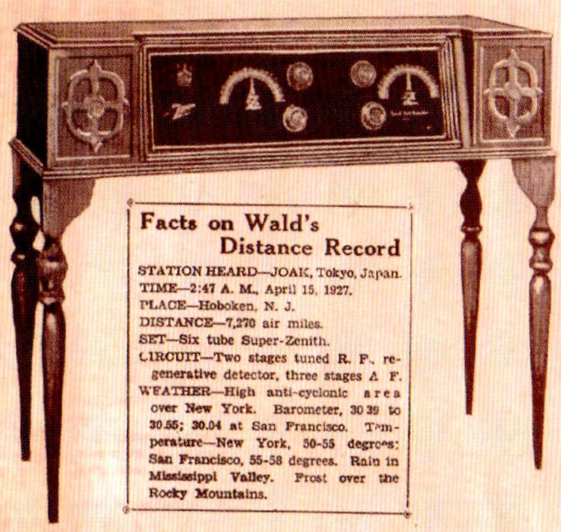

Facts on Wald's Distance Record

STATION HEARD—JOAK, Tokyo, Japan.
TIME—2:47 A. M., April 15, 1927.
PLACE—Hoboken, N. J.
DISTANCE—7,270 air miles.
SET—Six tube Super-Zenith.
CIRCUIT—Two stages tuned R. F., regenerative detector, three stages A. F.
WEATHER—High anti-cyclonic area over New York. Barometer, 30.39 to 30.55; 30.04 at San Francisco. Temperature—New York, 50-55 degrees; San Francisco, 55-58 degrees. Rain in Mississippi Valley. Frost over the Rocky Mountains.

Super-Zenith Six, was functioning perfectly, just having been overhauled by Anthony ("Doc") Izzo, the expert of the neighborhood.

Not a Special Set.

Matched tubes were employed in the radio frequency stages, but otherwise the tubes were ordinary.

A storage battery and two Layerbilt Eveready batteries supplied the motive power to the machine—the filament and plate current.

For an aerial Wald used a single wire spread 160 feet long, twenty-five feet above the roof of the building, which is two stories high. Ground connection was to a cold-water pipe under the sink, twenty feet away.

With these preparations, then, Wald was off.

The actual "flight" we are going to allow the radio "ace" to describe himself.

His Own Story.

"It was on Thursday evening, April 14, 1927, after finishing my day's duties, that I went upstairs to my home to open all of the windows for some fresh air, as our rooms had just been finished, cleaned and painted.

had bought the set only a few days previous, and it was functioning perfectly. Turning the dials, I suddenly heard a faint voice around 370 to 380 meters, as I figured it. This was about five minutes to three in the morning.

"The voice I heard talking sounded as though it were Chinese. As a matter of fact, I said, 'That's Chinese to me.' Nevertheless, I turned on all the volume and an announcer started talking in English. He announced that the call letters of the station were JOAK, but to me it sounded like a joke—JOKE—so I thought it might be a Western station broadcasting from a night club, and that one of the performers was imitating a Chinaman.

"Landed" in Tokyo.

"Following this, I heard the same unintelligible Oriental language. Then an organ solo was played, after which the announcer gave the call letters and the location. I began by that time to realize that it was not a dream.

"Really, I could not believe my hearing, and you can imagine the thrill I had for the moment.

"I listened to three more organ selections,

Extract reproduction of verification card received from Japan, confirming reception of station JOAK.
Left—The set which did the trick.

"I have had my set only since November 18, and actually sat up about fifteen or eighteen times to secure my selections of distant verifications.

Daylight Records.

"In addition to the foregoing list I have verification stamps for daylight reception of stations in Philadelphia, Wheeling, W. Va.; Wellesley Hills, Mass.; Hartford, Conn.; Schenectady, N. Y., and Pittsburg, Pa.

"Another noteworthy point about this reception is that the section where I live, on the main street in the very centre of congested Hoboken, is considered a dead spot, more or less. At frequent intervals I have tuned in and operated most all of the standard commercial receivers and some very good homemade circuits, but I must say that I consider my present set, a model 7 six-tube Super-Zenith superior to all of them. The set has two stages of tuned radio frequency, a regenerative detector and three stages of audio.

"Naturally, I have had many offers for the set, but it can't be bought. I'll keep it till it goes on the junk pile."

While, in a sense, Mr. Wald's reception was due to an accident, the reader should not get the impression that the DX record-holder is in any sense a novice.

He has owned and operated receiving sets for over three years and has learned all the "ins" and "outs" of his present set to a nicety.

Started with a "Neut."

His first radio set was one that he had built for him from a Fada kit, and with this he got the coast more than two years ago.

Although he has experimented with all sorts and shapes of antennae, he has found that he gets the best results from one about 160 feet long that extends in a northwesterly to southeasterly direction.

The house in which he lives is a two story brick building, located at 702 Washington street, Hoboken—not far from Stevens Point—a location where nearby residents experience great difficulty in getting DX over 500 miles.

Wald is a real fan who appreciates the relative values of out of town reception, getting more satisfaction out of logging a 100 watt station in Oil City, Pa., during day-

WNAC, Boston, Mass.; WSSH, Boston, Mass.; WJZ, Bound Brook, N. J.; WARS, Brooklyn, N. Y.; WBBC, WBKN, WBRS, Brooklyn, N. Y.; WPDQ, Buffalo, N. Y.; KYW, Chicago, Ill.; WBBM, WEBH, WENR, WHFC, WIBO, WLTS, WMAQ, WMBB, WMBI, WQJ, WSAX, WSBC, Chicago, Ill.; WKRC, WSAI, WLW, Cincinnati, Ohio; WTAM, Cleveland, Ohio; WPAP, Cliffside, N. J.; WQAO, Cliffside, N. J.; WAIU, WEAO, Columbus, Ohio; KOIL, Council Bluffs, Iowa; WRNY, Coytesville, N. J.; WOC, Davenport, Iowa; WHT, Deerfield, Ill.; WHO, Des Moines, Iowa; WMBC, WJR, Detroit, Mich.; KDKA, East Pittsburg, Pa.; WLIB, Chicago, Ill.; WOWO, Fort Wayne, Ind.; WOOD, Furnwood, Mich.; WASH, Grand Rapids, Mich.; PWX, Havana, Cuba; KFKX, Hastings, Neb.; WMCA, Hoboken, N. J.; WOK, Homewood, Ill.; KTHS, Hot Springs; WKBO, Jersey City, N. J.; WDAF, Kansas City, Mo.; KFI, KHJ, Los Angeles, Cal.; WMC, Memphis, Ill.; WIOD, WMBF, WQAM, Miami, Fla.; WJJD, Mooseheart, Ill.; WAAM, WDWM, WGCP, WNJ, WOR, Newark, N. J.; WEAF, WEBJ, WHAP, WHN, WLWL, WMSG, WNYC, WPCH, New York city; KGO, Oakland, Cal.; WLBW, Oil City, Pa.; CNRO, Ottawa, Canada; WODA, Paterson, N. J.; WFI, WIP, WLIT, WABQ, WRAX, Philadelphia, Pa.; KDKA, Pittsburg, Pa.; WCAE, Pittsburg, Pa.; WCSH, Portland, Me.; KGW, Portland, Ore.; WEAN, Providence, R. I.; WBBR, Rossville, N. Y.; KSD, St. Louis, Mo.; KPO, San Francisco, Cal.; KFOY, St. Paul, Minn.; WGY, Schenectady, N. Y.; WQAN, Scranton, Pa.; WBRL, Tilton, N. H.; CFCA, Toronto, Canada; WHK, Port au Prince, Hayti, West Indies; WOAX, Trenton, N. J.; WSBO, Wellesley Hills, Mass.; CNRV, Vancouver, B. C.; WJBV, Woodhaven, L. I.; WWRL, Woodside, L. I.; WCBD, Zion, Ill.; WWVA, Wheeling, W. Va.

Father Finn in Recital.

Father William J. Finn, director of the nationally known Paulist Choir, will be at the console of the Welte-Mignon organ during a program to be broadcast by the Red Network Sunday evening at 10 o'clock. Father Finn is a doctor of music from Notre Dame College and also wears the Palm of the French Academy, conferred by that body for his work in music. Pope Pius X. conferred upon him, after a concert at the Vatican, the title of "Magister Cantorum." WEAF will be the local broadcaster.

THE ZENITH FACTORY

The factory of the Zenith Radio Corporation is located in the heart of the famous Central Manufacturing District of Chicago. This immense factory houses the general offices of the company, the engineering and research laboratories, and all manufacturing departments. It produces the most complete line of radio apparatus in the world. It is a factory in the full sense of the word. It does not buy its major parts but manufactures them. Automatic machinery has been developed in this factory to maintain rigid standards and to keep the cost of production within reason. Every known labor saving device is employed. The system of testing is far beyond anything within our knowledge now in vogue in radio manufacturing. The checking and counter-checking makes it well nigh impossible for a defective Zenith to be shipped from the factory. It is a model radio plant fitted in every way to build the world's best radio set.

THE ZENITH LABORATORY

Since our radio products were first made Zenith has probably put more stress upon its laboratory than any other manufacturer. We are well aware of the tremendous cost of maintaining our laboratory, but we feel that this expenditure is wise because of the paramount importance of keeping Zenith in the forefront of the most modern development in the art.

Our laboratories are divided into four parts; the first consists of the research department. Here men eminently known in the radio art are reaching out into the future dealing with those little known and nebulous resources of nature which ultimately crystallize into concrete and definite ideas in the next department, which is known as the engineering department.

In the engineering department may be found men who are in the front rank of engineers and who apply themselves constantly to considering the all important task of keeping Zenith in the vanguard of the radio industry.

The next division is the model department. Here the work of the two former departments is put into practical operation. Experiments are conducted, data accumulated, and a vast store of information carefully collated and indexed, all of which forms an invaluable guide to check the work of the two former departments.

The fourth division is the drafting department. Here accepted constructions are carefully plotted, and blue prints are drawn upon which production is based.

The laboratory also has a staff of practical men, who under its direct supervision, are constantly circulating throughout the plant, examining the work coming through, and reporting any departure from the established standards. These are the liaison men between the laboratory and the factory. They do no inspection because that is done by a careful organization of skilled inspectors. Their sole purpose is to make sure that the production department adheres with great precision to the standards set up by the combined laboratories.

THE TRAFFIC DEPARTMENT

Facilities for delivering incoming and outgoing shipments are amply provided for on the first floor of the factory, where the Traffic Department holds forth. The traffic situation is really ideal. Freight tracks run into our yards and into the factory. Service is usually available on very short notice. Four freight cars can be loaded or unloaded simultaneously. The trucking service comprises means for loading or unloading many trucks at one time. These facilities include plenty of street room, loading platforms, and manual and mechanical help.

The Traffic Department studies the best routes for incoming raw materials and lays particular stress upon the best routes for shipping Zenith products to wholesale distributors in the most expeditious and economical manner. Located as the factory is, in the heart of the Central Manufacturing District, we have at our disposal, the prompt and efficient transportation service for which this district is famous the country over.

The Traffic Department is in charge of an experienced manager who devotes his entire time to the study of problems surrounding traffic.

ZENITH PACKING

The Zenith Radio Corporation takes extreme precautions to insure safe delivery of Zenith sets irrespective of the distance to destination. Our foreign shipments for instance, come to hand almost invariably without even a wire out of place. That is remarkable because the handling of foreign shipments is notoriously rough.

We know of no radio manufacturer that goes to the same expense and care that we do to guard against the wear and tear of shipping. First of all the Zenith set is wrapped in a protective covering of waxed paper to exclude moisture, and also a measure of protection against abrasions. Then it is put in a carton of heavy corrugated cardboard and sealed. This in turn is surrounded with three inches of excelsior to absorb any shocks resulting from fall or handling. Then the whole is inserted in a wooden case which is built with a view of being elastic, strong and puncture proof. It is in panel form of three-ply wood with wooden cleats or battens around all edges. The smaller cases are then wire bound. This is the ultimate today in packing.

As an additional precautionary measure, every shipping case bears "Handle With Care" and "Fragile" labels, besides one to indicate "Up" side of the instrument. But even though instructions are ignored, as so frequently happens, the method of packing is the best guarantee of safe arrival of the Zenith radio set.

WHY DOESN'T ZENITH SELL THE CHASSIS SEPARATELY?

This constantly recurring question can be easily answered. Buyers of the chassis alone are usually dealers or consumers who want to place a Zenith chassis in their own cabinets. Our position in this matter is that we object to our radios being installed in cabinets which are unsuited to them; where loud speakers of indifferent manufacture are used and where the whole ensemble is not a credit to Zenith. We take great pains and go to much expense to have Zenith cabinets consistent with Zenith quality and general standing. A Zenith chassis installed in a poorly designed cabinet would be a bad advertisement. Of equal importance is the loud speaker. Our speaker installations are all designed for the sets with which they are used. We match the impedances and take care of other factors which we know will give the best possible results.

ZENITH GUARANTEE

A guarantee of broad and liberal tenor is put inside of every Zenith set. This guarantee is reproduced below. The owner of a Zenith set may rest easy with the assurance of ample protection against possible defects in material or workmanship.

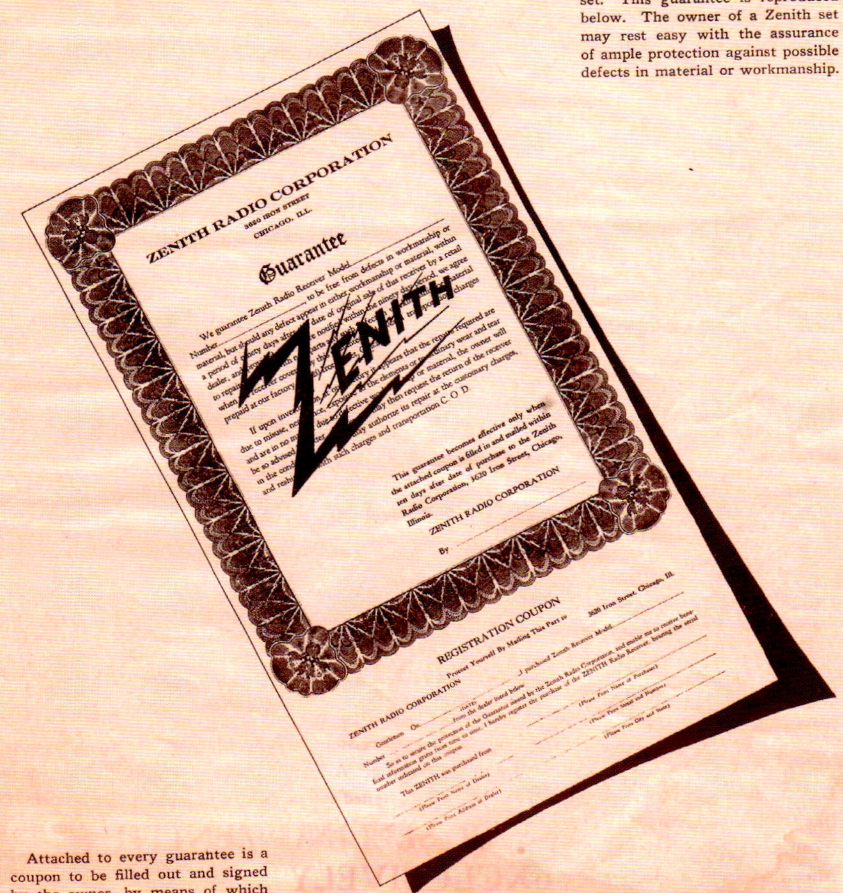

Attached to every guarantee is a coupon to be filled out and signed by the owner, by means of which he is registered at the factory in Chicago; the guarantee does not become effective unless the coupon is registered.

13

ZENITH ELECTRIC SETS

Our kaleidoscopic experience as pioneers in radio has been especially gratifying in the building of electric sets. We were the first to manufacture and market, practical, working electric radio sets. And when we describe them as electric, we are using the term in the full sense of the word. Many radio sets masquerade as electric, but in truth and in fact, operate by means of liquids or jellies, whereas in the Zenith electric sets, the rectification is entirely dry, there being no camouflaged batteries, accumulators or other subterfuge methods.

During the past year there has been much advertising of "Socket Power" sets. We realize that it is possible for a regular battery-operated set to be equipped with a small storage battery combined with a trickle charger in one case, and that this combination of old elements may be so treated as to give the prospective purchaser the impression that what is contained within the case is some sort of a real power unit. There are large manufacturers of batteries, who have put upon the market such devices, and when combined with an ordinary battery set, are thereby enabled to advertise their devices as "Socket Power." While there is no species of misrepresentation in the literal statement of "Socket Power" in connection with these devices, the effect is to mislead the intending purchaser into believing that he is really buying a truly electric set. Such devices however, are in no manner different from the old method of operating any radio set since radio started. Is it not a fact that a radio receiver made five years ago and actuated by an ordinary storage battery, could be said to be a "Socket Power" set, because the storage battery is renewed through the medium of a battery charger, which in turn receives its current from an electric light socket?

We see no difference in the so-called "Socket Power" set and any radio set made since the inception of the industry. In our opinion, when a dealer buys a battery set and equips it with some form of camouflage battery and trickle charger, and then proceeds to advertise such a combination as "Socket Power" or "Electric Power" set, he is misleading the public, because such a title causes the public to infer that they are buying a Simon pure modern electric set, utilizing the Zenith system. Salesmen should present this matter clearly to intending purchasers. They should put prospective buyers on their guard against assembled outfits that are really not batteryless. They should learn the difference between a so-called "Socket Power" set and a truly electric set, such as Zenith.

The vigilance committees of advertising clubs throughout the United States have had under consideration, the question of whether such "Socket Power" advertising is not misleading to the point of where it approaches the fraudulent. There are newspapers who refuse advertising of such "Socket Power" sets because they feel that when an advertiser states over his signature, that he is selling a "Socket Power" set, the customer has the right to believe he is buying a modern electric receiver, and not one employing the same system which has been used in all radio sets since the very beginning of the radio business.

We had long foreseen the day of the electric set. We knew it would be a big step in the evolution of radio. Our long experimentation finally bore fruit over two years ago, in the development of a set which greatly surprised radio technicians and engineers everywhere. It was an epoch-making set which created more business for us than we dreamed was possible. The public fairly leaped at the opportunity of buying a radio that could be operated without batteries, without the need of replenishing liquids or jelatinous substances of one kind or another. Of course, we have made progress since then, and while the first Zenith electric set still stands a marvel, and excells anything else that followed in its wake, the Zenith electric set for the coming season reflects an experience that no other manufacturer can boast.

REASONS FOR SELLING ZENITH EXCLUSIVELY

In handling the Zenith line exclusively, the dealer has the advantage of selling models ranging in price from $100.00 to $2500.00. In the matter of price he can fit the pocketbook of any prospective buyer crossing his threshold. In identifying himself solely with Zenith, the dealer becomes bracketed in the public mind with

14

the best in radio. Mention the dealer's name and the local public thinks of Zenith; mention Zenith and the local public associates it with the dealer. It needs no argument to perceive the value of advertising one make of radio and focusing selling effort on one make. The customer is impressed with the dealer's confidence. Whatever the dealer says on the subject of Zenith Radio is very effectually substantiated by the patent fact that he thinks so much of the line as to carry it in exclusion to all others. The customer is not confused by having to choose between different makes. He is more apt to buy on the spot. In a contrary situation, he would be apt to wait to think the matter over and clear up in his own mind, which of the different stories told by the dealer, about his different makes of radio, struck him the most. The selling arguments of the exclusive Zenith dealer strike home like a rifle ball, whereas those of the dealer handling a multiplicity of lines act like scatter shot.

ZENITH CIRCUITS

Every day announcements are being made of new radio circuits. One cannot pick up a newspaper without reading about some startling discovery that a certain writer is submitting to an anxiously waiting radio world. How many of these announcements however, are of serious importance? Hardly a day passes that we are not offered some new invention; some alleged improvement; some modification of a radio circuit. When these devices are examined in our laboratory under the critical gaze of our engineers, it is found that in 99 out of 100 cases, our records show that we have already conducted experiments along similar lines.

It is to be expected that a large manufacturing establishment like ours, maintaining extensive laboratories with a competent staff of engineers, working day after day to perfect radio, and with ample funds, facilities, and scientific apparatus, will make great headway on its own account. Many of the devices that are presented to us as brand new, were investigated in our laboratory a long time ago. It may be readily imagined that we have made countless experiments and built numerous pieces of apparatus which are not embodied, nor ever will be, in our circuits.

The three Zenith circuits are the result of years of thoughtful analysis and careful experimentation. Little by little we have improved our sets. The process of development has covered so many years that the result of our work of today stands as a real achievement. When improvements are made in radio, Zenith as a rule is first to make them. On another page we list some of the things that we were the first to use in the radio field. Some of these improvements represent milestones in the development of radio.

THE POWER SPEAKER

It was our good fortune a few years ago, to develop for use in the ZENITH DeLuxe Models, a dual speaker system which reproduced with very remarkable fidelity, the tones of the entire musical scale. The method consisted of two horns—one with a long tone column and an especially adapted unit to take care of the tones in the low and middle registers and the other, a smaller horn with a short air column and an especially adapted unit to take care of the tones in the high register. All audible frequencies were thus covered and the result was really astonishing. The wonderful reception of the ten-tube DeLuxe receiver was, of course, at its best with the help of this speaker system. An inferior speaker would have mutilated and marred and distorted the perfect reception of the DeLuxe. We never could have achieved our fine success with the ten-tube DeLuxe but for this efficient speaker system.

The dual speaker, however, has had its day. Wonderfully well though it has served, it is now supplanted by something better. We are equally fortunate today in being able to use a single reproducing device which accomplishes the same results and more. The new development is known as the "Power Speaker". It is an extremely sensitive mechanism, so perfectly conceived and executed that it not only performs the function of the dual speakers, but excels in reproduction and in other respects. The power speaker is more durable. It is equal to every possible strain. It will not only reproduce the entire musical scale but every possible shade of tone as well, from the merest whisper, the faintest pianissimo, to the loudest fortissimo—faithfully—fully, roundly, and with quality.

In a certain sense "Power Speaker" is a misnomer. The term suggests power. True, it produces large volume, but it is not in power only that this device excels, for it may be modulated to the softest rendition desired. Its forte lies in the fact that it faithfully reproduces ALL audible tones. How many people listening to radio have ever heard the slight hiss that is in every singer's or speaker's voice? How many have heard the kettle drums in a symphony orchestra, or the deep bass pipes of an organ. Zenith radio, equipped with the power speaker affords one the sensation of distinguishing between an "f" and an "s", or between an "m" or an "n" in song or speech; the sensation of "feeling" the heavy vibrations of the bass notes of a large pipe organ as is frequently experienced in a church. Critical musicians testify that the reproduction of the Zenith ten-tube DeLuxe receiver, equipped with power speaker is so true that it cannot be distinguished from the original performance.

All Zenith DeLuxe ten-tube receivers are equipped with the power speaker. The Model 16E is furnished with a power speaker at an advance of $50.00 over the regular models equipped with a cone speaker. The power speaker cannot be used on battery-operated sets.

15

ZENITH NATIONAL ADVERTISING

A STUPENDOUS CAMPAIGN FOR 1927-1928

16

A GREAT ADVERTISING CAMPAIGN

129,709,708

people will read Zenith magazine advertising between August and December 1927. This is more than the population of the United States.

15 great national publications will carry Zenith display copy.

9 outstanding trade magazines will carry the Zenith message to dealers.

54 Zenith wholesale distributors through 5126 Zenith retail dealers will present to consumers, through American newspapers, the advantages of Zenith radio and an open challenge to all competitors on performance.

ZENITH DEALER HELPS

This year we offer greater cooperation to a dealer than ever before as will be discovered upon a study of the following subjects:

Catalogs

The first edition of the 2,000,000 Zenith rotogravure folders showing the complete line in a convenient size for inserting in regular business envelopes, will be available to all authorized dealers.

Mats for advertisements.

A strong series of one- two- and three-column advertisements will be available. These will be furnished to authorized dealers in the form of mats, or in those rare cases where mats cannot be used, stereotypes may be had. A folder will be issued showing these advertisements, and dealers should order these mats by number only.

Direct by mail.

Last year's experience with a series of direct-by-mail letters was used to such good advantage by our dealers, that we have decided to continue this service another year. There is nothing more effective than well written letters mailed direct to a selected list of people who should own Zenith radio sets. These letters will be imprinted with the dealer's name and address, and will be sold to the dealer at a very nominal price.

Window displays.

Our authorized dealers will be entitled to receive copies of several different issues of handsomely lithographed window streamers which are designed to call the passer-by's attention to Zenith radio.

Bill board advertising.

We will furnish without charge, a reasonable number of 24-sheet posters to our authorized dealers. If desired, these can be furnished with the dealer's name imprinted thereon, at a charge not to exceed one dollar for each poster. Posting is to be done by the dealer and at his expense.

Authorized dealer signs.

After examining a large number of plans for authorized dealer signs, we finally decided upon a new process sign for placing on a glass of a door, a window, or a show case. The size is 7"x11", and consists of a photo-plated film. It is applied similar to a decalcomania transfer, but will not come off, and will stand washing. It will be furnished only through wholesale distributors to authorized dealers.

Electric window sign.

It is difficult to find a window sign of such attractiveness that it will arrest the attention of everyone looking in a window.

We have secured a very remarkable and attractive flashing window sign. The size is 14"x18". It consists of a base and a vertical sheet of plate glass with letters etched one-eighth of an inch deep. The light shining up from below, shows these letters in refracted silver. These signs are sold to dealers only through Zenith wholesale distributors. The price is $10.

ZENITH LIST PRICES

(Note—Those Models marked with (*) are electric.)

Model 12	6-tube—table model—metal chassis—fully shielded—single control—illuminated dial—for battery operation	$ 100.
Model 11	Same as Model 12 but larger cabinet	110.
*Model 11-E	Same as Model 11 but electrically operated	175.
Model 14	Has Model 12 chassis in fine butt and striped walnut floor console containing improved cone speaker—for battery operation	180.
*Model 14-E	Same as Model 14 but electrically operated	255.
Model 15	8-tube—loop set—table model—loop folds inside cabinet—metal chassis—fully shielded—single control—illuminated dial for battery operation	185.
*Model 15-E	Same as Model 15 but is electrically operated—price includes a Zenith AC power unit which comes separately	285.
Model 16	Has Model 15 chassis installed in beautiful cabinet with concealed loop and improved cone speaker—for battery operation	300.
*Model 16-E	Same as Model 16 but electrically operated	400.
*Model 16-EP	Same as Model 16-E but contains power speaker	450.
*Model 17-E	6-tube all electric receiver using the gas rectifier principle—this receiver is unsurpassed in tone quality and distance. (Spinet base $20 extra)	350.
*Colonial DeLuxe —	10 tubes all electric receiver—contains power speaker—six condensers on one shaft—every improvement known in radio—this receiver has no competition and is in a class by itself and is acknowledged by engineers to be the finest receiver built	650.
*English DeLuxe —	Same as Colonial but fine sample of period furniture	800.
*Italian DeLuxe —	Same as Colonial but has hand made cabinet by a famous builder of fine furniture	1250.
*Chinese DeLuxe —	Same as Colonial but cabinet is of rare beauty, hand made and authentic	1700.
*Spanish DeLuxe —	Same as Colonial but cabinet is the highest possible type of hand made furniture	2500.

No chassis sold separately. All sets shipped with tubes but not included in above listed prices, excepting with Model 17, one 350 milliampere gas rectifier is furnished.

WESTERN UNITED STATES PRICES SLIGHTLY HIGHER

TUBES REQUIRED FOR 1927-1928 ZENITH RECEIVERS

Note: Only Radiotron or Cunningham tubes may be used in Zenith receivers. The tube numbers given below are Radiotron numbers—every salesman knows the Cunningham equivalents.

Model 12—Requires six UX-201A. If a power tube is desired either UX-112 or UX-171 may be used in the socket marked "Power Tube."

Model 11—Same as Model 12.

Model 14—Same as Model 12.

Model 11E Requires four UX-226, one UX-171, one UY-227, and the power unit requires one UX-280 or UX-213.

Model 14E—Same as Model 11E.

Model 15—Requires seven UX-201A and one UX-171.

Model 16—Same as Model 15.

Model 15E—Requires six UX-226, one UY-227, one UX-210 and the power unit requires one UX-281 or UX-216B.

Model 16E—Same as Model 15E.

Model 17E—Requires five UX-201A and one UX-171. The power unit requires one 350 MIL gas rectifying tube, which is furnished gratis with the set.

DeLuxe Models require eight UX-226, one UY-227, one UX-210 and the power unit requires one UX-281 or UX-216B.

All the above listed Zenith models will be shipped from our factory completely equipped with the necessary Radiotron or Cunningham tubes, and we shall charge for the tubes at the current list prices authorized by the manufacturers.

THE ZENITH MODEL 12 CHASSIS

FRONT VIEW

This is Zenith's latest contribution to the radio art. It is a high grade, sensitive, selective receiver of unusual tone quality at a popular price. Unlike some so-called "6-tube" receivers which use one tube for antenna coupling and only five effective working tubes, this is a bonafide 6-tube receiver, with every tube a working tube.

The Model 12 contains a well devised antenna compensator located at the rear of the chassis. Judged by the standards of many other 6-tube sets, the Model 12 is the equivalent of 7 tubes. The transformers used are made by Zenith and are of a construction and ratio such as to produce surprising tone quality. The dial is calibrated in both kilocycles and meters, thus being in step with the intention of the Federal Radio Commission to use kilocycles only. An electric light is placed behind the dial giving a soft glow and making easy reading. The switch is combined with the volume control which is of the roller type giving smooth and gradual modulation. Output filters are provided for loud speaker reproduction.

Four Zenith condensers and control wheel are accurately and rigidly secured to one shaft, giving permanent balance and freedom from service. The construction is all metal and fully shielded. The Model 12 is economical in the use of batteries. It is wired for a power tube.

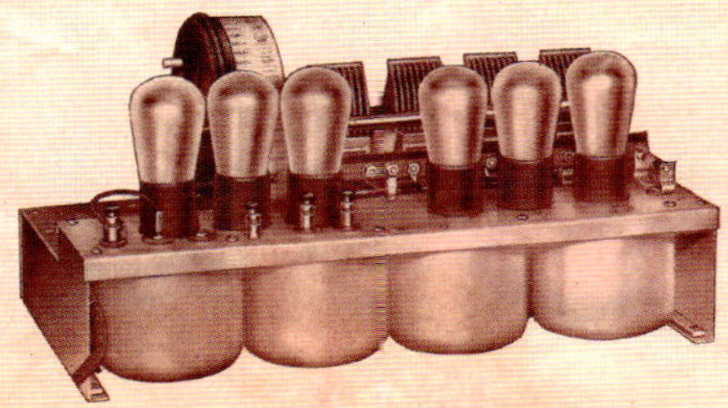

REAR VIEW

ZENITH MODELS 11 and 12

MODEL 12

Model 12. Beautiful brown mahogany cabinet, well finished. The front panel is of selected veneer with two-toned routed motif in center, and gracefully shaped mouldings.

Dimensions: 19 inches long, 13⅝ inches deep, 10½ inches high.

Model 11. This cabinet is made of selected butt matched walnut veneer in semi-matte finish. Front panel beautifully routed and two-toned with appropriate mouldings.

Dimensions: 25 inches long, 13⅝ inches deep, 10½ inches high.

MODEL 11

REAR VIEW MODEL 11E

Model 11-E. Same cabinet as Model 11. This rear view shows the Zenith ABC Power Unit installed, ready for operation.

ZENITH MODEL 14

The cabinet of Model 14 is of selected walnut veneers with ornamental overlays and panels of beautifully figured maple. The finish is black walnut and high-lighted to bring out the beauty of the grain with a rich reddish color on the maple embellishments. The lower compartment is for batteries. The doors when open do not stand out like wings, but fold back against the sides of the cabinet out of view. This means much to a person who is concerned about the appearance of the home, because radio sets are open 90% of the time, and if the doors do not swing out of the way, they are a disfigurement.

The upper front doors folded back disclose an ornamental metal grille treated to harmonize with the cabinet.

Model 14-E is the same as Model 14 excepting that it is electrically operated.

Dimensions: 48 inches high, 26 inches wide, 15½ inches deep.

AT LEFT
MODEL 14
OPEN

AT RIGHT
MODEL 14
CLOSED

THE ZENITH MODEL 15 CHASSIS
8 TUBES—LOOP

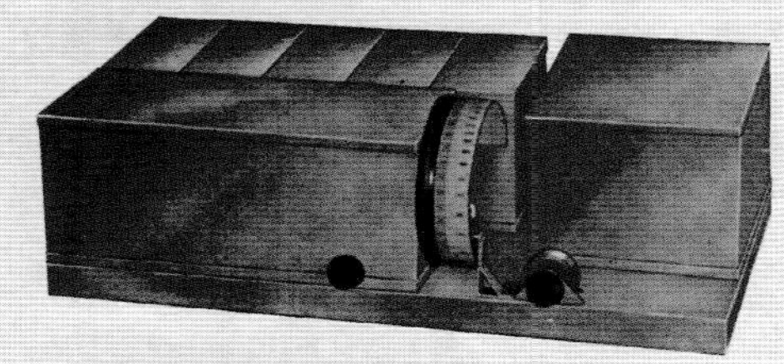

FRONT VIEW

REAR VIEW

The eight tube loop set employs four stages of tuned radio frequency amplification, a tuned detector, and three stages of audio frequency amplification. The four radio stages and the detector are tuned by a five gang condenser. This condenser is die cast throughout and is of very high accuracy. It is made so that the tuning of the loop will follow the tuning of the other circuits very closely throughout the entire wave length range. A small vernier adjustment is provided which enables the operator to adjust the loop circuit to exact resonance with the remaining circuits when extreme distant reception is desired. For ordinary purposes it is not necessary to adjust the vernier each time a station is tuned in. This special construction gives a single controlled eight tube loop receiver which we believe to be unique in its accuracy and simplicity of operation. The volume control and sensitivity control are combined in one knob which also serves to turn the set off and on. This set is completely and totally shielded and the turning of the loop does not affect the operation of the set.

The audio amplification of this receiver is of the latest design. The system employed is a special tuned double impedance amplifier and the frequency character of this amplifier gives a rising curve on the lower frequency. This means that the lower frequencies are amplified to a greater extent than the intermediate and high frequencies. This special feature accomplishes two things. First, it makes up for the deficiencies of the average loud speaker in reproducing the lower frequencies, and it also gives sufficient amplification of the lower frequencies to allow the volume to be turned down to a very low point without losing the low note reproduction. This feature is not found in other radio sets and is one of the outstanding achievements of radio reproduction.

The loop employed with this receiver is arranged to fold inside of the cabinet when not in use. The dial is calibrated in both wave lengths and kilocycles and is also illuminated.

ZENITH MODELS 15 and 16

At the right is shown Model 16 closed. Spanish Renaissance design, in selected butt matched walnut, and blister maple overlays on the pilasters and decorative front instrument panel. Semi-matte lacquer finish, beautifully high-lighted and toned in black walnut and Rembrandt brown. The loop, concealed from view, is controlled from outside the cabinet. An ornamental colored metal grille is disclosed when doors are folded back.

Dimensions: 43½ inches high, 31½ inches wide, 17¼ inches deep.

At the left is shown the Model 16 open. A power speaker at an advance in price may be had in the Model 16 when it is electrically operated. The power speaker cannot be used with battery operated sets.

Model 16-E is the same as Model 16 excepting that it is electrically operated.

At the right is shown Model 15. Constructed of a combination of selected veneers, butt matched walnut top and center front panel, and figured maple side front panels, with ornamental wood mouldings, lacquer finished in black walnut, high-lighted with a reddish cast on the maple panels. A beautiful bronze plate sets off the cabinet to best advantage. The ornamental loop is in an upright position when in use and folds into the cabinet when not in use.

Model 15-E is the same as Model 15 excepting that it is electrically operated and comes with a separate Zenith A-B-C Power Unit.

Dimensions: 33½ inches long, 15⅜ inches deep, 11⅝ inches high.

ZENITH MODEL 17 ELECTRIC

This year's Model 17 is the last word in electric sets. There is no way in which we can make it better. It reflects our best thought and the most extensive experience of any manufacturer of electric radio receivers. There are higher priced Zenith sets, but none are better value relatively than the Model 17. It will grace any home no matter how luxuriously appointed, and it will not be out of place in the home of modest surroundings. For the man who wants the best in radio, without the extra cost of furniture, his every wish will be fulfilled with this Model. The demand for it the coming year promises to be large. The reason for this is not far to seek because it is a set that appeals to those of affluence, as well as those of moderate means.

Model 17 contains a new single tube 400 milliampere power unit. It is noteworthy for its full, deep tone of surpassing quality. It is a six-tube set and incorporates the latest development in the application of a gas rectifier to a power supply. The list price includes one 400 milliampere tube.

For distance reception this receiver cannot be excelled by any other six-tube receiver and it is seldom equalled. The control is simplicity itself. One main knob does the tuning while a volume control regulator provides means for modulating which is a delight to those whose musical ear requires sonorous reproduction.

The advantages of the gas rectifier system as compared with the AC tube system of electrification are:

1—The working tubes are the standard 201A tubes which are so well known for their stability and reliability as to need no endorsement by us.

2—An electric power unit comprised of transformers, condensers, chokes, resistances and rectifying tube is better known in the art than any other form of current supply.

The Model 17 is a time tried and tested receiver which will give consistent service of highest quality at all times.

MODEL 17E
SHOWN WITH SPINET BASE

ZENITH 10-TUBE ELECTRIC DE LUXE RECEIVERS

There is no radio easier and more profitable to sell than the Zenith DeLuxe ten-tube sets, ranging in price from $650.00 to $2500.00. The exclusive features of this radio put it in a class by itself. In truth, and in fact, it has no competition. It appeals to most people because most people even those of limited means, want the best of everything and buy the best of everything.

The most direct route to a sale of a Zenith DeLuxe is the demonstration. The wonderful performance of the set will do more to interest a prospect than anything that can be said or written. It has the maximum distance and power. It is highly selective. There is positively no more wonderful tone in radio than that produced by the Zenith DeLuxe. Only a demonstration can give the prospective buyer full realization of its exquisite qualities.

The distinguishing features of the DeLuxe are briefly classified as follows:

1. The only set functioning successfully without loop or antenna. It operates from self-contained receiving plates.
2. The only regularly manufactured set having ten tubes.
3. The only set with six gang condensers. Therefore the unusual selectivity. These condensers are the most accurate to be found in any radio set.
4. The only set with a wave range of 80 to 550 meters. At present broadcasting is confined to meters from 200 to 550. There is no doubt in our minds but sooner or later the broadcasting range will be lowered. With one of these DeLuxe receivers the owner is assured of having a wider range of reception than can be built in the low-price receivers.
5. Employs complete double shielding.
6. Illuminated dial.
7. Single control.
8. All DeLuxe models are equipped with the latest development in the power cone speaker. This speaker is so sensitive and powerful that it faithfully reproduces all the colors and effects of the original. The actual low notes of the tuba, the kettle drums, and the pedal notes of the organ are brought out with amazing fidelity.

Note: Colonial, English and Spanish Models are carried in stock. Italian and Chinese are custom built requiring 60 days delivery.

THE 10-TUBE ZENITH DE LUXE CHASSIS

The ten-tube Zenith set consists of five stages of tuned radio frequency amplification, a tuned detector, and four stages of audio frequency amplification. The radio frequency stages and the detector are all tuned by means of a six gang condenser which is the most accurate variable condenser on the market. This condenser is entirely of die cast construction and has seven bearings for the steel shaft which carries the rotor plates. After being assembled, it is carefully checked and balanced so that error is less than one-half of one percent.

The audio end of this set employs one stage of resistance coupling and three stages of transformer coupling with a ratio of 1 to 1. By using four stages of amplification with low amplification per stage, a high over-all amplification is obtained with excellent quality. A power tube of the UX-210 type is used in the last audio stage.

This set is made entirely electric by employing AC tubes of the UX-226 and UY-227 type. The B eliminator employs the UX-281 rectifying tube. The ballast resistance is employed in the power supply unit which compensates for line voltage changes, and it is only necessary for the operator to turn the set off and on. All voltages are automatically adjusted to the proper value.

RIGHT—
FRONT VIEW

LEFT—
REAR VIEW
WITH SHIELDING
REMOVED

ZENITH 10-TUBE ELECTRIC DE LUXE MODELS

COLONIAL
(OPEN)

At the left is shown the Zenith ten-tube DeLuxe Colonial Model. There is a quiet romantic sentiment expressed in the woods and design of the Colonial Model, which conveys a patriotic closeness to the days of its historic period. It is design of this character that creates a background for those who persist in their search for a true record of culture and refinement in the home.

(Carried in stock).

COLONIAL
(CLOSED)

At the right is shown the Zenith ten-tube DeLuxe English Model. The English Model conveys a dignity with an abundance of characteristic interest in the chaste, severe lines known to this particular class of exclusive furniture and expresses the spirit of the present day in the application of simple mouldings and a few curved lines, which have an historical background.

(Carried in stock).

ENGLISH (CLOSED)

ENGLISH
(OPEN)

27

ZENITH 10-TUBE ELECTRIC DE LUXE MODELS

ITALIAN

At the left is shown the Zenith ten-tube DeLuxe Italian Model. Such exceptional design will have a special appeal to those who prefer the unusual. Concentrated splendor is offered in the grille and panel detail. There is a "feel of the wood" that the eye caresses, as the fingers do old velvet or fine silk. Observe the harmony of every feature—the fine proportions.

(Custom build—60 days delivery).

At the right is shown the Zenith ten-tube DeLuxe Chinese Model. In this we have one of those extraordinary accomplishments which has succeeded in producing a thing of great beauty and interest, the mechanisms of which can only be realized when skillfully disposed among other decorative styles.

(Custom build—60 days delivery).

CHINESE

28

ZENITH 10-TUBE ELECTRIC DE LUXE MODELS

SPANISH

SPANISH (CLOSED)

Above is shown the Zenith ten-tube DeLuxe Spanish Model. Wisdom is as necessary a quality in art as in life itself. For after the novelty wears off, too much action becomes a source of irritation to the eye and the thought. The Spanish Model—a piece of old cabinet work, portrays vividly the treasured elegance of olden Spain.

(Carried in stock).

HOW TO SELL ZENITH RADIO

This handbook will be read and used by young salesmen who are seeking more efficient ways of selling radio as well as by experienced salesmen. If some of the suggestions made herein seem elementary to the oldsters, they must take into account the younger element.

First in importance is to display radio correctly. A dirty, dusty set is an abomination. Good furniture stores keep people dusting all day long. Besides this, their furniture is "oiled off" whenever a blue coat appears on the finish. Varnish on wood is simply a transparent covering which, like window glass, becomes dirty. How long ought windows to go without washing? Apply the same time element to wood finishes. Five cents worth of lemon oil and a few minutes will often produce more business than two hours of talk. How to use lemon oil? Simply rub on with a small cloth and then wipe dry with a CLEAN rag. The cleaner the rag, the cleaner the surface.

Take your favorite Zenith Model—keep it in the pink of condition. Old tubes get fuzzy. A demonstrator should use only the best of accessories.

When possible have a special room or space for your Zenith Model. Concentrate upon the set you think will fit the pocketbook of the customer. Then start the Zenith story and don't stop until you have the sale.

Perhaps your customer has visited another store where he has stood up and become tired. Perhaps a lady has been shopping and "can't go another foot." Give your customer a chair and ask, "Are you comfortable?" Simple as it is, it starts the deal right and frequently undoes much that the former salesman has done.

Cultivate a courteous, respectful and attentive manner. Ascertain the name of your customer at the earliest moment and use it frequently. Never call your customer "lady" but say, "Are you comfortable, Mrs. Brown?" Using the buyer's name creates friendliness and moves closer to the sale.

It is fatal to talk a more expensive set than the buyer can afford. If you are an exceptional salesman you might sell a $1250 Zenith Italian Model DeLuxe 10-tube electric set to a buyer who ought to spend $300. But what is the result? The buyer says he won't have anything but the $1250 set and will wait until he can afford it. Wait! That is the danger sign to be watched at all times. Most buyers want to "wait." The great purpose of the salesman is to eliminate the thought of "wait" and put before the buyer's mind the all important thought of "now."

Another, and still stronger reason for talking the set which fits the buyer's purse is the installment angle. Long terms on anything are undesirable. Radio is no exception. When you sell a set which fits a customer's purse, the terms are easier arranged.

On the other hand, it would be just as fatal to talk a Zenith $100 Model to a customer who could buy a $1250 Model. The first effort should, therefore, be to obtain a line on the customer's actual buying power. This may be achieved through a few leading questions designed to bring out their ideas. Don't say, "How much do you want to invest in a set?" It is much better to lead with the graceful question, "Mrs. Brown, have you given any thought as to the type of radio set you would like to own? Do you prefer a floor model or one which is placed on a table? Our Zenith radio sets start at $100 for the table models and $180 for the floor consoles." Similar questions will give a good index of the buyer's ability to pay.

Now start building confidence. That is the foundation of every sale. Convince the buyer of your sincere effort to serve him. Go over the Zenith story from its beginning. Explain to him that Zenith is one of the oldest and earliest builders of radio (P2). Bear down upon the point that experience is necessary to build a good radio. Mention the Berengaria test. Impress MacMillan's use of Zenith when frozen in within 11 degrees of the North Pole as the best guarantee of Zenith dependability. MacMillan could take no chances. HIS radio set HAD to work. So he selected Zenith (P3). Tell your buyer that not every dealer can handle Zenith. Only selected dealers who are authorized by the factory (Par. 1—P3). Show him your authorized dealer sign.

If your buyer is considering sets retailing for less than $100, use the material in paragraph numbered 2 on page 3. Does a radio buyer want to own a radio set which may be cut and slashed in price at a later date? Certainly not. The reason for upsets in the radio market are given in paragraph 3 on page 3 and paragraph 4 on page 4.

Does it not stand to reason that a manufacturer who is licensed under the basic patents, who can make what he wants, can produce better radio than one who must go around the bush? Some manufacturers will be put out of business because of patent difficulties. Zenith is licensed —its position is secure. The buyer should take this into consideration (P4). Argue with your buyer that this is the day of strong institutions. Zenith has over a million in assets and owes less than ten thousand. This financial strength means permanency. The buyer of a Zenith is making a lasting purchase of which he will be proud (P4).

If necessary—prove Zenith's supremacy by showing what it has been first in doing (P5). Emphasize the ramifications of Zenith by describing our train control demonstrations (P6). Put a punch in Zenith sales argument by the use of Zenith slogans (P7). If they are interested in long distance reception, display the double page spread (P 8 and 9) showing reception in Hoboken, New Jersey from Japan.

Display a picture of the factory (P10). Every buyer likes to see where and how his radio set is made. Our immense daylight factory is impressive and real.

Touch upon the laboratory (P11). This department is the heart of any radio manufacturing business. There are not more than three other manufacturers who equal our expenditures on research and development.

If your buyer is from out of town and wants his set shipped, don't fail to show our packing methods (P12). What about the guarantee? This interests all buyers? Show him Page 13.

Now comes the salesman's most important point. If your buyer has an electric current in his house, sell him an electric set. Study the arguments on Page 14. Zenith leads in electric sets. It is the first and oldest manufacturer. If anyone can build good, dependable electric sets, Zenith is certainly entitled to first place from the standpoint of experience and ability.

When the buyer can afford the best in radio without regard to price, you have very impelling material to draw to his notice in the considerations outlined on page 26. In addition, stress the advantage of the power speaker, which is a very remarkable adjunct to the De Luxe. See page 15. As you know, this power speaker is also furnished with model 16-E at an extra cost of $50.

There are few people who have not heard of Zenith. If one of these rare individuals appear as a buyer, you can give him an idea of Zenith publicity by showing Pages 16 and 17. The burden is then on him to say why he has not heard of Zenith. If he can read, he must see our advertising. If he can hear he must listen to our station WJAZ at some time.

Don't fail to impress the buyer with Zenith tone quality, Zenith careful workmanship, and Zenith best quality materials. The purchase of a radio set is an important event to any buyer. He must live with his set for a long time. A radio set which has a nasal, thin tone soon becomes tiresome. Day in and day out the sonorous tone of Zenith radio sets is a constant pleasure to owners.

Finally, give the buyer the famous Zenith challenge printed on the rear cover of the booklet and close with the impressive statement that "Zenith is the largest manufacturer of high grade radio in the world."

In closing these suggestions, may we remark that they have been written by a man who uses the arguments contained herein every day and who has been, in our opinion, more successful in selling Zenith Radio than any other man with whom we have come in contact. These suggestions are not, therefore, mere vaporings but are the result of practical experience.

31

Zenith Famous Challenge to Combat

Under identical circumstances and conditions, and taking our various models in their proper grade, we challenge any make of radio receiver to equal or surpass Zenith performance in quality or quantity of tone; selectivity; sensitivity; distance; ease of operation; precision of manufacture or general all around performance.

Zenith Radio Corporation

Notice:
Our receivers are licensed only for radio amateur, experimental and broadcast reception.
Zenith Radio Corporation.

Printed in U.S.A.

Zenith Trade Letters

(Reprinted with permission of Zenith Electronics LLC)

No. 6 - Page 1

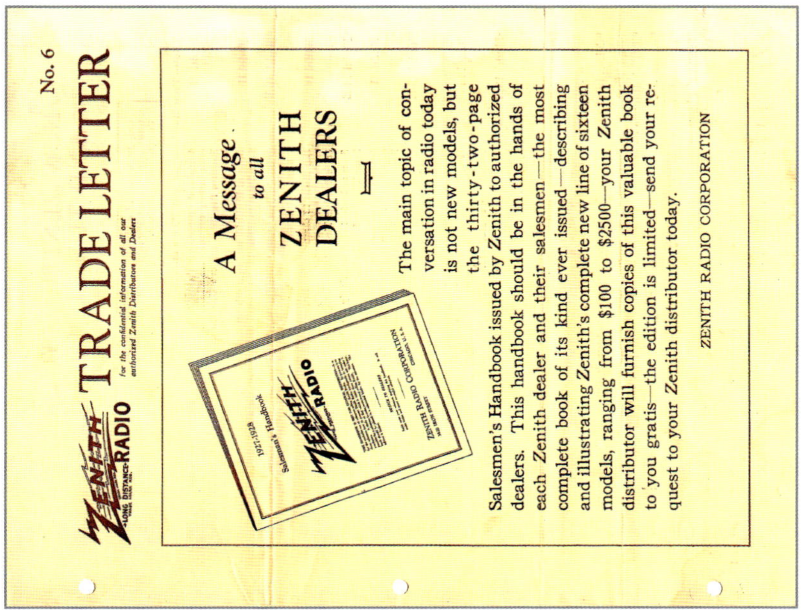

No. 6

ZENITH LONG DISTANCE RADIO **TRADE LETTER**

For the confidential information of all our authorized Zenith Distributors and Dealers

A Message to all ZENITH DEALERS

The main topic of conversation in radio today is not new models, but the thirty-two-page Salesmen's Handbook issued by Zenith to authorized dealers. This handbook should be in the hands of each Zenith dealer and their salesmen—the most complete book of its kind ever issued—describing and illustrating Zenith's complete new line of sixteen models, ranging from $100 to $2500—your Zenith distributor will furnish copies of this valuable book to you gratis—the edition is limited—send your request to your Zenith distributor today.

ZENITH RADIO CORPORATION

No. 6 - Page 2

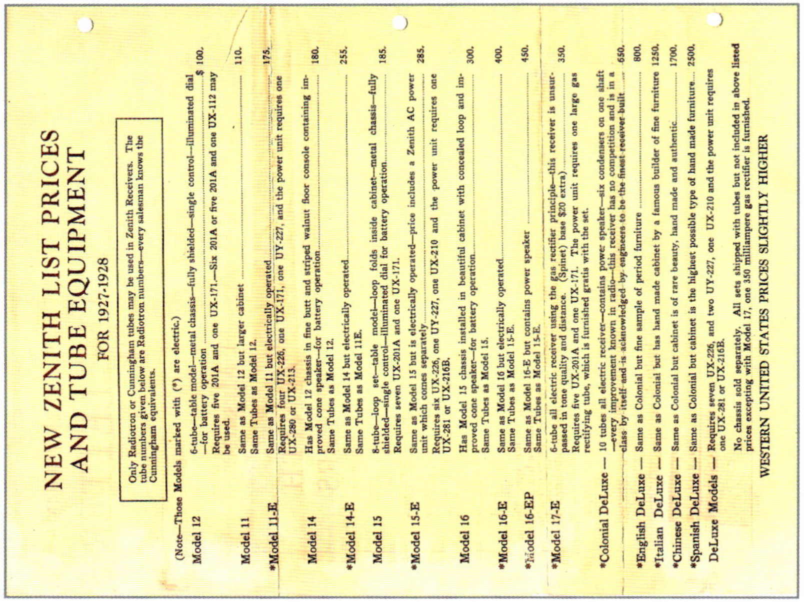

NEW ZENITH LIST PRICES AND TUBE EQUIPMENT

FOR 1927-1928

Only Radiotron or Cunningham tubes may be used in Zenith Receivers. The tube numbers given below are Radiotron numbers—every salesman knows the Cunningham equivalents.

(Note—Those Models marked with (*) are electric.)

Model 12	6-tube—table model—metal chassis—fully shielded—single control—illuminated dial —for battery operation. Requires five 201A and one UX-171.—Six 201A or five 201A and one UX-112 may be used.	$ 100.
Model 11	Same as Model 12 but larger cabinet. Same Tubes as Model 12.	110.
*Model 11-E	Same as Model 11 but electrically operated. Requires four UX-226, one UX-171, one UY-227, and the power unit requires one UX-280 or UX-213.	175.
Model 14	Has Model 12 chassis in fine butt and striped walnut floor console containing improved cone speaker—for battery operation. Same Tubes as Model 12.	180.
*Model 14-E	Same as Model 14 but electrically operated. Same Tubes as Model 11E.	255.
Model 15	8-tube—loop set—table model—loop folds inside cabinet—metal chassis—fully shielded—single control—illuminated dial for battery operation. Requires seven UX-201A and one UX-171.	185.
*Model 15-E	Same as Model 15 but is electrically operated—price includes a Zenith AC power unit which comes separately. Requires six UX-226, one UY-227, one UX-210 and the power unit requires one UX-281 or UX-216B.	285.
Model 16	Has Model 15 chassis installed in beautiful cabinet with concealed loop and improved cone speaker—for battery operation. Same Tubes as Model 15.	300.
*Model 16-E	Same as Model 16 but electrically operated. Same Tubes as Model 15-E.	400.
*Model 16-EP	Same as Model 16-E but contains power speaker. Same Tubes as Model 15-E.	450.
*Model 17-E	6-tube all electric receiver using the gas rectifier principle—this receiver is unsurpassed in tone quality and distance. (Spinet) base $20 extra). Requires five UX-201A and one UX-171. The power unit requires one large gas rectifying tube, which is furnished gratis with the set.	350.
*Colonial DeLuxe —	10 tubes all electric receiver—contains power speaker—six condensers on one shaft —every improvement known in radio—this receiver has no competition and is in a class by itself and is acknowledged by engineers to be the finest receiver built.	650.
*English DeLuxe —	Same as Colonial but fine sample of period furniture	800.
*Italian DeLuxe —	Same as Colonial but has hand made cabinet by a famous builder of fine furniture	1250.
*Chinese DeLuxe —	Same as Colonial but cabinet is of rare beauty, hand made and authentic	1700.
*Spanish DeLuxe —	Same as Colonial but cabinet is the highest possible type of hand made furniture	2500.
DeLuxe Models —	Requires seven UX-226, and two UY-227, one UX-210 and the power unit requires one UX-281 or UX-216B.	

No chassis sold separately. All sets shipped with tubes but not included in above listed prices excepting with Model 17, one 350 milliampere gas rectifier is furnished.

WESTERN UNITED STATES PRICES SLIGHTLY HIGHER

No. 6 - Page 3

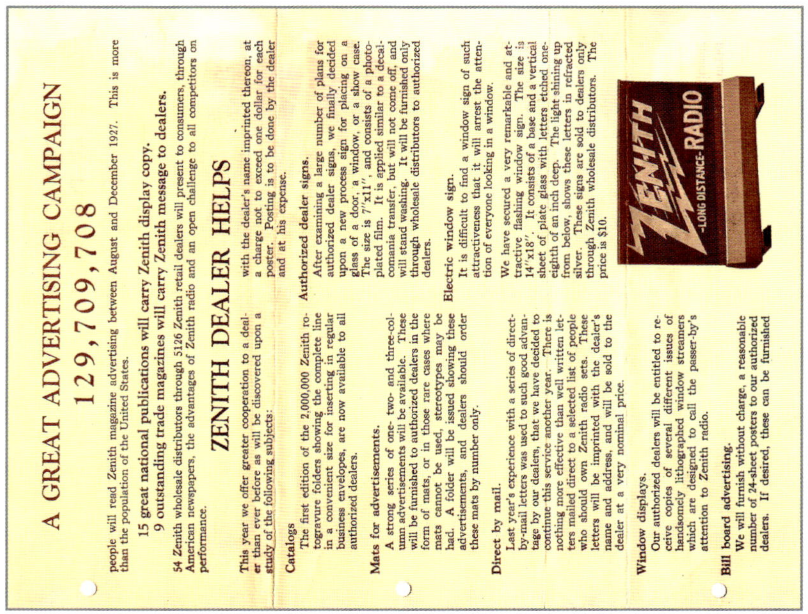

A GREAT ADVERTISING CAMPAIGN

129,709,708

people will read Zenith magazine advertising between August and December 1927. This is more than the population of the United States.

15 great national publications will carry Zenith display copy.
9 outstanding trade magazines will carry Zenith message to dealers.

54 Zenith wholesale distributors through 5126 Zenith retail dealers will present to consumers, through American newspapers, the advantages of Zenith radio and an open challenge to all competitors on performance.

ZENITH DEALER HELPS

This year we offer greater cooperation to a dealer than ever before as will be discovered upon a study of the following subjects:

Catalogs

The first edition of the 2,000,000 Zenith rotogravure folders showing the complete line in a convenient size for inserting in regular business envelopes, are now available to all authorized dealers.

Mats for advertisements.

A strong series of one- two- and three-column advertisements will be available. These will be furnished to authorized dealers in the form of mats, or in those rare cases where mats cannot be used, stereotypes may be had. A folder will be issued showing these advertisements, and dealers should order these mats by number only.

Direct by mail.

Last year's experience with a series of direct-by-mail letters was used to such good advantage by our dealers, that we have decided to continue this service another year. There is nothing more effective than well written letters mailed direct to a selected list of people who should own Zenith radio sets. These letters will be imprinted with the dealer's name and address, and will be sold to the dealer at a very nominal price.

Window displays.

Our authorized dealers will be entitled to receive copies of several different issues of handsomely lithographed window streamers which are designed to call the passer-by's attention to Zenith radio.

Bill board advertising.

We will furnish without charge, a reasonable number of 24-sheet posters to our authorized dealers. If desired, these can be furnished with the dealer's name imprinted thereon, at a charge not to exceed one dollar for each poster. Posting is to be done by the dealer and at his expense.

Authorized dealer signs.

After examining a large number of plans for authorized dealer signs, we finally decided upon a new process sign for placing on a glass of a door, a window, or a show case. The size is 7"x11", and consists of a photo-plated film. It is applied similar to a decalcomania transfer, but will not come off, and will stand washing. It will be furnished only through wholesale distributors to authorized dealers.

Electric window sign.

It is difficult to find a window sign of such attractiveness that it will arrest the attention of everyone looking in a window.

We have secured a very remarkable and attractive flashing window sign. The size is 14"x18". It consists of a base and a vertical sheet of plate glass with letters etched one-eighth of an inch deep. The light shining up from below, shows these letters in refracted silver. These signs are sold to dealers only through Zenith wholesale distributors. The price is $10.

No. 6 - Page 4

ZENITH NATIONAL ADVERTISING

A STUPENDOUS CAMPAIGN FOR

1927-1928

No. 9 - Page 1

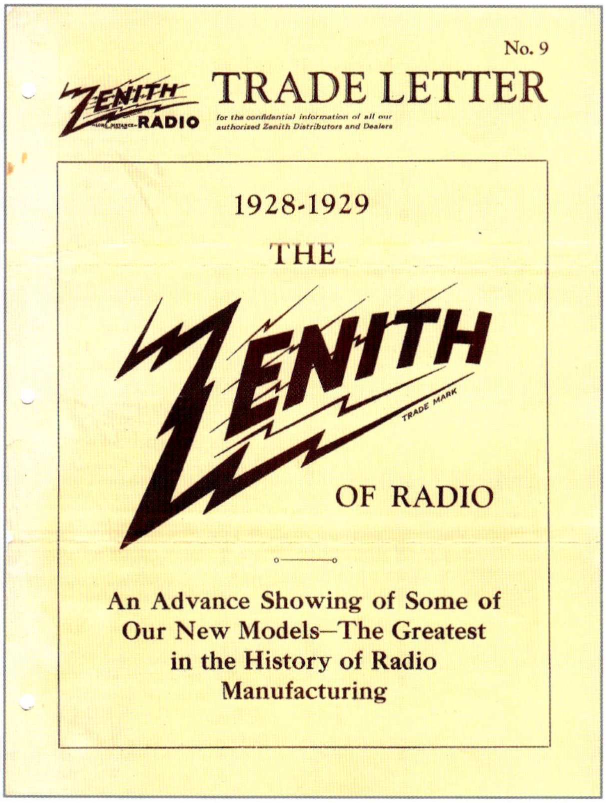

No. 9

ZENITH RADIO TRADE LETTER

for the confidential information of all our authorized Zenith Distributors and Dealers

1928-1929

THE

ZENITH

TRADE MARK

OF RADIO

An Advance Showing of Some of Our New Models—The Greatest in the History of Radio Manufacturing

No. 9 - Page 2

Models 31 and 33

The cabinet of dark walnut veneer is beautifully shaded and hand rubbed and further embellished with diagonally matched satin-wood overlay on the vertical front panel. A dark brown transparent illuminated dial, calibrated in both meters and kilocycles is recessed in a cove in top center of the cabinet. The appearance is further enhanced by an embossed escutcheon plate. An all metal, effectively shielded chassis, with each coil separately and individually shielded employs four Zenith die-cast condensers accurately and rigidly secured to one shaft, assuring permanent balance and freedom from variation. The volume is controlled *instantaneously* through an improved roller type control. A highly sensitive, selective receiver assuring quality of reproduction found only in Zenith Radio. Electric Model employs the following seven tubes—

5 UY 227 or 5 C 327 — 1 UX 171 or 1 CX 371 — 1 UX 280 or 1 CX 380

For battery operation—The following six tubes—

5 UX 201 or 5 CX 301 — 1 UX 171 or 1 C 371

List Price, Less Tubes

Model 31 For Battery Operation .... $100 Model 33 Completely Electric $150

Models 32 and 34

A low-boy console cabinet of highly figured walnut veneer with dark dull rubbed finish, shaded with a handsomely figured overlay. An arched opening reveals the slanting panel and illuminated transparent dial. Cross bar between legs approximately eight inches from floor. Cone type speaker mounted behind gracefully carved grille, this speaker faithfully reproduces the justly famous Zenith tone quality.

Circuit and Chassis are identical with those described under Models 31 and 33 (above). Electric Model employs the following seven tubes—

5 UY 227 or 5 C 327 — 1 UX 171 or 1 CX 371
1 UX 280 or 1 CX 380

For battery Operation the following six tubes—

5 UX 201 or 5 CX 301 — 1 UX 171 or 1 CX 371

List Price, Less Tubes

Model 32 For Battery Operation $180
Model 34 Completely Electric $230

MODEL 34

WORLDS LARGEST MAKERS OF HIGH GRADE RADIO

No. 9 - Page 3

Models 35 and 35P

An authentic English design cabinet with richly figured walnut veneers, solid walnut legs and cross bar. Doors swing completely back to sides of cabinet when closed, the doors are embellished with handsome linenfold carvings. The legs are fluted and beautifully carved. Inside front panel designed in the form of an arched decoration with split carvings and figured overlay design correctly toned and colored to match the cabinet. This Model must be seen to be appreciated, the meager description herewith does not do justice to the eye value of this receiver. Cone type speaker same as described in Models 32 and 34. Chassis same as described in Models 31 and 33.

This receiver (Model 35) uses the following seven tubes—

5 UY 227 or 5 C 327
1 UX 171 or 1 CX 371
1 UX 280 or 1 CX 380

List Price, Less Tubes $270

MODEL 35 CLOSED

Model 35P

Cabinet identical with Model 35 chassis, same in appearance except special construction for use with Dynamic cone speaker. The Dynamic cone type speaker brings a new era in radio reception. The amazing fidelity of tone and greater volume without distortion all go toward making the Dynamic speaker most favored where true perfection of reproduction is desired. This Model uses the following eight tubes—

5 UY 227 or 5 C 327
1 UX 210 or 1 CX 310
2 UX 281 or 2 CX 381

List Price, Less Tubes $330

MODEL 35 OPEN

WORLDS LARGEST MAKERS OF HIGH GRADE RADIO

No. 9 - Page 4

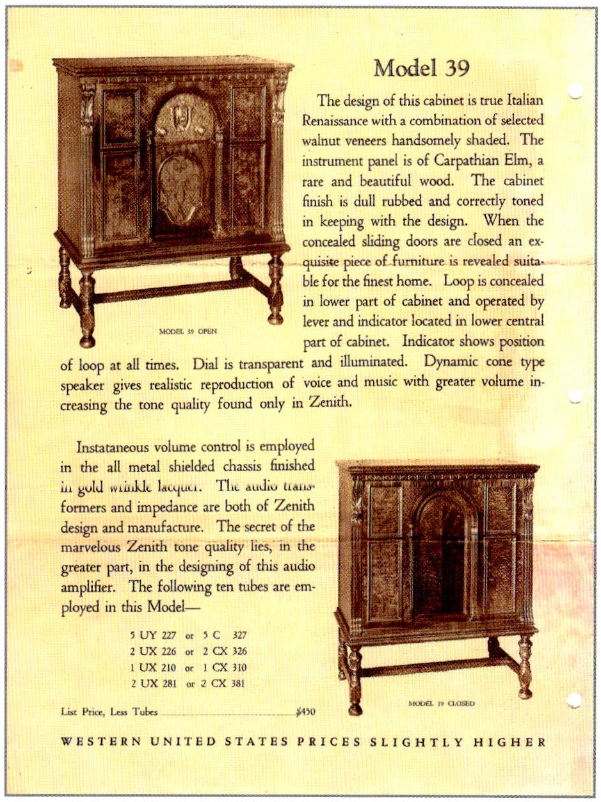

Model 39

The design of this cabinet is true Italian Renaissance with a combination of selected walnut veneers handsomely shaded. The instrument panel is of Carpathian Elm, a rare and beautiful wood. The cabinet finish is dull rubbed and correctly toned in keeping with the design. When the concealed sliding doors are closed an exquisite piece of furniture is revealed suitable for the finest home. Loop is concealed in lower part of cabinet and operated by lever and indicator located in lower central part of cabinet. Indicator shows position of loop at all times. Dial is transparent and illuminated. Dynamic cone type speaker gives realistic reproduction of voice and music with greater volume increasing the tone quality found only in Zenith.

MODEL 39 OPEN

Instataneous volume control is employed in the all metal shielded chassis finished in gold wrinkle lacquer. The audio transformers and impedance are both of Zenith design and manufacture. The secret of the marvelous Zenith tone quality lies, in the greater part, in the designing of this audio amplifier. The following ten tubes are employed in this Model—

5 UY 227 or 5 C 327
2 UX 226 or 2 CX 326
1 UX 210 or 1 CX 310
2 UX 281 or 2 CX 381

List Price, Less Tubes $450

MODEL 39 CLOSED

WESTERN UNITED STATES PRICES SLIGHTLY HIGHER

Section III

Radio Restoration

Chapter 15: Zenith Radio Restoration Guide and Tips

Radio Restoration in Five Easy Steps

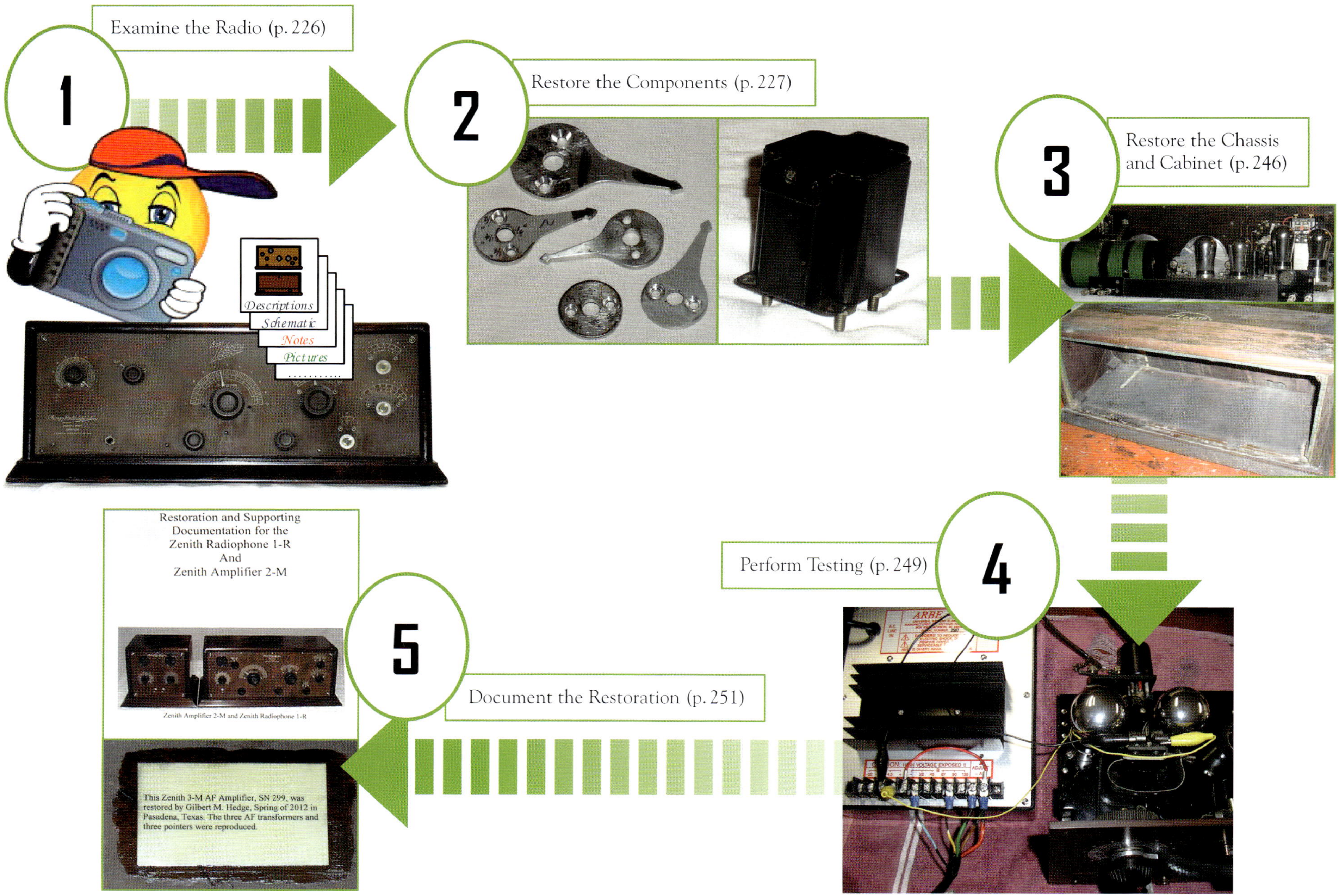

Restoration Notes

The scope of any restoration text cannot possibly cover every possible situation. This text is compiled to addresses the restoration of the Zenith equipment covered in this book. Most of these repair techniques can also be applied to the restoration of equipment that was built by other radio manufacturers.

During the restoration of your radio, you may encounter problems or situations that could not possibly be anticipated and therefore are not covered in this book; that's the nature of radio restoration. Be prepared to be creative and think out of the box when addressing these previously-unencountered problems.

Some radios do not require an extensive restoration similar to the 3-R discussed in this chapter. The models 11 and 12 would typically only need a good cleaning, and maybe the occasional replacement of a transformer. Be selective and use only those restoration techniques that are applicable to your radio.

Each restoration technique is followed by the specific Zenith models that are applicable to that technique, e.g., "Pulley Construction (applicable to the 3-R and VII)." Sometimes a suggestion is made whenever the restoration technique could be appropriate for the restoration of other manufactured radios (all radios, portables, battery radios, etc.).

The dimensions and measurements of the parts presented here may not be the same as the part in your set, so always double check all the characteristics of a part to make sure it will fit. It's always a good idea to build a test piece and perform a fit check. Most dimensions are not critical; however, be sure everything looks and functions as originally intended.

Prior to starting the restoration of the radio, review the radio description chapter for that radio, so that you are familiar with the layout and location of the various components and assemblies. Keep it close at hand throughout the entire restoration process. Compare your work to the drawings, diagrams and pictures in this book or other documentation that you may have. This will help make certain that the originality of the radio is maintained. It can also help in the identification of alterations that a previous owner may have made. If you do find any alterations, try to return the radio to its original configuration. Alterations can usually be distinguished from manufacturing changes by inspecting the wiring and general workmanship. If it looks neat and professional, it was probably a manufacturing change, but use caution when making this determination. A second or third opinion may be worth the effort.

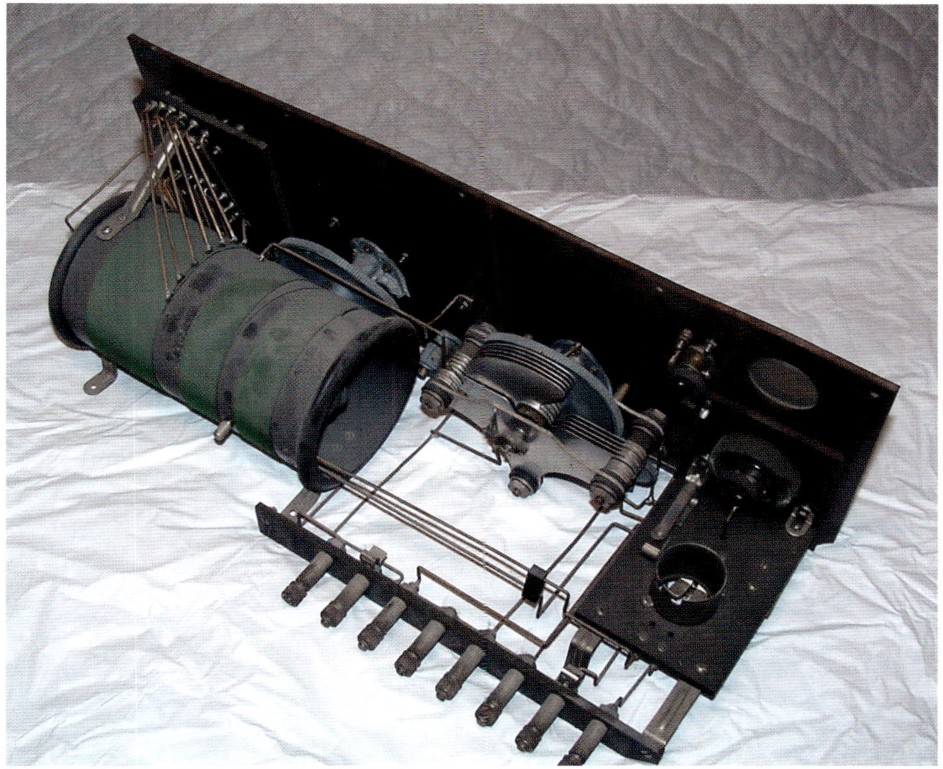

Unrestored 1-R Chassis

Step 1: Examine the Radio

During the examination take plenty of photographs and take numerous notes on what you find or don't find. Should you suspect that something is missing, use the radio description or other documentation to make comparisons. This information will be used during the assembly of the radio and can also be used as reference material for future restorations.

The value of visual and tactile inspections cannot be overstated. As a visual inspection example, a bright (75- to 100-watt) incandescent lamp illuminates an area in such a manner as to expose small cracks in components, and aids in visually locating shorts between wires. The tactile inspection involves the gentle movement of controls or components and defines the functional integrity of that control or component and any related mechanical or electrical assemblies. Make sure you mark the exact spot of any loose or broken connections so that you can locate and fix them later, or fix them now.

Perform a visual inspection of the exterior and interior of the radio. Remove any internal replaceable or loose components (screws, "C" battery, power cable, etc.) Always be on the lookout for loose mechanical (nuts, screws, rivets, etc.) or solder joints. Record or make notes about anything that is missing or needs restoration, and pay special attention the each of the following:

Outside Inspection

- Knobs, pointers and shafts
- Front panel condition (screws, raised washer)
- Cabinet condition (cracks, scratches, dings, trim molding, etc.)
- Decals (missing or damaged)
- Back door, spring hinges, underneath chassis mounting screws

Inside Inspection

- Loose parts or hardware
- Manufacturer's labels
- Power cable
- "C" battery
- Tubes

A final comment with respect to the tubes. Always remove a tube by holding on to the base and not the envelope. Pull the tubes and test each one. Then segregate the good tubes from the bad tubes (do the same for any other component that is bad). One bad or weak tube can degrade the performance of the radio. Later, once the radio is playing, the best-performing tubes will be selected.

Chassis Inspection. Remove the chassis assembly from the cabinet. See *Removal of Chassis from Cabinet* instructions in the description for this radio model.

The following is offered as an example of the type of data that needs to be recorded during this phase of the restoration. This particular process is applicable to the 1-R and 3-R type radios that have pulleys or gears, and may not apply to your particular set.

IMPORTANT—This activity must be completed before any disassembly of the chassis. The orientation or clocking of several components (vario-coupler, variable capacitor, etc.) needs to be determined to make sure they are reassembled in the current position. Correct orientation is essential to proper radio tuning. Make a special note of the orientation and direction of movement of the tickler coil located inside the main coil of the vario-coupler, in relation to the "T" pointer on the front panel. Make another special note of the orientation and direction of the movement of the rotating plates, located on the variable capacitor shaft, in relation to the "C" pointer on the front panel. Make sure you have enough information to reassemble them in their current configuration. This information will be used during the reinstallation of the pulleys in the *Chassis and Cabinet Restoration* section.

Perform a visual inspection of the chassis and make a note of any missing or broken or damaged parts. Frequently parts are missing or alterations have been made by a previous owner. Use the pictures and schematics in the descriptions chapter for your radio to help identify any missing parts or alterations. Pay special attention to the physical condition of the each of the following.

- Cabinet condition inside (cracks, broken wood corner supports)
- Circuit board (check for loose connections and rivets)
- Grid leak resistor
- Filament resistors (one for each tube)
- Battery mounting hardware and terminals
- Rheostat wiper
- Tap switch wipers
- Transformers
- Stage selection switch mechanism
- Bus wires
- External power socket
- Variable Capacitor (pulley, plates, vernier)
- Vario-Coupler (pulley, tickler, vernier)
- External terminals
- Underneath chassis (capacitors, wires, rivets)

Remove the grid leak resistor and filament resistors. Check each for proper resistance. Refer to the description of the radio for the proper values.

Connections Inspection. Broken solder joints and loose mechanical connections are a common problem in these early Zenith radios. Pay special attention to the connections on the vario-coupler. A little effort in performing this inspection now will save a lot of time and aggravation later on when power is applied to the set. Perform a tactile inspection of all mechanical and solder connections by gently moving each wire, rivet or component, while watching the connections. During this phase of the restoration, if a loose or bad connection is located, either make a written note of its location or fix it immediately.

AF Transformer Testing. In order to test the typical transformer, the windings must be isolated from the rest of the circuits, which usually requires physically removing the transformer from the set. The Zenith 3-R (L) AF transformers can be checked without removing them from the chassis, if several actions are performed before making the measurements. Remove all the tubes, disconnect (unplug) the battery cable and turn the switch to the OFF position. This opens all the circuits to the transformer primary and secondary windings, so that they can be checked without removal. Use an ohm meter to measure the DC resistance of the primary and secondary of each transformer (T1, T2, T3). Note: Some of the transformer connections may be partially obscured from view by a bypass capacitor, but can be reached with the meter probes. Compare these measurements to the *Transformer DC Resistance* values given for the transformers for your specific radio. If the measurements indicate an open circuit or are significantly different (2 or 3 times) from those suggested, then the transformer may need to be repaired or replaced.

Make sure that each part (pointers, pulleys, cable, batteries, etc.) that needs to be reproduced, repaired or modified has been identified.

At this point you should have a good idea of what it is going to take, in terms of resources and expertise, to restore your radio.

Outsourcing. After the radio examination has been completed, you may want to outsource some of the reproduction work; especially for tasks that you may not be comfortable performing, or for which you don't have the time or tools to complete the job (cabinet refinishing, making a pulley or pointer, etc.) If you do outsource some of the work, make sure you communicate exactly what it is that you want to have done before any agreements. For instance, if you are going to have someone build a pulley, provide them with the dimensions, show them the old pulley, show them the pictures in this book of these pulleys and then show them where and how the pulley would fit into the radio. That way they fully understand what you are asking them to do and why. It is recommended that you not start the actual restoration process until all the replacement parts have been built and acquired. Otherwise you run the risk of having to do some of the work over again, or not having the parts at all, or having parts that don't fit. If you do outsource something, always do a complete inspection, as soon as you receive it. Perform a fit-check to make sure that it fits.

Step 2: Restore the Components

Pointer Construction (applicable to the 1-R, 2-M, 3-M, 3-R and 4-R). The pointers are small, medium and large. The vernier spacer is similar to the small pointer, but without a pointer. Notice that the large pointers for Variable Capacitor and Vario-Coupler are the same, but have different center hole sizes.

The templates on page 229 can be used to trace the replacement pointers. Measure the screw hole dimensions of your knobs and compare those dimensions to the template screw holes. Make appropriate adjustments to the templates in order to make sure that the pointers will fit to your knobs properly.

Cut the templates from a copy of the Pointer Template page. Place the templates on a 0.125" aluminum bar and trace around each with a thin tip black marker. Center-punch the shaft and screw holes. Make as many traces as needed.

Use a band saw with a narrow blade (0.125" wide) to rough cut each of the pointers.

Rough Cut Pointers

Utilize a small bench vise to hold the pointer. For each pointer, file a small amount off of one edge then turn it over and file a small amount off of the other edge. Keep rotating the pointer and filing, taking off only a small amount of material at a time. As the pointers are filed, compare them to each other in order to make sure that they all match when they are completed.

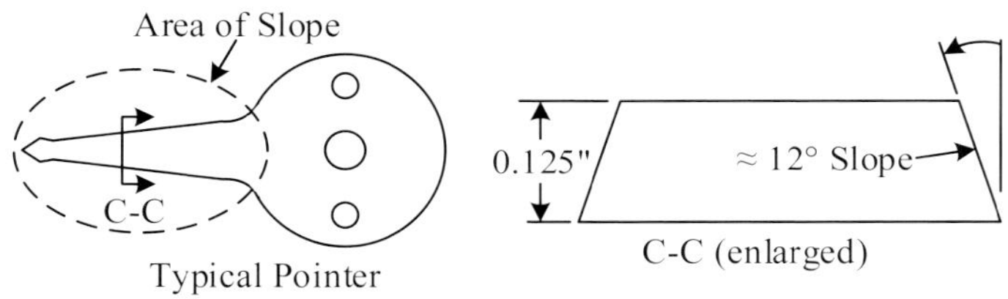

Area of Slope

Pointer Slope (original)

The variable capacitor pointer uses an adaptor that allows the pointer to be mounted onto the shaft. Machine the adaptor, shown below. The 6-32 threaded holes are for the set screws. Use a small amount of J-B Weld to attach the adaptor to the pointer on the countersink side, as shown.

Add the slope to the edge of each pointer, as shown in the Area of Slope diagram above.

Drill the holes, with the appropriate size bit, at the center punch marks. Countersink the knob screw holes on the opposite side from the slope. Use a very fine file to smooth the edges and then use fine steel wool or very fine sandpaper for additional smoothing. Perform a final polish with very fine "0000" steel wool.

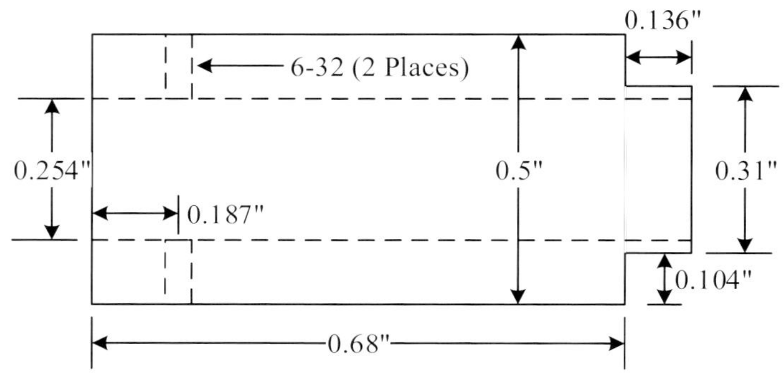

Variable Capacitor Pointer Adaptor Dimensions

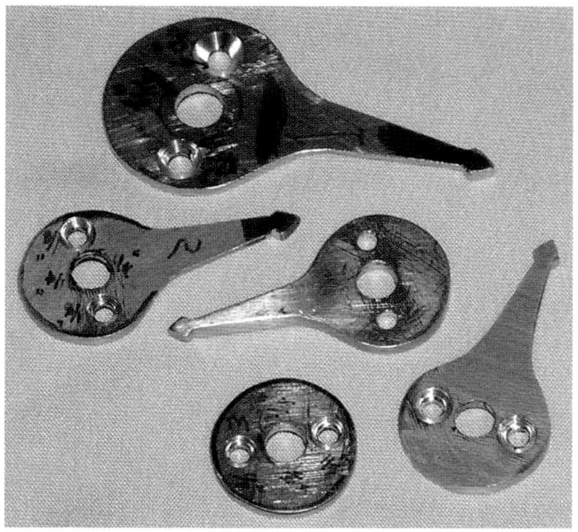

Finished Pointers

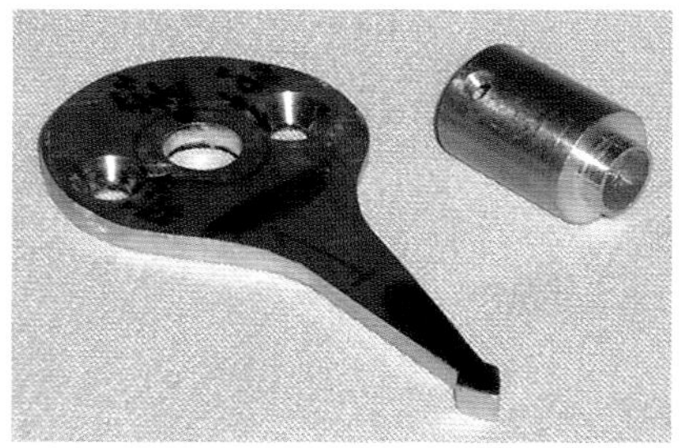

Variable Capacitor Pointer and Adaptor

Pointer Templates

Do not cut the Templates from this page; you may need them in the future. Make a photocopy of this page. Cut out the templates from the photocopy and compare them to the Templates on this page, to make sure they are the correct size. Follow the instructions on the previous page to make the pointers.

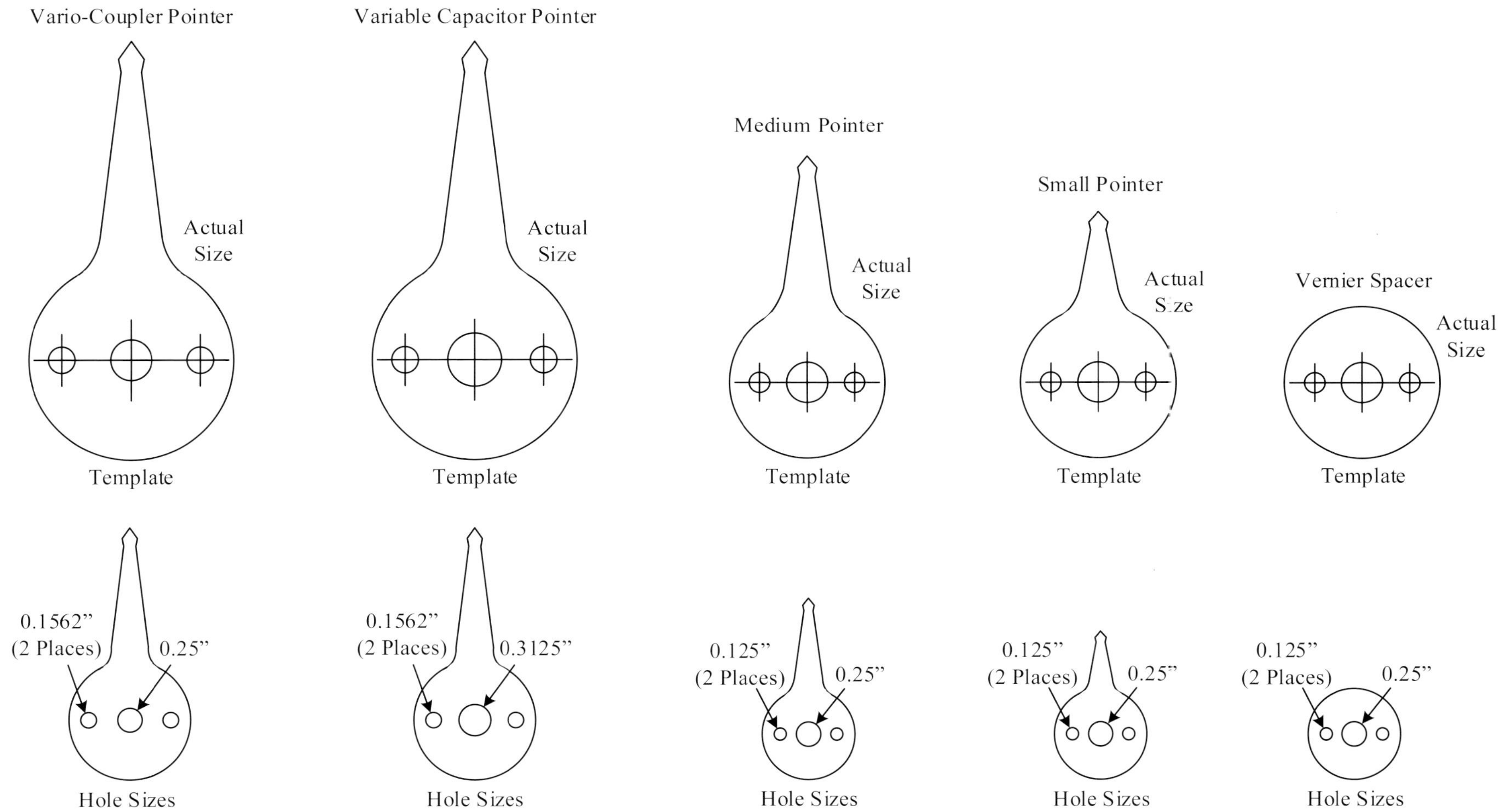

From *The Early Zenith Radios: The Battery Powered Table Sets 1922–1927*, © 2014 by Gilbert M. Hedge with Durell M. Roth. ISBN 978-0-7643-4674-3. Schiffer Publishing Ltd. www.schifferbooks.com

Vario-Coupler Pulleys and Partially Completed Blanks

Variable Capacitor Pulleys and Partially Completed Blanks

Pulley Construction (applicable to the 3-R and VII). The large pulleys are made from 0.25" thick by 4" wide aluminum flat plate. An insert is added in order to provide holes for the set screws that hold the pulleys to the shafts. See the Pulley Detail on the next page.

The first step is to machine the pulley inserts shown below. Both inserts are identical except for the hole in the center. The hole in the center of the Vario-Coupler Pulley Insert is 0.25" and the hole in the center of the Variable Capacitor Pulley Insert is 0.375".

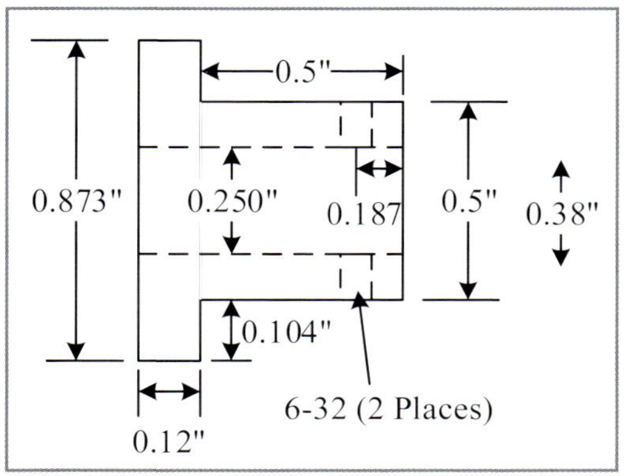

Vario-Coupler Pulley Insert Detail

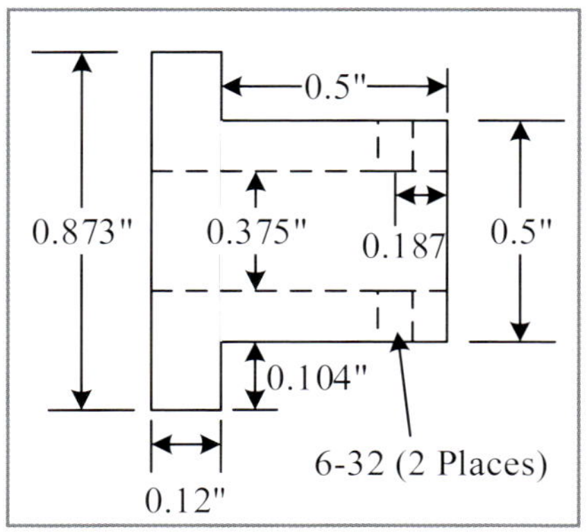

Variable Capacitor Pulley Insert Detail

The 2 large pulleys are made by center punching and then making 3.75" circles on the 0.25" aluminum plate. Rough cut each on a band saw. Drill a 0.5" hole in the center of each for the inserts. Glue each insert into each pulley with a small amount of J-B Weld. Use a lathe to machine a V groove into the outside edge of each pulley. The outside diameter is not critical but should be approximately 3.65". Machine the face of each pulley.

Mark and drill two holes for the dial string and spring screws. Thread each hole with a 6-32 tap. Mark and drill the dial string access holes on the outside V groove, making sure they are aligned with the dial string and spring screw holes. File these holes completely smooth so that the dial string will not chafe. Perform a final polish with steel wool.

Both Sides of Pulley

The variable capacitor shaft is made of 0.375" brass round bar and holds the pulley, rotating plates, spacers and locknut. The large end is threaded as shown with a 3/8"-24 NF die. The small end is 0.25" thick and receives the variable capacitor adaptor that holds the pointer and knob.

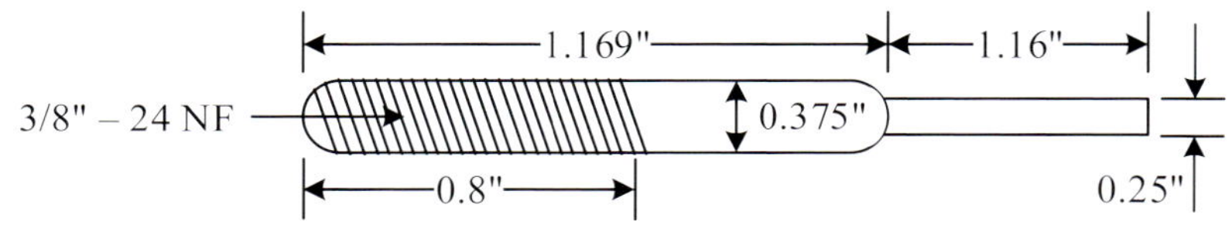

Variable Capacitor Shaft Detail

The two small vernier pulleys were made from 0.5" brass round bar and machined on a lathe. They measure 0.4375" long and 0.5" diameter. Each has a 0.25" shaft hole in the center and a 6-32 threaded hole for a set screw. The dial string groove is approximately 0.1875" wide, 0.25" deep and near the end.

The reassembly of the variable capacitor assembly is described on page 241.

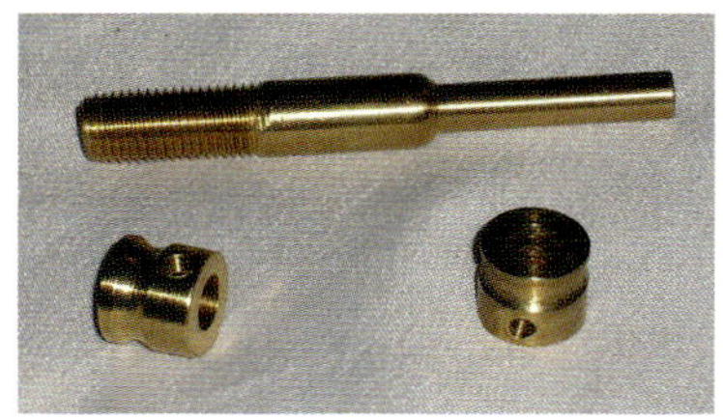

Variable Capacitor Shaft and Small Pulleys

Pulley Detail

Note: The same Variable Capacitor pulley and Vario-Coupler pulley is used in both versions (Early and Late) of the 3-R.

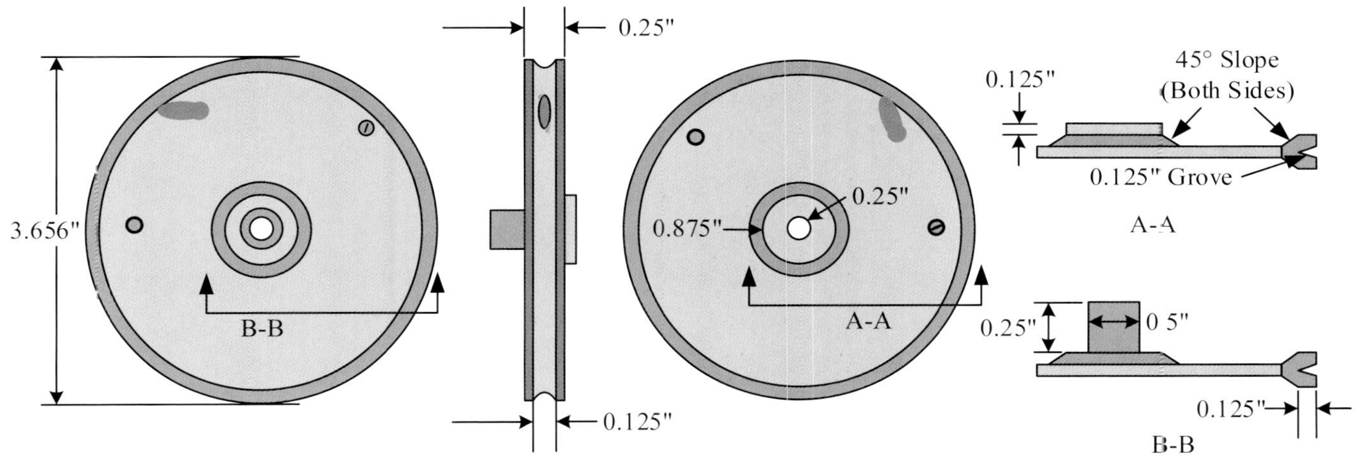

Vario-Coupler Pulley

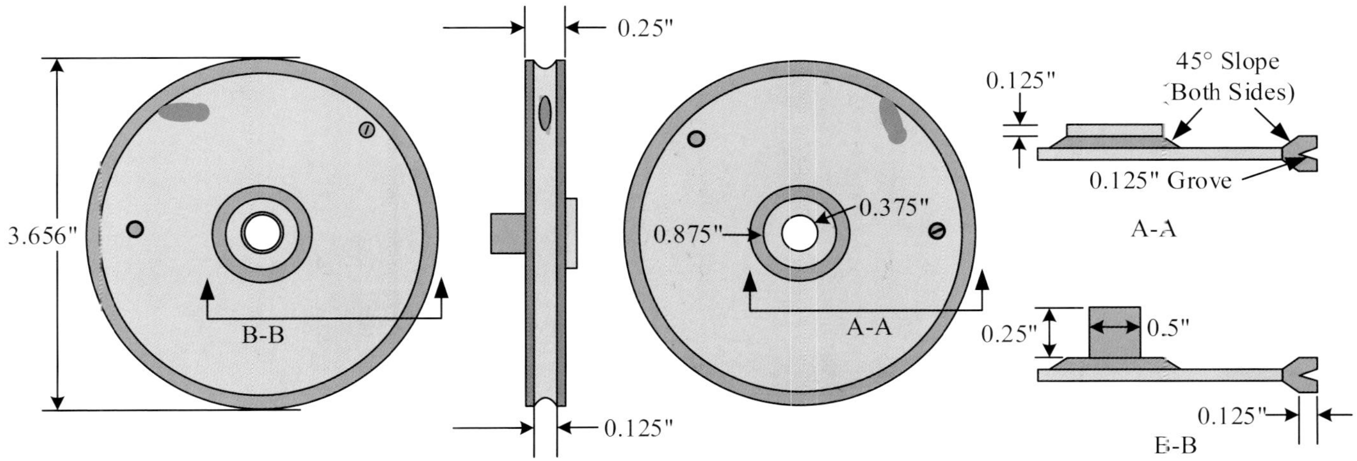

Variable Capacitor Pulley

3-R 4-R

Trim Molding Construction (applicable to the 3-R and 4-R). These radios have trim molding strips on the front panel that are attached to the bottom of the cabinet, shown above. They are frequently missing and some radios may not have had them. There were some inconsistencies with the number of screws required and the distance between the mounting holes. The trim molding for the 3-R is 22" long and the 4-R is 19.5" long. They are made from 0.3125" square strips of mahogany that are 2 feet long (the 3-R trim molding requires an 80° backside angle to compensate for the front panel slope.) Each piece is then shaped with a 3/16" Roman Ogee router bit and cut to the proper length. Trim to fit and shape the ends. Drill the holes to match your set and countersink each. Apply stain and several coats of lacquer.

Reproduced 3-R and 4-R Trim Molding

Battery Bracket Construction (applicable to the 1-R, 2-M, 3-M, 3-R, 4-R, and other battery radios). The construction of two different common battery brackets is discussed here. Because there is so much variation in the bending process, some details have been omitted and the actual dimensions should be determined on a trial and error basis, using the appropriate battery.

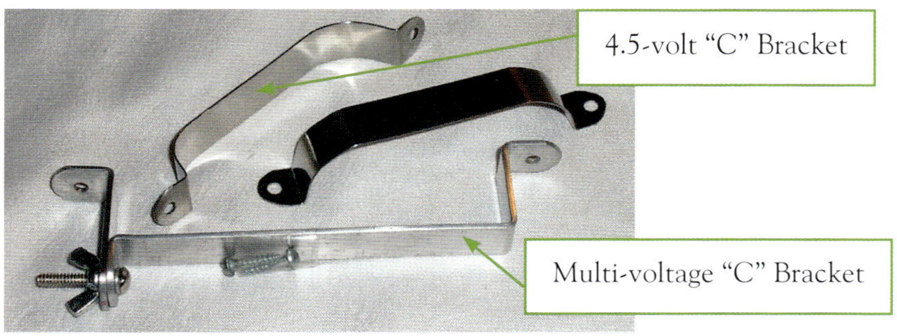

Battery Bracket Examples

The multi-voltage "C" battery bracket, shown below, for the 3-R (E) and 4-R (E) is made from 0.0625" thick by 0.5" wide aluminum flat bar. The dimensions of the two pieces can be determined fairly easily by trial and error. Cut the required strips, drill the mounting holes, and smooth with light sandpaper or steel wool. Bend as appropriate.

The 4.5-volt "C" battery bracket for the 3-R (L), 4-R (L), 2-M, and 3-M is made of 0.0312" thick by 0.5" wide stainless steel bar. Cut a strip of stainless 4.375" long and drill 0.125" mounting holes 0.3125" from each end. File the ends smooth. Bend the ends 90° about 0.3125" from each end, in the same direction. Now use a 0.5" round object to make the large bends starting at 0.375" from each end.

Multi-voltage "C" Battery and Bracket

"C" Battery and Bracket

Multi-Plug and Power Cable Construction (applicable to the 3-R and 4-R). The original Jones Multi-Plug is disassembled by unscrewing the internal platform out of the case. The multi-plug reproduced here uses a small no. 2 flat head screw to secure the pin holder to the case, after it has been slid into place. All other aspects of this reproduced multi-plug are very similar to the Jones Multi-Plug. There are three main parts to the multi-plug: case, pin holder, clocking post.

The case and the pin holder are made of 1.5" solid black Acetal Rod. The post and pins are made of brass.

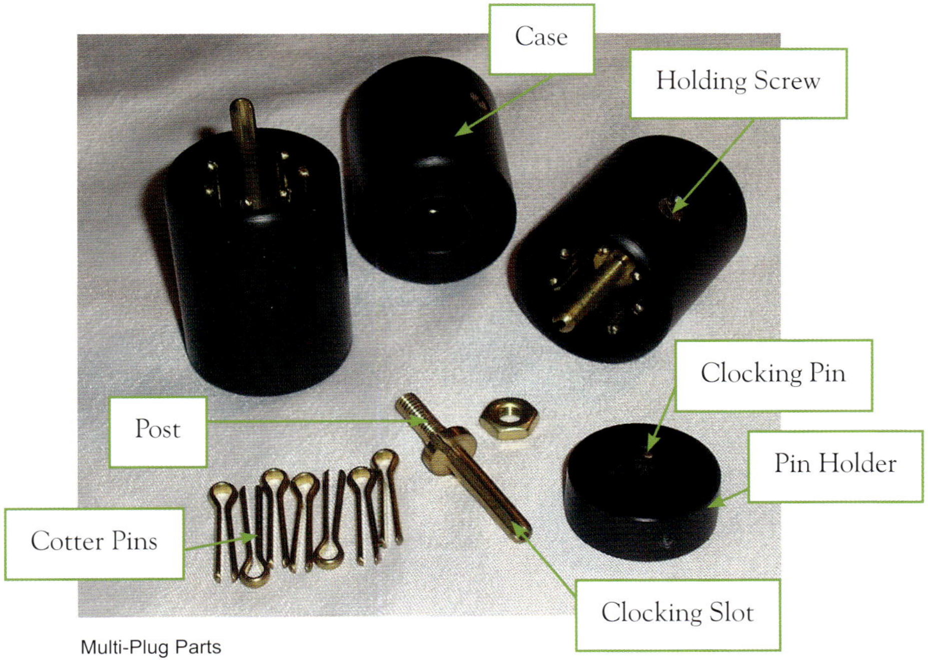

Multi-Plug Parts

The multi-plug is assembled by inserting the post that has a shoulder with a clocking slot, into the center of the pin holder. The clocking slot slips over the clocking pin in the pin holder. The clocking slot ensures the multi-plug can only go into the power cable socket one way. A brass nut is then installed on the large end of the post to hold it in place. The pins are then inserted into the pin holder. The pin holder assembly is then inserted into the case. The pin heads rest on a shoulder inside the case. The hole in the case is aligned with a hole in the pin holder and the no. 2 flat head is installed to hold everything together.

Specific dimensions and reproduction techniques are not provided due to the complexities involved in the construction of this multi-plug. If you are interested in reproducing this multi-plug, obtain an original Jones Multi-Plug and use it as a guide, along with the techniques described here, to create a reproduced multi-plug.

Power Cable Assembly. The Zenith power cable is assembled as follows. Slide the case over the cable, small end first. Then solder a brass pin to each wire. Adhere to the color codes (wire, pin holes) that are shown in the *Power Cable and Multi-plug* diagram in the description for this radio model. Insert each pin, by wire color, into the corresponding hole in the pin holder. Put a small tie wrap where the wires meet at the cable and tighten, for support. Slide the case down the cable until the pin heads rest on the shoulder inside the case. Turn the case until the holes line up and insert the screw and tighten. Install suitable lugs on the end of each loose wire. Perform a continuity test on each wire and double check the color code for each wire and pin connection.

The cable can be either a 7 wire cable (used in the early version) or a 5 wire cable (used in the late version). An 8 wire (part MCC-8) cable can be obtained from Radio Daze at www.radiodaze.com. Another cable option, which is used here, is the very flexible Selector Cable (Otis Elevator part VP-730550). See the 3-R and 4-R Multi-plug information on page 188.

See the individual radio descriptions for specific details related to external power for a particular set.

Pins Connected to Wires (notice color coding of wires and pin holder)

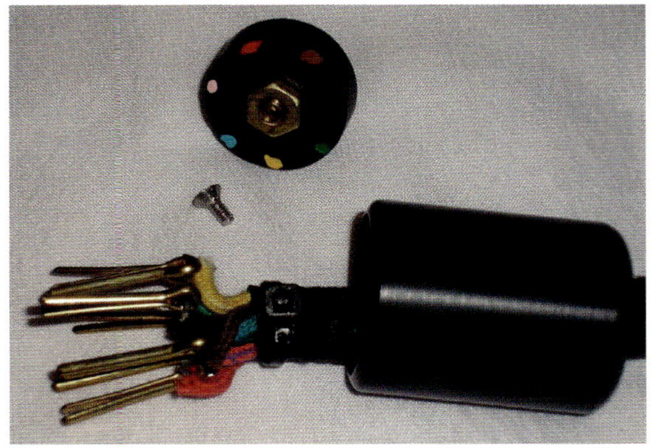

Completed Zenith Power Cable with Reproduced Multi-Plug

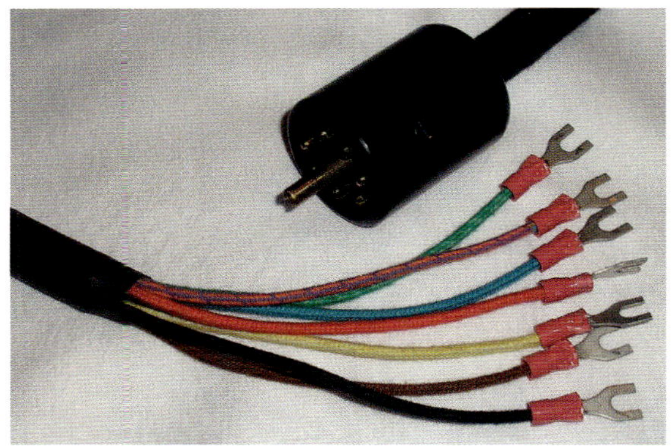

Paper Mâché Horn Speaker Repair or Construction (applicable to the Super-Portable and other radios that have similar horn speakers). Many of the radios that were produced in the early 1920s had built-in horn speakers of various shapes, sizes and colors. They were usually made of wood, plastic or paper mâché. The damaged ones that were made of paper mâché can be repaired fairly easily. Those that are missing, maybe made of some unknown material, can possibly be replaced with a reproduced paper-mâché horn speaker, provided the dimensions and construction method can be determined.

Repairing the broken or damaged paper mâché speaker is pretty straightforward, as long as it is not too extensive. Mix enough paper mâché to complete the job (always follow the manufacture's recommendations). Apply the mix to the area and form as needed. Let it dry completely, sand, and paint. The thicker the paper mâché application, the longer it will take it to dry.

Paper Mâché Repair

The Zenith Super-Portable paper mâché horn speaker, or other similar horn speakers, can be reproduced using the following steps: determine the dimensions and create a drawing, construct the wood mold, create the paper mâché horn speaker, and make the appropriate cuts to form the 90° turn. Each of these steps will be discussed in greater detail.

The shape and size of the horn speaker can be obtained by copying an existing speaker or by measuring the dimensions of the void where the horn speaker was supposed to be, or by trial and error. One way to get the exact angles and dimensions is to make an exact life-size drawing of the horn speaker, laid out on a large piece of paper or poster board. Since a mold is being created, the life-size drawing needs to define the inside surfaces of the speaker. Then the lengths, widths and angles can be measured and transferred to a normal-size drawing. Chapter 6 includes a drawing of the Super-Portable horn speaker (page 106), with all dimensions and angles. The drawing also includes some additional notes about the construction of the horn speaker.

- Construct a wood mold, utilizing the inside dimensions and angles from the drawing (this requires a wood lathe). Or have a local wood shop make the mold. Each end should have a little extra length, which will be trimmed to fit later.
- Wrap the mold in clear plastic wrap (food wrapping) in order to make sure that the paper mâché does not stick to the wood (depends on the type of wood) and assure a fairly smooth inside surface.
- Mix the paper mâché according to the manufacturer's instructions and then apply it to the entire mold, starting at the bottom and working toward the top. Take your

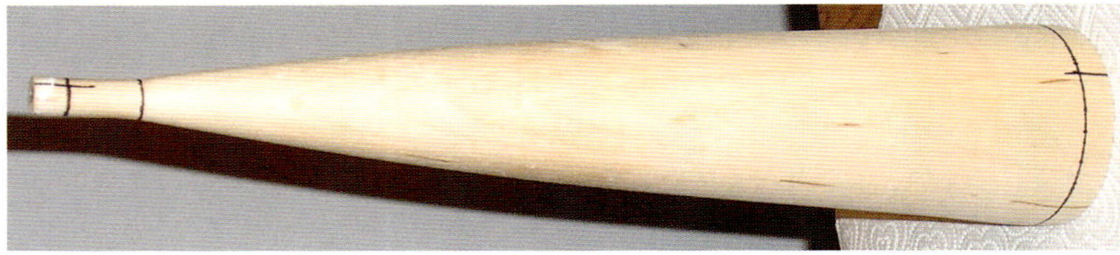

Horn Speaker Mold

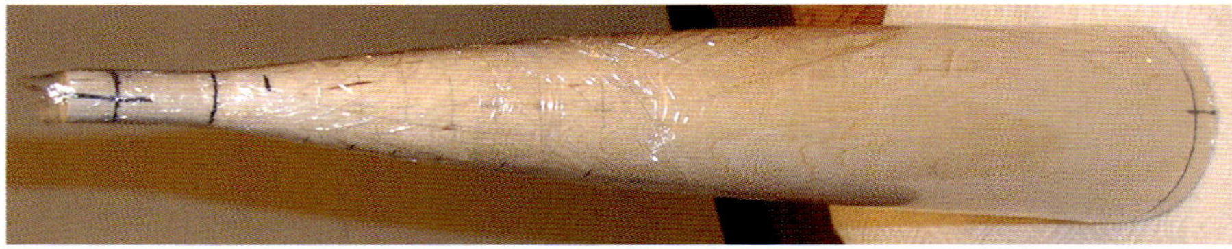

Mold Wrapped in Clear Plastic Wrap

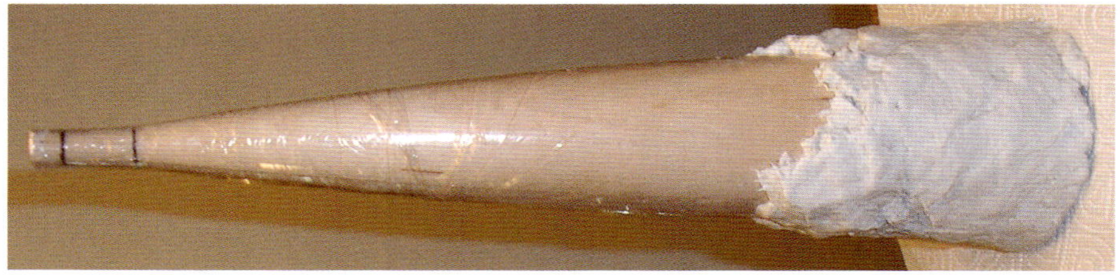

Application of Paper Mâché

Wet Paper Mâché

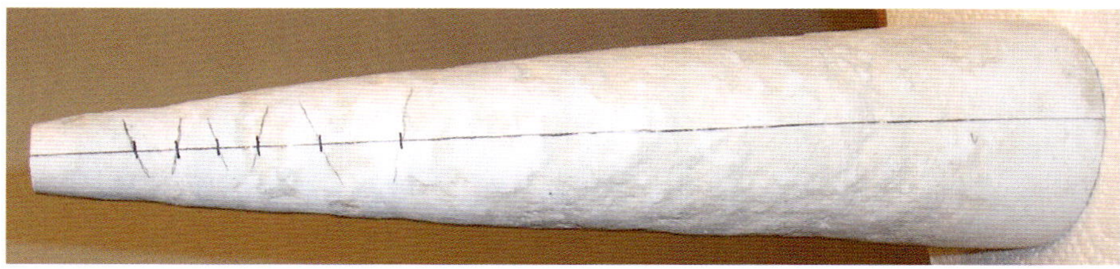

Markings

time because the paper mâché dries slowly. The thickness of the application should be as uniform as possible, about 0.125" to 0.1875" thick. Allow it to dry, which may take several days. Once dry, remove the newly formed horn speaker from the wood mold.

Trim each end, just enough to square them up. They will be trimmed to fit, later.

• Draw a long straight line, center to center, from the small end to the large end. Draw another line on the other side that is 180° from the first line. These lines will be used later to help align the individual pieces after they are cut and then rotated during reassembly.

• On the small end, apply 6 marks, at the locations shown in the drawing (page 106). Each of the marks is for one of six cuts that will be made later. Draw a line at each mark, in the directions indicated. These lines are about 7.5° and show the angle and direction of the cut. These lines just provide an indication of the direction of the cut, but the actual true angle of 7.5° will be measured on the band saw relative to an imaginary straight line through the center of the horn speaker.

On the band saw, keep the imaginary center of the speaker level with respect to the saw table and at an angle of 7.5° in relation to the blade. The 7.5° cut should go through the mark (cross) in the center, so band saw alignment is critical. Make the six cuts completely through on the band saw.

Prior to gluing, perform a fit check by reassembling the pieces in the same order that they were originally in, except turn each piece 180°. Use some tape to temporarily hold the pieces in place. The small top piece should be at a right angle to the largest bottom piece. If not, rotate some of the individual pieces to get the exact 90° angle. There will be a slight mismatch of the inside at each intersection of the pieces, due to the rotation of the pieces. File each inside intersection prior to reassembly as appropriate. If everything looks good, mark each intersection and number each piece, so that they can be put back together in the current order and position. Remove the pieces, apply white glue, reassemble and let dry overnight.

Perform a fit check and trim the ends to fit. Sand, paint and install.

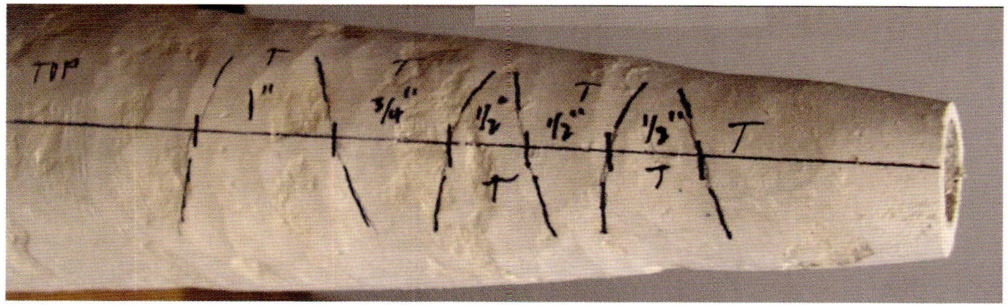

Pre-Cut Markings

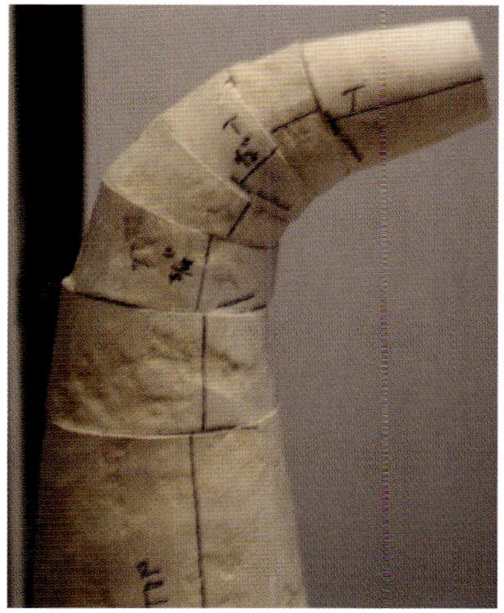

Flipped Pieces

Original Horn Speaker

Installed Reproduced Horn Speaker

General Battery Construction Considerations (applicable to all battery radios). Reproduced batteries should always look and function just like the original batteries when they are installed into the radio. The only difference is the use of modern internal batteries that are removable and easily replaced.

There are four basic criteria that need to be determined in order to design a replacement battery: the voltages required, the style of the case, the dimensions of the case, and the type of terminals. Once these have been identified there are several options available for building the battery. Any types or styles of modern batteries that are available can be used internally in the reproduced battery, as long as they meet the voltage requirements for the radio and will fit inside the case.

It's always best if you have an old battery that can be copied, but if not, then there are several ways that the requirements for the reproduced battery can be determined.

The voltage requirements can be determined from the radio or from a schematic of the radio. Some batteries provide multiple voltages such as "C" battery −1.5v, −3v, −4.5v, −6v, −7.5v, or "B" battery +22.5v, +45v. The number of terminals and their style, the voltage or voltages and their polarity (+, −) must be identified.

The size of the reproduced battery case can be determined in a couple of different ways. If you have an existing battery, use the dimensions from that battery. Or use a pamphlet or other literature that the various battery manufacturers distributed in the 1920s. Another method is to measure the void where the battery was placed inside the radio. Sometimes the radio had a battery bracket that can also be used to help determine the dimensions. Always use care when making these types of measurements, because the reproduced battery needs to fit and look like it belongs inside the radio when completed.

Next determine what type of terminals is needed, where the terminals will be located on the case, and how they will be mounted. Most of the batteries of the 1920s used either Fahnestock clips or knurled nuts and some even had unterminated wires for making the connections to the radio.

Any material that is available can be used to make the case, but make sure that glue and paint will adhere to the surface.

A removable door also needs to be a part of the case so that the batteries can be easily replaced. One option is to add a door to the bottom of the case that would be held in place with small screws, which is the method that was used for the example "B" and "C" batteries. Another method is to create a door that is held in place by friction, which is the method used for the "A" battery example. The friction type door requires a precise construction method.

The reproduced battery needs a label that gives it that old style battery look. Use a computer and any good graphics software package to create the label. The design can be obtained from any old battery, or just create your own style that mimics the old style batteries. Once the label has been created, print it out, cut out the label and perform a test fit. Make adjustments to the computer version and retest. Print out the final label on a color laser printer, if possible, and glue it to the case.

Various Reproduced Batteries

"A" Battery Construction (applicable to the Super-Portable and other portable radios). The 1.5-volt, No. 6 "A" battery (2.5625" by 6.125") was made from a carpet freshener container that was obtained from a local dollar store. This particular container has the proper width and length to make an original-looking reproduced "A" battery.

The metal bottom of the tube needs to be removed and replaced with a 0.0625" piece of plywood. Cut the tube at both ends such that the total length is 6.125". Cut a 0.25" ring from the excess tube for later use to make the stop for the lid.

Tube

Lid Stop Ring

Lay out the lid on a piece of 0.375" plywood that measures the same as the inside diameter of the tube. Lay out a similar piece of 0.0625" plywood for the bottom. Drill and tap 6-32 holes in the center of each piece and an additional 6-32 hole in the lid that is 0.3125" from the edge for the negative (−) terminal. Rough cut the circles with a band saw. Machine each to the inside diameter of the tube. Each piece should fit tightly into the ends of the tube. The lid should be slightly tapered to ensure a snug fit. Sand and paint before final assembly, as the paint affects the fit.

Lid Layout

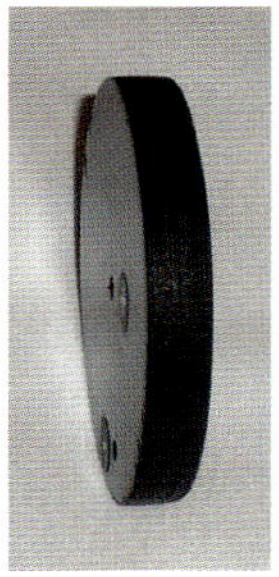
Completed Lid

Top View of Lid

Trim the "D" cell battery holder wires to length and connect the red wires together and the black wires together and attach two lugs. This puts two 1.5-volt batteries in parallel and supplies twice the current of one cell.

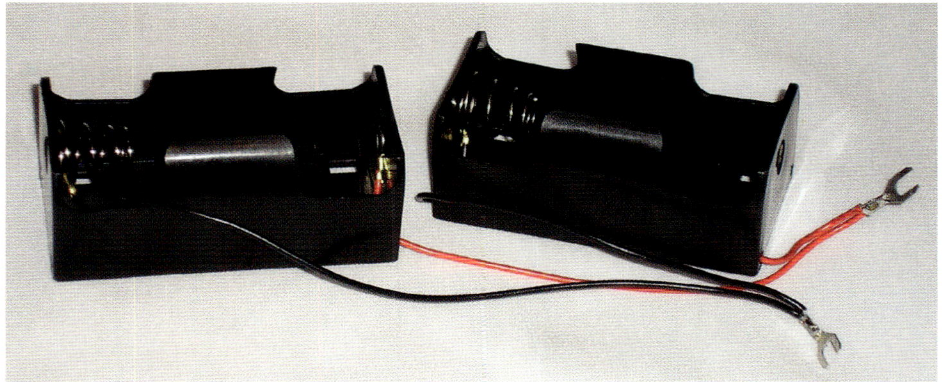
"D" Battery Holders

Remove 0.125" of tin foil from the inside bottom of the tube. Apply white glue and insert the 0.0625" plywood to enclose the bottom.

Starting at 0.375" from the top of the container, remove 0.25" of the foil from around the inside of the tube. Apply white glue all the way around the outside of the stop ring. Place the stop ring around the inside of the tube where the tin foil was previously removed. Use the tapered lid to set the stop to the proper depth. Make sure that the top of the lid is even with the top of the tube. Remove the lid and let the stop ring dry. Once it is dry, you should be able to push the lid into the tube and it will stop where it is even with the top of the tube. If everything fits properly, the lid will hold in place with friction.

Completed Bottom

Lid partially inserted into tube.

Screw two 0.75" by 6-32 brass screws through the underside of the lid. Attach the battery holder wires to the terminals on the lid with the black going to the outside terminal and the red going to the center terminal.

Install batteries and check the voltage for +1.5-volts on the center terminal with the negative terminal as reference. Slide the batteries and lid into the tube. One word of caution: the length of the tube as described here (6.125") works correctly with a 0.375" lid, the length of the two "D" cell battery holders and the thickness of the bottom. Changes in any of these may result in the lid not closing properly.

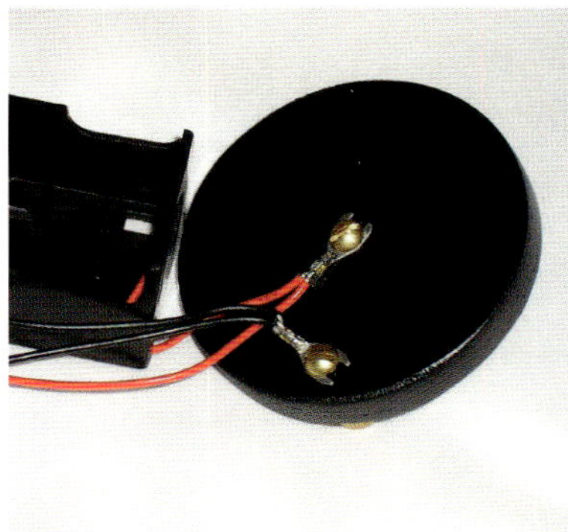
Battery Connection to Lid

Final Assembled Battery Holders and Lid

Paint the top and bottom of the case in preparation for the label. Create a battery label on the computer with any graphics software package. Also include a + and negative indicator (with black background). Print the label and cut it out. Perform a fit test by wrapping the label around the case. Make adjustments to the computer version. Print the final version, with a laser printer if possible. Glue the label to the case.

Glue a positive (+) indicator next to the center terminal and a negative (−) indicator next to the terminal near the edge. Install a brass washer and knurled nut onto each screw on top of the lid.

Install the battery assembly (batteries, holders, lid) into the tube. Check for proper voltage.

"A" Battery Top and Bottom

"B" Battery Construction (applicable to the 1-R, 2-M, 3-M, Super-Portable and other battery radios). There are many different sizes of "B" batteries with various terminals for the voltages ranging from 22.5 volts to 120 volts. The external dimensions of the "B" battery described here measure 4" long by 2.625" wide by 3" high and has three terminals: negative (−), +22.5 and +45.

The case is made of 0.125" plywood. Draw (allow 0.125" depending on how you fit the corners together) and cut the 6 pieces (2 sides, 2 ends, top and removable bottom). Glue the 2 sides and 2 ends to the top. Measure the inside from the bottom of the top to the bottom edge of the case and cut four corner-support 0.25" by 0.25" pieces to that length. Insert and glue the four pieces into the corners. Make sure these corner supports run the full height of the battery and are flush at the bottom. Sand all sides smooth.

"B" Battery

"B" Battery Pre-cut Pieces

Place the removable bottom in place on the bottom of the case and drill the four corner holes. Use a bit size appropriate for the screws that are available. Countersink the holes in the removable bottom. Install the removable bottom with the screws and sand smooth.

Mark the location for the three terminals on the top of the case. Drill and tap the three 6-32 terminal holes. Paint the case with black paint.

The batteries use five 9-volt snap on connectors and one 4.5-volt battery holder. Solder the connectors in series with the 4.5-volt holder tapped between the third and forth 9-volt connectors. Use a 6-32 screw of the proper length to connect the wires through the bottom of the case. Connect the black wire to the negative terminal. Connect the last red wire to the 45-volt terminal. Connect the red 4.5-volt wire to the 22.5-volt terminal. Install the batteries and check the voltages and polarities. Adjust if necessary. Install one washer and knurled nut on each terminal. Should another style of terminal be used, proceed appropriately with the installation.

The battery label can be made in two identical pieces for easy installation. Create, print and install the label and voltage indicator labels.

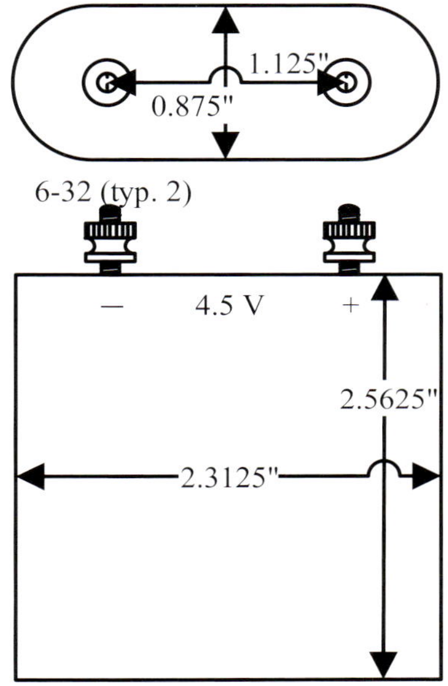

"C" Battery Outer Dimensions

"C" Battery Construction (applicable to the 2-M, 3-M, 3-R, 4-R, VII and most other battery radios). The 4.5-volt "C" battery was a very common battery that was used to provide a negative voltage on the grid of the tube and controls the flow of electrons from the cathode to the plate.

The case is built with six pieces: 2 sides, 2 ends, top and bottom. Determine dimensions for each piece using trial and error.

Cut the individual pieces and lightly sand each. Glue the sides to the ends with white glue. Clamp until dry.

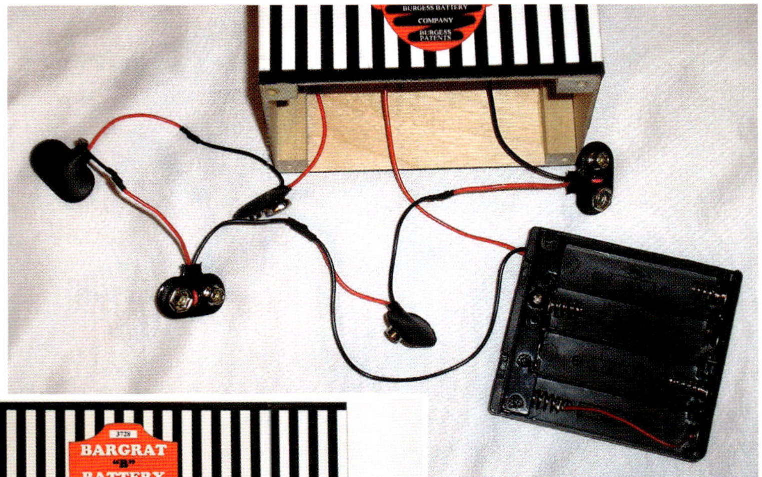
Battery Holders

Battery Labels and Voltage Indicators, before cutting.

Clamped Pieces

Case Pieces

Removable Door

Corner Marks on Top

Four Corners Trimmed

Now glue the top to one end of the box. The top will be where the terminal posts will be installed later. Clamp and let dry.

The removable bottom will be attached to the box bottom with two small brass wood screws. Place the bottom on the bottom end of the box and drill two pilot holes for the brass wood screws. Countersink the holes and install the screws.

Lightly sand just enough to smooth all sides and edges of the box.

The original "C" battery had rounded ends. The rounded ends of this box will be simulated by trimming the corners at 45°. Mark the top end with two even 45° lines (approximately 0.25"). Use a table saw to cut the four 45° corners. Carefully align the saw blade to one of the marks. Use a test piece of wood and compare it to the box. If it looks close, cut each of the 4 corners of the box. Sand all edges.

Drill and tap the two 6-32 terminal holes at the locations shown.

Paint the top and bottom of the case in preparation for the label. Print the label and cut it out. Perform a fit test by wrapping the label around the case. Make adjustments to the computer version. Print the final version, with a laser printer if possible. Glue the label to the case.

Obtain a 3-AAA battery holder, available from suppliers like Radio Shack. Use two brass 6-32 screws that are 0.5" long. Strip each wire and wrap the ends around each of the screws, or use lugs. Insert the screws from the inside of the bottom, into the holes of the top. Snug them but do not overtighten.

Install three AAA batteries and check the 4.5 volts. Check the positive and negative polarity. If everything is correct install the bottom with the two wood screws.

Install a brass washer and knurled nut on each terminal.

Typical "C" Battery Label

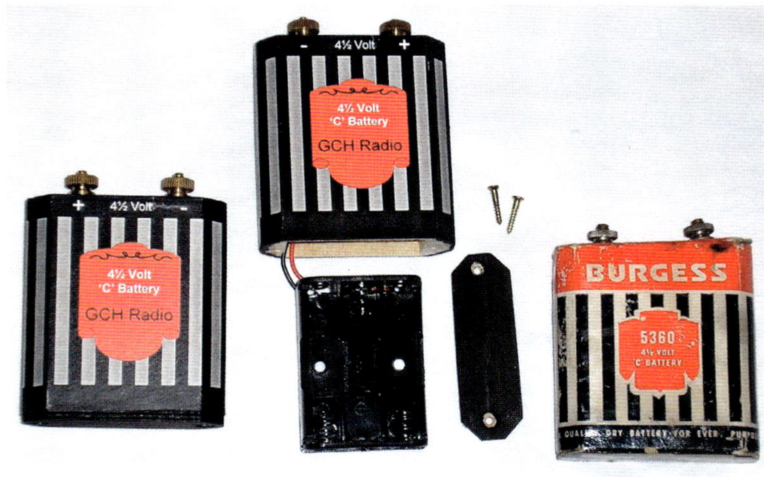

Reproduced and Original "C" Batteries

Variable Capacitor Assembly Repair (applicable to the 1-R and 3-R). The variable capacitor will need to be removed from the chassis. The repair of the variable capacitor requires that it be disassembled, a new pulley and shaft be installed, and then that the unit be reassembled. Adjustments will need to be made so that the plates mesh properly. Several quick checks will be made to help determine if it has been assembled and adjusted correctly. Perform the following steps to repair the Variable Capacitor.

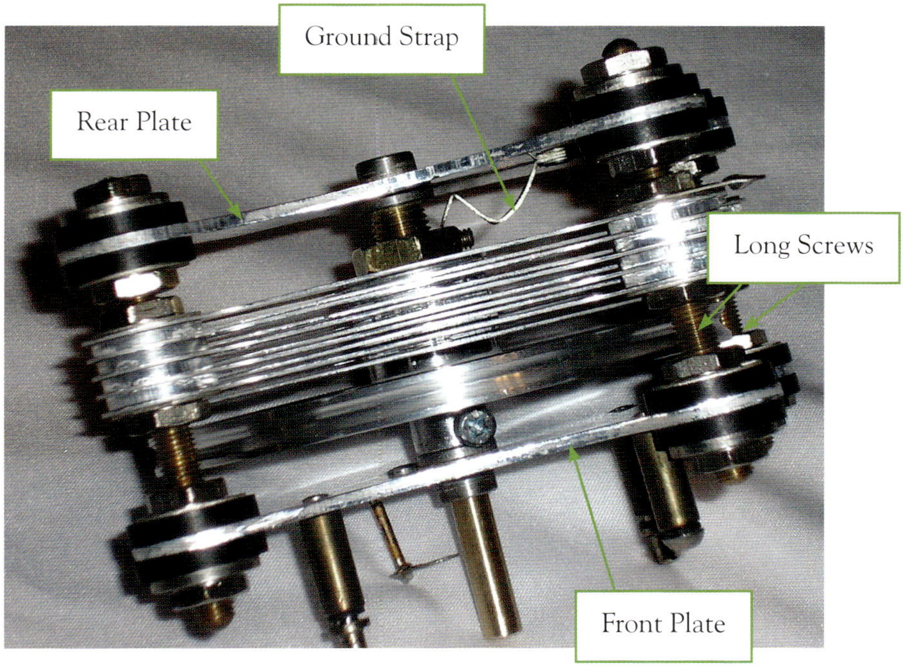

Variable Capacitor Assembly

Original Variable Capacitor

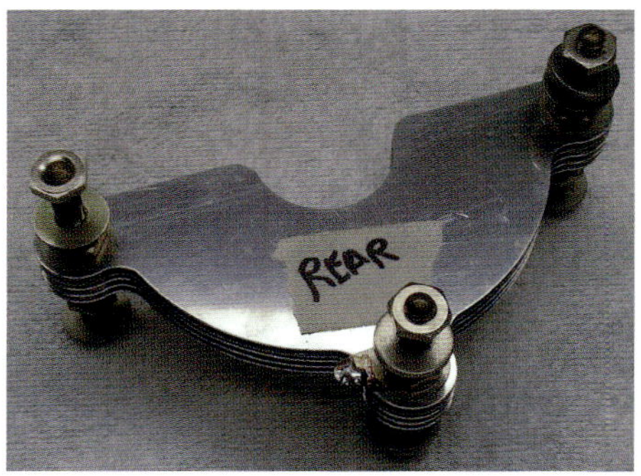

Stationary Plate Mark

1. Remove the Variable Capacitor from the chassis. See *Removal of Variable Capacitor from Chassis* procedure for this radio model.

2. Remove the front plate from the Variable Capacitor assembly.

3. Mark the stationary plate next to the rear plate (for later assembly). Remove the rear plate from the Variable Capacitor assembly.

4. Remove the shaft assembly from the stationary plate assembly.

5. Disassemble the Variable Capacitor shaft. Remove the pot metal material from the plates and washers with a screwdriver and hammer (tap lightly and do not damage the aluminum or brass parts). Remove pot metal from round holes (e.g., nut) by drilling the center with a small bit (e.g., 0.0625"), then gradually step up the drill bit size until most of the pot metal is removed. Do no damage to the parts or threads. Use a small sharp object or punch to knock out the remaining pot metal.

6. Straighten each individual rotating plate, by laying it on a flat surface and lightly tapping high spots with a small hammer. Check both sides until the plate is perfectly flat. Clean all parts.

Parts Containing Pot Metal

Drill out pot metal material.

Straighten rotating plate.

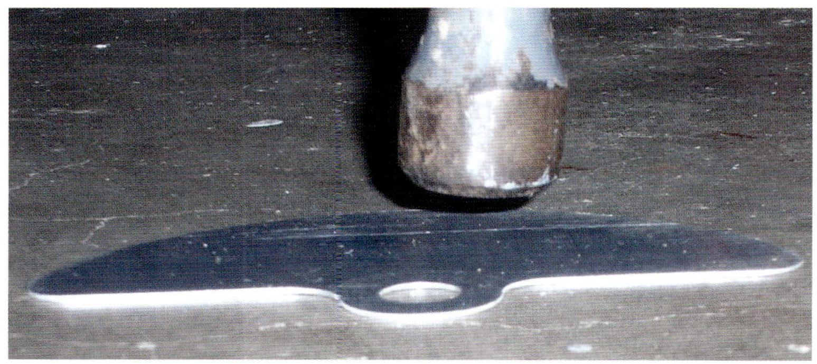

Two Views of Assembled Shaft

7. Place the pulley on the shaft with the set screw toward the small end. Slide the plates and spacers onto the shaft. Install the 0.375" brass nut. Lightly tighten the nut and all set screws.

8. Install the shaft assembly into the stationary plate assembly. Install the front and rear plates. Hand tighten all nuts.

9. Center the three bolts through the front and rear plates (use a double nut on one end). Adjust the pressure on the ends of the shaft with the front and rear plates, so that the 0.25" shaft will be hard to rotate by hand.

10. Adjust the fixed plate spacers so that the rotating plates turn without touching the fixed plates. Position the ground strap as shown.

11. Make any final adjustments required to get the rotating plates to mesh with the stationary plates, without touching. Patience and attention to detail are required at this point. The following checks can be made to make sure the Variable Capacitor operates correctly.

12. The rotating plates should rotate 360° and never touch any of the fixed plates.

Repaired Variable Capacitor

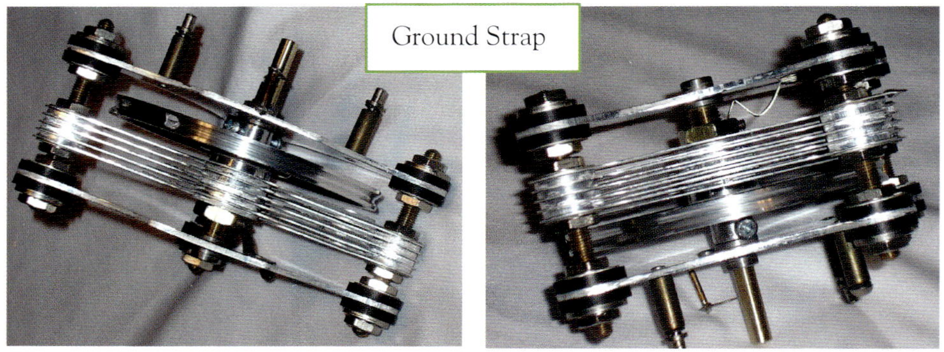

Top View

Bottom View

Front View

Rear View

13. The shaft should rotate freely with no horizontal or vertical movement. If it is too loose the weight of the plates will cause it to rotate on its own. If it is too tight it will not rotate. Adjust tension on the front and rear plates so that the shaft will be snug but will rotate smoothly.

14. If any of the criteria is not met, make adjustments, as appropriate, and go through steps 11–13 again.

Reinstall the variable capacitor into the chassis.

Rear View

Acme RF Transformer Repair (applicable to the Super-Portable and other battery radios with similar Acme transformers). The Super-Portable radio has three Acme RF transformers that have questionable reliability. If they are open, they need to be replaced or repaired. Instructions for repairing the R-2 and R-3 RF transformers are included here to address this problem. Drawings of each transformer are provided in the description of the Super-Portable, pages 104 and 105. Use these drawings as a guide to rewind the transformers. The original wire diameter is 40-gauge, but 30- to 38-gauge lacquer-covered wire should work.

Remove the bad transformer from the radio. Remove the four corner screws from the top and open the transformer. They have been potted with wax that can be removed by heating them in an oven. Stand the transformer up on a small pan, positioned so that the wax can drain out. Put the pan in the oven and set the temperature at 170°F or the lowest heat setting. As the wax is heated it will drain out of the transformer. When all the wax has drained out, remove the transformer, let it cool, and then remove any residual wax.

Disconnect the transformer wires from the lid. Unwrap and count the outer windings. Gently remove the wax paper that separates the secondary (outer winding) from the primary (inner windings); this paper will be reused. Unwrap and count the inner winding. Leave the paper that covers the core in place. Using the count or the information provided in the detail drawing, rewind the inner winding at the location shown in the detail on the page for this transformer. Leave excess wire on the ends for connecting to the terminals. Wrap the inner winding with the wax paper that was removed earlier. Rewind the outer winding in the similar manner as before, with the required number of turns and location. Solder the secondary and primary wires to the proper terminals on the lid. Test for continuity and reassemble the transformer. Reinstall the transformer into the radio.

Three Acme RF Transformers

Open RF Transformer Case

Internal RF Transformer Removed

Rewound Acme RF Transformers

Thordarson AF Transformer Modification (applicable to the 3-R, 4-R, Super-Portable, VII and other battery radios). Remove the AF transformer that needs to be replaced. If a spare transformer is unavailable, the original transformer can be disassembled and the internal windings can be replaced with a new transformer such as the P-T156 or A53-C interstage transformer available from Antique Electronic Supply (www.tubesandmore.com).

Use a small screwdriver to bend the tabs, at the top, straight up. Remove the top. Remove the sides by grasping them with a pair of pliers and pulling them up. Cut the leads and remove the old transformer.

Audio Transformer

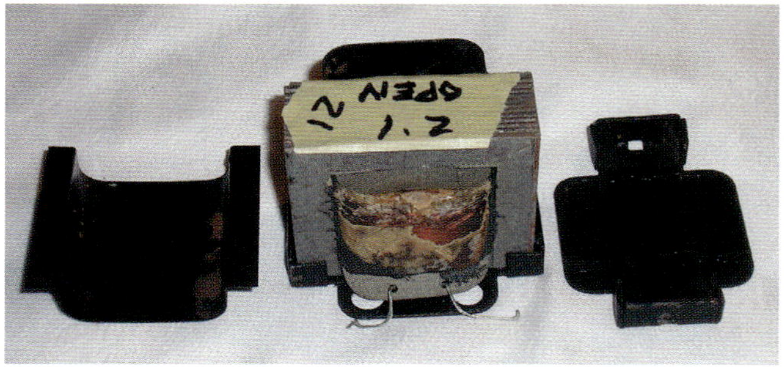

Sides Removed

Install sides.

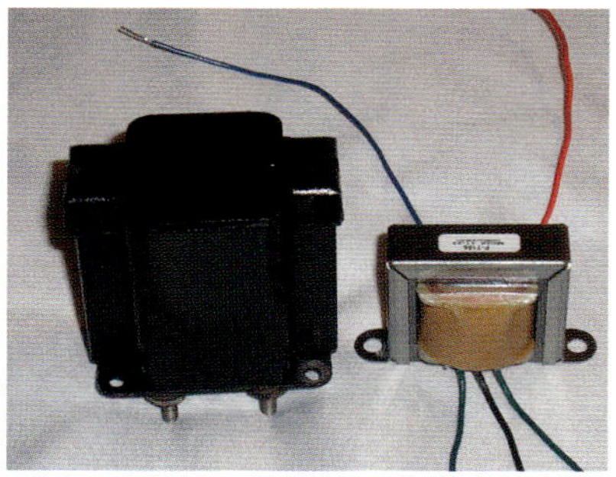

Modified Transformer and P-T156

Install the new transformer into the case and solder the terminals. Install wood shims, as appropriate, to keep the new internal transformer from moving inside the case.

Slide the sides into place. Install the top and bend the tabs over to hold it into place.

Reinstall the transformer in the radio while maintaining proper primary and secondary orientation.

"D" Battery Adaptor Modification (applicable to the Super-Portable, VII and other battery radios). The Zenith Super-Portable and Super-VII radios have a "D" cell battery that provides the negative bias to the grid of the first RF tube. In the early 1920s the "D" battery was the same as those today, except that the case was the negative terminal. The old style battery holder had a band to hold the battery in place, about two-thirds of the way up the length of the battery. This band was the negative terminal. Today's "D" batteries have an insulating material for the outer covering; therefore, an adapter is required to allow them to function in the manner of the older units. The adapter provides a connection from the negative end of the modern battery to the band that holds the battery in place.

Use the "D" Battery Adapter Template diagram, below, to draw the outline of the adaptor on a 0.0156" brass sheet. Cut out the template with a heavy pair of scissors or tin snips. Smooth the edges with a small flat file or light sandpaper. Form the adaptor by rolling the template around a "D" cell battery as tightly as possible, then put several strong rubber bands around the outside to hold it in place. Remove the "D" cell battery and spot solder the seam in several places. Remove the rubber bands and apply more solder, if needed. Bend the three tabs to the center and solder them where they meet.

Install the "D" cell battery into the adaptor, and then install both into the battery holder of the radio in the normal manner.

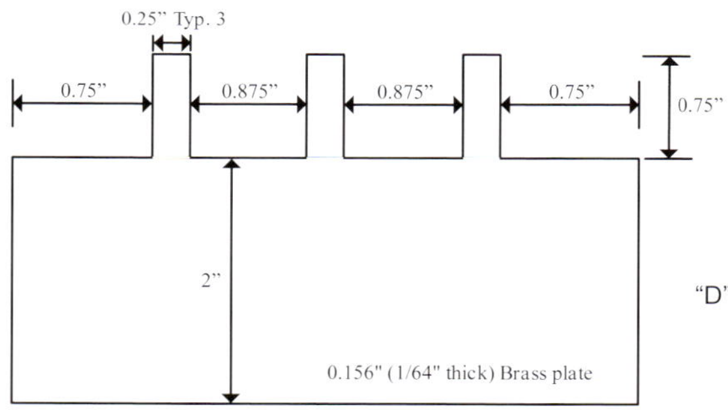

"D" Battery Adapter Template

Completed "D" Battery Adapter

1-R Gear Replacement Modification (applicable to the 1-R). The 1-R radio uses gears for the vernier mechanisms. Fine tuning is accomplished with small gears (c or t) that engage the larger gears on the variable capacitor and tickler (C or T) shafts. If those gears are bad, this procedure explains how to replace them with the same pulleys that are used to restore the 3-R. Warning: This modification does alter the original configuration of the radio and cannot be easily reversed. **Do not implement this procedure if you want to maintain the originality of the radio or if there is a possibility of obtaining the original style replacement gears.**

Follow the process for reproducing these pulleys on page 230.

The restoration of the 1-R is very similar to the restoration of the 3-R, except for the following: While the appropriate parts (gears, variable capacitor, etc.) are removed from the set, use a fine-tooth cutting tool (small saw, small grinder, etc.) to cut each vernier shaft housing at a point 0.75" from the rear side of the front panel as shown in the Chassis Underside photograph below. File any sharp edges that may remain on the shaft housing. During the installation of the pulleys, small adjustments may be required (additional grinding or the inserting of a spacer) in order to assure correct alignment of the large and small pulleys. See the Tuning Capacitor Underside and Vario-Coupler Underside photographs for examples of both pulleys installed in the 1-R receiver.

Continue with the restoration techniques described for the 3-R.

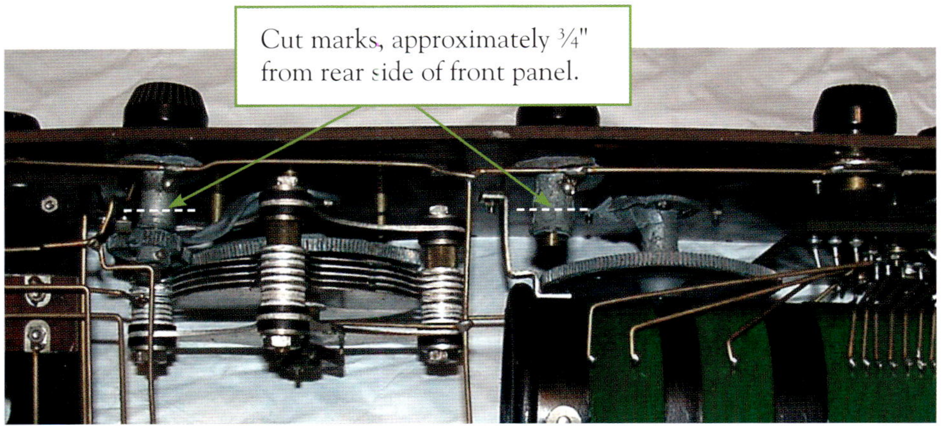

Chassis Underside. Notice the damage to the original 1-R gears.

Tuning Capacitor Underside. Reproduced replacement pulley installed.

Vario-Coupler Underside. Reproduced replacement pulley installed.

Step 3: Restore the Chassis and Cabinet

This section discusses the restoration of a 3-R receiver in order to demonstrate various restoration techniques. All these techniques are typically applicable to restoring the 1-R, 2-M, 3-M, 3-R, 4-R and VII. Additionally, you will find that many of these techniques are helpful in working on the Zenith 11, 12, and other equipment in general. When restoring equipment having fabric covering on the cabinet, such as the Zenith Super-Portable, refer to Fabric Reconditioning on page 248.

It is assumed that the construction, repairs or modifications to parts or assemblies have already been completed prior to starting this restoration. The restorer does have the option of pausing during this process to address a particular component, then returning to this section. Refer to the *Component Restorations* section for additional information related to construction, repairs, or modifications to parts.

Control Removal (applicable to the 1-R, 2-M, 3-M, 3-R and 4-R). The chassis should have already been removed from the cabinet during the *Radio Examination* step. After the chassis has been removed from the cabinet the knobs, pointers and shafts need to be removed. On the rear side of the front panel remove all wipers or collars that are holding the shafts in place and then pull each shaft out from the front. Frequently the pot metal shafts are "swollen" and will not turn. For the stuck shafts, break the knob and pointer assembly off (see *Original Condition of Front Panel*). Now, drill a 0.0625" hole in the center of each broken shaft. Now drill with a successively larger bit until the pot metal can be removed from the hole. Be careful not to damage any internal parts or the panel. Use a small round file or piece of rolled sandpaper to smooth the inside of each shaft hole, so that the new shafts will turn freely when installed.

Swollen Shafts

Chassis Cleaning (applicable to all battery radios). Gently wipe the vario-coupler, circuit board, wiring, sub-panels and back side to the front panel with a damp cloth. Then go over the circuit board and the back side of the front panel with crème polish (Bristol, Paragon, etc.). Furniture crème polish does wonders on cleaning the plastic surfaces. Use a dampened toothbrush or small stiff paintbrush to lightly clean the underside of the circuit board or other hard-to-reach areas.

Clean all mechanical switches, i.e., taps, wipers, and jack. The taps and wipers' contacts points need to be cleaned by lightly sanding them with fine sandpaper (200 to 400 grit). For the jack, hold the contacts open with a screwdriver and insert a flat piece of 200 to 400 grit sandpaper between the contacts and gently move it back and forth.

Panel Renewal (applicable to all battery radios with similar panels). Clean the front panel with diluted Murphy Oil Soap (follow the manufacturer's instructions). You may need to repeat this cleaning procedure as the panels usually have tobacco stains and dirt that have accumulated over the decades. Allow the panel to dry. Do not attempt to remove any of the surfaces from the front panel. The markings are cut into the panel and the removal of any of the surface material may eliminate them. During this process pay special attention to the markings on the front panel, and stop if they look like they are being negatively affected.

Original Condition of Front Panel

The luster of the brown faux wood grain front panel can be restored fairly closely to its original look by cleaning and then applying several coats of crème polish. Apply a small amount to a folded cloth. Gently rub in a circular motion over a small area. Continue over the entire surface. Repeat the polishing several times.

The engraved markings on the front panel are occasionally worn down. They were cut in at the factory during manufacturing with an engraving machine and then filled in with white paint. Sometimes you can use a white crayon or white Wipe-On Paint Stick (available from Antique Electronic Supply) to fill the voids, but at times this does not work and no other satisfactory fix has been identified. Use caution when applying crayon or Wipe-On Paint Stick because it can smear on to the panel and is sometimes hard to remove, so try a small test spot first.

Apply Novus #1 (available from Antique Electronic Supply, www.tubesandmore.com) to the surface of the front panel for additional protection and also to provide anti-static protection.

Renewed Front Panel

Control Installation (applicable to the 1-R, 2-M, 3-M, 3-R and 4-R). Thoroughly clean all parts (shafts, pointer, knobs, pulleys) with soap and warm water. An old toothbrush can be used to get into the voids and notches of the knobs. Apply Novus #1 to the knobs to protect them and to provide anti-static protection.

All the following shafts are made from 0.25" brass rod. Always double check all measurements before applying J-B Weld or other suitable metal glue. Make sure that the pointer is straight and flush with the front panel and/or shaft before gluing.

Reproduced Pointers and Shaft

Refer to the pictures and drawings for this radio (3-R) when installing the shafts, pointers, knobs and pulleys. Pay special attention to the clocking or alignment of the variable capacitor rotating plates and vario-coupler tickler coil in relation to the pointers. Use the clocking orientation information that was recorded during the *Radio Examination*.

Each individual 0.25" brass shaft will be measured and cut in the following manner. Make sure you have the correct hardware (knob, pointer, wiper, pulley, switch, etc.). Before making any cuts always double check the measurements. Measure the length of the shafts and include the thickness of the pointer and the width of the collar (tap, wiper, etc.) plus a little extra (approximately 0.125"). Cut the brass rod and file the ends. Use a small amount of J-B Weld to fasten the pointer to the shaft. Allow to dry overnight. Attach the knob to the pointer and install the knob, pointer and shaft assembly into the panel.

The variable capacitor shaft has already been installed during the variable capacitor repair (see page 241). Attach the large knob to the large variable capacitor pointer. Slide the variable capacitor knob and pointer assembly onto the variable capacitor shaft and tighten the set screw. Attach the dial string and spring. See the dial string routing in the description for this radio model. Align as recorded during the *Radio Examination*.

The original 0.25" vario-coupler shaft is rarely damaged and can usually be reused. Attach a large knob to the vario-coupler pointer. Apply a small amount of J-B Weld to the vario-coupler pointer and shaft and then slip the vario-coupler pointer onto the end of the shaft. Make sure it is flush and straight on the shaft. Let it dry overnight. Install the knob onto the pointer. Install the shaft into the vario-coupler through the pulley and tickler. Attach the dial string and spring. See the dial string routing in the description for this radio. Align as recorded during the *Radio Examination*. Tighten all set screws.

Cabinet Refinishing (applicable to all radios with wood cabinets). Never strip a radio cabinet, unless it is absolutely necessary. The original dark red mahogany look of most early Zenith cabinets may be difficult or impossible to obtain if the cabinet is stripped. Too, if there is a decal it will most likely be lost. If the cabinet does not need to be stripped then clean it with diluted Murphy Oil Soap (follow the manufacturer's instructions) or other suitable furniture cleaner and apply several coats of crème polish.

Replace any missing corner support blocks, and pay special attention to the blocks that support the front panel via a front panel mounting screw. The front panel support blocks need a 10-degree slope (1-R, 3-R only); see Corner Support Blocks (page 248).

If stripping is required, perform a test (stripper and stain) on an inconspicuous location or on another piece of wood, in order to make sure that the desired results can be achieved. Remove the back door. Cover and protect all manufacturing labels. Strip the cabinet and door (only the side that faces out) per the manufacturer's instructions. Make sure the cabinet and door are completely clean and let them dry thoroughly.

Prior to staining, fix any problems with the wood. If there are any large cracks or dents you can fill them with mahogany type wood filler. Allow it to dry, then sand as needed. Stain the cabinet and door with red mahogany stain, following the manufacturer's instructions. Sometimes you may need to apply multiple coats. If the

Original Cabinet

Corner Support Blocks

cabinet is not dark enough then apply a coat of black (ebony) stain, wipe, and allow to dry. Now apply the red mahogany stain in order to get that deep dark red tone. Getting the correct color and tone may require some trial and error.

After the cabinet has thoroughly dried, apply several coats of semi-gloss clear lacquer, per the manufacturer's instructions. If you have a reproduced decal, apply the decal and then apply another coat of semi-gloss clear lacquer spray over the entire cabinet. This makes it look uniform all over and helps protect the decal.

Perform a thorough cleaning of the inside of the cabinet with a damp cloth. Do not damage any of the labels.

Replace any missing green felt. Attach the felt with general purpose white glue. The felt helps protect the back side of the front panel.

Remove any covers protecting the manufacturing labels.

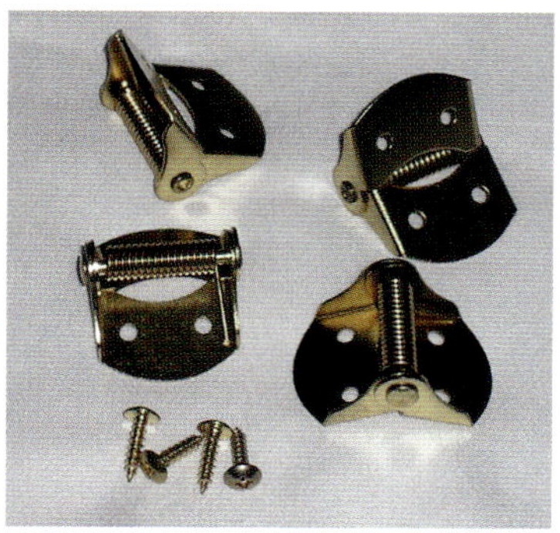
Spring Hinges

Clean the hinges and install the door. The brass spring hinges can be cleaned with a good metal polish. Replacement brass-plated spring hinges may be sourced on the Internet. After the lacquer has had several days to dry, apply several coats of crème polish to the whole cabinet.

Refinished Cabinet

Fabric Reconditioning (applicable to the Super-Portable and other portable radios). Most of the portable radios of the 1920s had fabric covering over the cabinets and the Zenith Super-Portable was no exception. Usually this fabric covering can be reconditioned close to its original appearance. The results of the following techniques depend on the current condition of the fabric. The methods described here assume that the covering is complete and intact with only some small tears, rips, and minimal fraying. When the covering is in bad condition and/or has pieces missing, new matching material can be obtained and the cabinet completely recovered. The techniques described here, however, only address the reconditioning of the fabric, striving to match the original color as closely as possible. All products mentioned are available from local hardware or merchandise stores. In using them, follow the manufacturer's instructions.

Remove all hardware and clean, paint, or polish, as appropriate.

The edges, corners, and other ripped areas usually have fraying. Remove any excess fraying by cutting with scissors, but only if absolutely necessary. Further fraying or ripping can be stopped by gluing these areas with Elmer's Multi-Purpose Glue-All; it dries clear. Use only a small amount of glue and allow to dry.

Clean the whole cabinet with water or use diluted Murphy Oil Soap. Wipe lightly to avoid damaging the fabric material. Let it dry.

Use a cotton swab to apply a small amount of shoe dye: black, brown, red, etc., to blend in the blemishes or other problem areas. Try to make the problem area blend in with the surrounding area as closely as possible. Be careful not to make the area stand out. Allow it to dry.

Depending on the condition of the material, here are two additional repair options. First, if the fabric is in fairly good condition, an application of a couple of coats of neutral shoe polish may work wonders. In cases where the fabric needs just a little help, apply a coat of colored shoe polish (black, brown, etc.) Second, in cases where the fabric needs a lot of help, apply a coat of shoe dye, then a coat or two of shoe polish. Always start on the conservative side with your approach and progress to the more extreme. The object is to restore the fabric to its original color and luster, which usually had very little shine.

Reinstall the hardware.

Examples of Shoe Dye and Polish

Original Fabric-Covered Cabinet

Reconditioned Fabric-Covered Cabinet

Step 4: Perform Testing

Preliminary Test Information (applicable to all battery radios). Your radio restoration has now been completed and all that remains is to test the radio and then document the restoration. The testing procedures outlined in this section test a 3-R (L) radio; however, the process is typically applicable to the testing of any battery radio.

The A and B power for this test set (3R) is provided by a battery eliminator and the C power is provided by an internal reproduced "C" battery. Any combination of batteries and/or battery eliminator can be used as long as the correct voltages are applied to the proper circuits.

WARNING: Never remove or insert any components (tubes, resistors, etc.) from or into the radio while the power is applied. Such action could possibly result in damage to components inside the radio.

During the testing, one important, but not essential, test instrument is the "variable AC transformer" available from Jameco Electronics at www.jameco.com. The variable AC transformer allows you to slowly increase the AC voltage from 0 to 120 VAC. This voltage powers the battery eliminator for your radio and allows you to slowly increase all voltages to the radio. Such a procedure allows you to monitor your radio for excessive current while the set has less than full power applied, which helps prevent component damage. If a variable transformer is not available, then full power must be applied.

In this test example of a 3-R (L), the battery eliminator is powered from a variable isolated AC supply, shown in the photo. This unit allows the slow increase of AC voltage from 0 to 120 VAC and the panel meters monitor both AC voltage and AC current. These meters provide a good indication of what is happening inside the radio. As the voltage is increased, if a short exists in the set, the current meter will indicate high current and the power can be turned off and the set examined to locate and correct the short or other reason for the excessive current.

Variable Isolated AC Supply

Preparation for Testing (applicable to all battery radios). Refer to the drawings, pictures and schematics in the description chapter for your radio model. This will help refresh your memory as to the location of specific components which may be important during testing. Perform the following set-up steps:

- Install the filament resistors and the grid leak resistor.
- Install the "C" battery.
- Attach the antenna and ground.
- Insert the headphone plug into the jack.
- Connect the power cable to the power supply (see *Preparation for Operation* for your set for this radio).
- Plug the battery eliminator power cord into the 120-volt receptacle or to the variable isolated AC supply, if used.

DO NOT install the tubes at this time.

Set the "S" switch, on front panel of radio, to position 3 (this applies power to all pins of the tube sockets).

3-R Bench Test Connections

Trial Test (applicable to all battery radios). If a variable AC supply is used, plug the battery eliminator into the isolated receptacle, setting the voltage control to zero volts. If the variable AC supply is not used, plug the battery eliminator into the AC outlet. Turn all power switches to ON (radio switch, battery eliminator, and any variable AC supply).

Use a voltmeter to measure the voltage on each tube socket element (pin) to make sure the correct voltage is applied. The appropriate voltage can be obtained from the radio descriptions for this radio model. If any voltage seems to be abnormal, proceed to **Troubleshooting**. If everything seems to be normal, turn all switches to OFF and continue with the **Final Test.**

Final Test (applicable to all battery radios). Install the tubes. Put on the headphones and turn the power switch to ON. Follow the Operation instructions for this radio model. If any problems are encountered, proceed to **Troubleshooting**. If it seems to be working fairly well then proceed to **Maximizing Performance**.

Troubleshooting (applicable to all battery radios). Select the appropriate problem description below, identify the problem, and resolve it as appropriate. Use the drawings, pictures and schematics for your radio to verify the configuration and wiring. **Note:** Don't make more than one alteration (fix) without retesting or you will risk not being able to determine exactly what you did to fix the problem (loss of lesson learned).

Excessive current or heat—Perform resistance checks to determine possible shorts or open circuits. Check all electrical connections. Check battery connections. Look for shorts.

Incorrect voltage on tube terminals—Perform resistance checks to determine possible shorts or open circuits. Check power connections for proper voltage. Check for proper resistance of filament resistors. Check for proper voltage from power supply and batteries, as applicable.

No sound—Check all connections (antenna, ground, power, headphones, etc.). Check batteries ("A", "B", "C", etc.). Verify that all components have been installed (tubes, resistors, etc.). Test the tubes. Check all mechanical connections (switch, taps, wipers, jack, rheostats, etc.). Verify that the AF transformers are not open. Look for broken or faulty wiring or connections. Check all Vario-Coupler solder connections.

Low volume—Make sure you have a good antenna and ground. Check batteries. Check the settings of the tap switches and review *Operation* instructions for this radio model. Check clocking or orientation of the Variable Capacitor, Vario-Coupler, switch contacts, etc. in relation to the pointers. See *Maximizing Performance*.

Tuning problem—Check the settings of the tap switches and review the *Operation* instructions for this radio model. Check clocking or orientation of the Variable Capacitor and Vario-Coupler in relation to the pointers.

Return to **Trial Test** to continue the testing.

Maximize Performance (applicable to all battery radios). Never remove or insert a tube or any other component with the power applied or the switch ON.

There are not a lot of things that can be done to these early radios that will increase their performance; "it is what it is." However, there is something that can be done to make sure you are using the best tubes that are available to you. Some tubes seem to work in some radios better than others. The performance of the radio can be maximized by selecting the tubes that work the best in the particular radio. This technique does require having a pool of extra tubes.

Turn the radio ON and tune it to a clear station. Now turn the radio OFF. Take out a tube and replace it with another similar tube from the pool. Turn the radio back ON and see if it plays any better than the one that was just removed. Use the tube that played the best. Repeat this process for each stage of the radio. Repeat the process until there is no improvement. The radio now contains the best tubes that are currently available.

Final Assembly (applicable to all battery radios). Install the chassis into the cabinet using the Removal of Chassis from Cabinet process in reverse order. Hook up the power and perform an operational test.

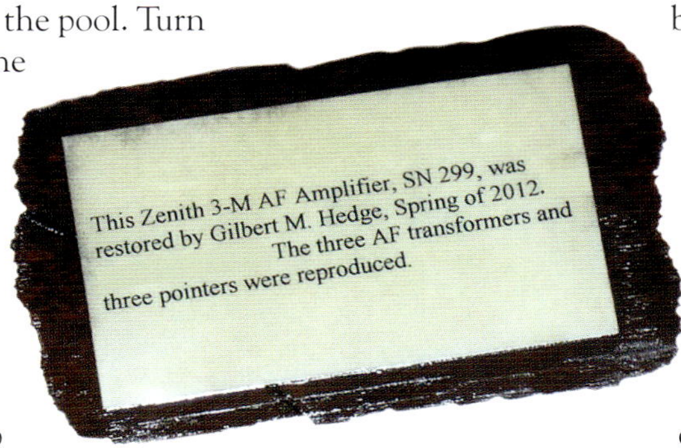

3-M Restoration Label (positioned underneath cabinet)

Typical Show Display

Step 5: Document the Restoration

Radio Identification (applicable to all radio restorations). Now that you are finished, the restoration information needs to be suitably, and permanently, identified somewhere on the radio. It could be located on the inside, but the preferred placement is underneath, on the bottom where it is inconspicuous, easily viewed and does not diminish the look of the radio. The marking should include model number, manufacturer, who restored it, when and where it was restored, and any significant restoration information, e.g., "The gears were replaced with reproduced pulleys." It can now be enjoyed by future generations and everyone will know who restored this beautiful radio.

The 3-M Restoration Label was created with Microsoft Word on Avery Business Card stock and printed on an ink jet printer. The card was then glued to the bottom of the radio with Elmer's multi-purpose Glue-All, and allowed to dry. Then it was covered with three coats of lacquer.

Restoration Record (applicable to all radio restorations). Create a record of the restoration; now, before you forget anything. Try to capture everything that was done to restore this radio: parts replaced, parts reproduced, cabinet refinishing, cleaning, problems encountered, condition of radio before restoration, etc. Include plenty of before-and-after photos. Assemble the documentation, notes, photos, comments, etc. in a binder or with some other suitable method. Make sure this document stays with this radio, so that the historical facts can be accurately interpreted in the future.

Save replaced parts for future reference or display. All this information can be used to put together an informative and entertaining backdrop for presenting your set at one of the local or national antique radio shows and conventions. Use your imagination, and don't forget to enjoy your newly restored radio.

Typical Restoration Binders

Glossary

"A" BATTERY—A battery that furnishes electrical current to heat the filaments of a vacuum tube.

ALTERNATING CURRENT (AC)—Current that changes direction, measured in cycles per second or hertz.

AUDIO FREQUENCY (AF)—An audible frequency approximately 20 to 20,000 cycles (hertz) per second.

"B" BATTERY—A battery that provides sufficient voltage to the plate of a vacuum tube.

BATTERY ELIMINATOR—Provides A, B and C power to the receiver and gets its energy from an AC source (eliminates batteries).

BIAS—A negative voltage applied to the grid of a tube to reduce current flow between the cathode and plate.

BUS BAR—A heavy gauge wire used in making radio connections.

BYPASS CAPACITOR—A capacitor used to provide a path for alternating current.

"C" BATTERY—A battery that has one or more voltages usually less than 10 volts. Applied to the grid circuit to control the flow of electrons between the cathode and plate.

CAMSHAFT—A shaft fitted with cams that open or close various switches.

CAPACITOR—A Condenser (vintage term). A device that stores electrical energy. It passes AC and blocks DC.

CHASSIS—The base on which a radio is built.

CHOKE—A coil, usually wrapped around an iron core that opposes alternating current while allowing direct current to pass.

CLOCKING—The orientation of one object to another, e.g., rotatable fins in relation to the dial pointer.

COMPONENT—An individual part of a circuit (i.e., resistor, transformer, capacitor, tube, speaker).

CURRENT—The flow of electrons.

DETECTOR—A device (tube, crystal, etc.) that removes the radio frequency part of a signal, leaving only the audio portion.

DIRECT CURRENT (DC)—Current (electrons) that flows in one direction.

DRY BATTERY—A battery not employing a liquid electrolyte, formed with a paste.

DRY CELL—See **Dry Battery**.

FEEDBACK (REGENERATION)—The transfer of a portion of the energy from the output (plate) to the input (grid).

FILAMENT—An element of the vacuum tube that is heated to emit electrons.

FREQUENCY—The number of cycles per second (hertz) of an alternating current.

FRONT PANEL—The surface on which most of the receiver controls are located.

GRID—A wire coil or screen between the cathode and plate of the vacuum tube. It controls the flow of electrons from the cathode to the plate.

GRID CAPACITOR—A capacitor used in the detector circuit of a vacuum tube, located between the tuning coil and the grid.

GRID LEAK RESISTOR—A resistor in the grid circuit of a tube that allows electrons to leak off.

HEATER—The element of the vacuum tube that heats up to allow the cathode to emit electrons.

INDUCTION—The magnetic opposition to any change in current flow, measured in henrys.

JACK—A socket that accepts a **Plug** at one end and attaches to electric circuitry at the other.

LOUDSPEAKER—A device that converts audio frequencies into sound.

NEGATIVE—Any potential that is lower than another (i.e., −4.5 volts is negative to −3 volts).

OPEN—Not electrically connected.

PIN DESIGNATION—The identification of the various tube electrodes or elements.

PLATE—Collector for electrons in a vacuum tube. Has a positive charge related to the cathode.

PLATE VOLTAGE—The positive B voltage between the plate and cathode of a vacuum tube.

PLUG—A device, usually connected to wires at one end, that inserts into a socket (**Jack**).

POST—See **Terminal**.

POTENTIAL—The excess or depletion of electrons on a component that produces electrical pressure or voltage difference.

POT METAL—Slang term used to refer to cheap metal alloys with a low melting point. There is no formal definition of pot metal, so it can be hard to determine its contents. Some common metals included in such alloys include zinc, lead, copper and tin, among others.
PRIMARY—The input winding of a transformer which inductively induces energy into the **Secondary** winding (output).
RADIO FREQUENCY (RF)—Frequencies higher than audible sound waves.
REGENERATION—See *feedback*.
REGENERATIVE CIRCUIT—The method use to transfer the output signal from the plate to grid using induction.
RESISTANCE—The opposition to the flow of current. Measured in ohms (Ω).
RHEOSTAT—A variable resistance.
SECONDARY—A output winding of the transformer that obtains its energy by induction from the **Primary** (input).
SELECTIVITY—The ability of a receiver to tune to a particular frequency while rejecting all others.
SENSITIVITY—The degree to which a receiver responds to a given signal.
STAGE—A single tube and its associated circuit components.
STAGE SELECTION—The ability to select any number of the available stages.
TERMINAL—A binding post or other fitting that provides a connecting point between components.
TICKLER, TICKLER COIL—A coil placed in the plate circuit of a regenerative receiver and coupled inductively in the grid coil in order to obtain regeneration.
TRANSFORMER—Two windings that transfer power by induction, from one winding (primary) to another (secondary). May be used to step up or step down voltage, or for isolation.
TRANSFORMER DC RESISTANCE—A measurement that provides an indication that the transformer primary or secondary winding has continuity.
TROUBLESHOOTING—To investigate and locate a malfunction in equipment and then repair it.
TUNED RADIO FREQUENCY (TRF)—A radio employing multiple ganged RF amplifier circuits that are collectively tuned to the desired wave length by varying the variable capacitors in parallel with the secondary of the RF transformers.
TUNING—Selecting a desired frequency (resonate) and blocking all others.
TUNING CAPACITOR—A variable capacitor that has stationary and rotating plates. The amount of engagement of the plates determines the capacitance. Works in conjunction with an inductive coil that resonates at the desired frequency.
VACUUM TUBE—A bulb with a vacuum containing a filament, plate and grid.
VARIABLE CAPACITOR—See **Tuning Capacitor.**
VARIO COUPLER—A tuner having a primary and a secondary coil inductively coupled, with the secondary arranged so as to be rotatable within the primary. The amount of coupling depends upon the variation of the angle between the axes of the two coils.
VERNIER DIAL—A device by which a complete turn of the control knob turns a shaft only part of a turn.
VOLT—The electrical pressure that will cause one ampere of current to flow through a resistance of one ohm.
VOLUME—The degree of loudness of sound produced by a loudspeaker.
WINDING—The wire forming a coil.

Notes

Chapter 1, Zenith 1-R

1. Zenith Radio: The Early Years 1919–1935. (Atglen, PA: Schiffer Publishing Ltd., 1997), pgs. 15 and 18.

Chapter 2, Zenith 2-M

1. *Antique Radio Restoration Guide: 2nd Edition*. (Iola, WI: Krause Publications, 1992), pg. 130.

Chapter 3, Zenith 3-M

1. *Antique Radio Restoration Guide: 2nd Edition*. (Iola, WI.: Krause Publications, 1992), pg. 130.

Chapter 6, Zenith Super-Portable

1. *The Zenith Story: A History From 1919*. (Chicago, Illinois: Zenith Radio Corporation, 1955), pg. 7.

Chapter 12, Notes on the Early Zenith Radios

1. *Zenith Radio: The Early Years 1919–1935*. (Atglen, PA: Schiffer Publishing Ltd., 1997), pg. 22.
2. Ibid., pg. 109.
3. *Collecting Old Radios and Crystal Sets*. (Des Moines, IA: Wallace-Homestead Book Company, 1977), pg. 35.
4. *The Portable Radio in American Life*. (Tucson, AZ and London, England: The University of Arizona Press, 1991), pg. 74.
5. *Guide to Old Radios: Pointers, Pictures and Prices: Second Edition*. (Iola, WI: Kruse Publications, 1995), pg. 35.
6. *Zenith Radio: The Early Years 1919–1935*. (Atglen, PA: Schiffer Publishing Ltd., 1997), pgs. 24 and 32.
7. Ibid., pg. 109.

Chapter 13, Successes and Failures

1. This material was previously published as an article in *Sound Waves*, Irving, Texas: VRPS, July 2011; in *The Grid Leak*, Houston, Texas: HVRA, Sept.–Oct. 2010; and in *Antique Radio Classified*, Port Washington, NY: Vintage Radio Publishing, LLC, Volume 28, Number 6, June 2011.
2. *Zenith Radio: The Early Years 1919–1935*. (Atglen, PA: Schiffer Publishing Ltd., 1997), pg. 216 n. 21.
3. Ibid., pg. 56.
4. Ibid.
5. *Radio News*, "Portable Broadcast Stations." (New York, NY: Experimenter Publishing Company, Inc.), January 1925, pg. 1149.
6. *Zenith Radio: The Early Years 1919–1935*. (Atglen, PA: Schiffer Publishing Ltd., 1997), pg. 56.
7. *The Zenith Story: A History from 1919*. (Chicago, IL: Zenith Radio Corporation, 1955), pg. 7.
8. *Radio Manufacturers of the 1920s: Vol. 3: RCA to Zenith*. (Chandler, AZ: Sonoran Publishing, LLC., 1999), pg. 255.
9. *Zenith Radio: The Early Years 1919–1935*. (Atglen, PA: Schiffer Publishing Ltd., 1997), pg. 209 n. 61.
10. Ibid.
11. *Radio Manufacturers of the 1920s: Vol. 3: RCA to Zenith*. (Chandler, AZ: Sonoran Publishing, LLC., 1999), pg. 274.

Bibliography

Alth, Max. *Collecting Old Radios and Crystal Sets* (Des Moines, IA: Wallace-Homestead Book Company, 1977).

Cones, Harold N., John H. Bryant. *Zenith Radio: The Early Years 1919–1935* (Atglen, PA: Schiffer Publishing Ltd., 1997).

———. *Zenith: The Glory Years 1936–1945: History and Products* (Atglen, PA: Schiffer Publishing Ltd., 2003).

Douglas, Alan. *Radio Manufacturers of the 1920s: Vol. 1: A-C Dayton to J. B. Ferguson, Inc.* (Vestal, NY: The Vestal Press, Ltd., 1988).

———. *Radio Manufacturers of the 1920s: Vol. 2: Freed-Eisemann to Priess* (Chandler, AZ: Sonoran Publishing, LLC., 1995).

———. *Radio Manufacturers of the 1920s: Vol. 3: RCA to Zenith* (Chandler, AZ: Sonoran Publishing, LLC., 1999).

Gernsback, Sidney. *S. Gernsback's Radio Encyclopedia 1927: Replica Reprint* (Palos Verdes Peninsula, CA: Vintage Radio, 1974).

Grinder, Robert E. and George H. Fathauer. *The Radio Collector's Directory and Price Guide* (Scottsdale, AZ: Ironwood Press, 1986).

Hallmark, Clayton L. *How to Repair Old-Time Radios* (Blue Ridge Summit, PA: Tab Books, 1979).

Johnson, David. *Antique Radio Restoration Guide: 2nd Edition* (Iola, WI: Krause Publications, 1992).

Johnson, David and Betty Johnson. *Antique Radios: Restoration and Price Guide* (Des Moines, IA: Wallace-Homestead Book Company, 1982).

———. *Guide to Old Radios: Pointers, Pictures and Prices: Second Edition* (Iola, WI: Krause Publications, 1995).

McMahon, Morgan E. *1887–1929: Vintage Radio: Second Edition* (Palos Verdes Peninsula, CA: Vintage Radio, 1973).

———. *1921–1932: Radio Collectors Guide: Revised Edition* (Palos Verdes Peninsula, CA: Vintage Radio, 1981).

———. *A Flick of the Switch 1930–1950* (Palos Verdes Peninsula, CA: Vintage Radio, 1976).

Middleton, H. A. *Receiving Tube Substitution Guide Book* (New York, NY: John F. Rider Publishing, Inc., 1953).

RCA. *RCA Receiving Tube Manual* (Camden, NJ: RCA Corporation, 1975).

Roth, Durell M. *Oscilloscopes and Antique Radios* (Self-published, 1994).

Schiffer, Michael Brian. *The Portable Radio in American Life* (Tucson, AZ and London, England: The University of Arizona Press, 1991).

Gilbert M. Hedge has been collecting and restoring radios for 35 years. He holds an AAS in Electronics Technology and BS Engineering: Quality Control and Quality Assurance, and is retired from The Boeing Company where he worked for 25 years in various quality engineering disciplines.

After seeing several antique radios at local flea markets he acquired a small wood cabinet set to see if he could get it working. That was all it took; he was hooked. The visits to the flea markets led to meeting other collectors and joining the Houston Vintage Radio Association. Over the years his interest evolved into collecting Zenith battery sets, Trans-Oceanic radios and portable radios of the 1920s. Today the main focus of his collection is the early Zenith battery sets of the 1920s. Because of the nature of these early Zenith radios and the lack of available replacement parts, Hedge developed a special set of methods and skills in order to make the replacement parts that are usually required for their restoration.